LARGE-SCALE GEOLOGIC STRUCTURES

LARGE-SCALE GEOLOGIC STRUCTURES

Jacques DEBELMAS
Professor Emeritus
Joseph Fourier University
Grenoble

Georges MASCLE
Professor
Joseph Fourier University
and CNRS
Grenoble

Translated from French
by

N. Venkat Rao
Professor
Centre of Exploration Geophysics
Osmania University
Hyderabad

A.A. BALKEMA/ROTTERDAM/BROOKFIELD/1998

Published by arrangement with Masson Éditeur, Paris.

Aidé par le ministère français chargé de la culture.
Published with Financial Aid from the French Ministry of Culture.

Translation of: *Les grandes structures géologiques*, 3e, July 1997, Masson Éditeur.

Cover Photograph: Mont Blanc Massif (French-Italian Alps). Aerial view. Courtsey: authors

A.A. Balkema, P.O. Box 1675, 3000 BR Rotterdam, Netherlands
Fax: +31.10.4135947; E-mail: balkema@balkema.nl;
Internet site: http://www.balkema.nl

Distributed in USA and Canada by
A.A. Balkema Publishers, Old Post Road, Brookfield, VT 05036-9704, USA
Fax: 802.276.3837; E-mail: Info@ashgate.com

ISBN 90 5410 776 6

PREFACE TO THE FIRST EDITION

The need for a book devoted to major geological structures has long been felt. It may be surprising that to date not a single book has touched upon this topic. The reasons are obvious.

Firstly, the volume of data to be compiled is enormous, which becomes especially difficult in view of the need to integrate quite diverse aspects of the geological sciences.

Secondly, there is the problem of according a level of confidence to the innumerable published explanatory models, most of which are mere working hypotheses and much too simplified

Finally, given the rapid accumulation of a large mass of data constantly being gathered by increasingly extensive explorations of the sea bottom and the refinements of geological and geophysical methods for analysing the continental structures, a book attempting to incorporate such data faces the risk of becoming obsolete all too soon.

These difficulties notwithstanding, however, we have undertaken the task at the behest of students, especially those preparing for competitive examinations, and professors of teaching institutions.

Clarity demands conciseness. Hence, this volume does not constitute an exhaustive treatise but rather a simple presentation of some examples of large structures of our planet and their dynamic evolution. In this presentation we have assumed that the basic concepts (such as stratigraphic, tectonic or petrographic terminology, plate tectonics) are familiar to the reader and thus have excluded the tectonophysical problems dealt with in other publications.

The simplicity and the conciseness desired by our publishers, as well as by the readers to whom this book is addressed, obviously necessitated very deliberate planning. Such an approach is always fraught with danger and we cannot escape from the classic dilemma expressed by P. Valery: Whether to keep it simple knowing full well that exactitude would suffer somewhat, or to offer complex material that would be more precise but less comprehensible. Eventually we decided that the former would permit a more global perception of the phenomena and would better conform to the present trends in geology.

Since the previous edition, significant progress has been achieved in various domains and several regions. Excellent books have appeard on different aspects of major geological structures. It hence became necessary to revise some sections and to add new ones. The oceanic region, continental margins and collision chains have more particularly benefited from these new concepts. Gravitational spreading (collapse) of orogenies has become a specially developed subject. On the other hand, in order to avoid too much enhancement in the volume of the book, some sections less directly related to the description of major types of structures have been deleted (Caucasus, for example).

This volume is devoted to the study of major geological features such as oceanic ridges, continental margins, sedimentary basins, mountain chains etc. These are studied in a dynamic or evolutionary perspective wherein we have attempted to correlate the various phenomena that operate simultaneously (sedimentation, deformation, volcanism, metamorphism etc.). The book thus represents an integrated approach with obvious applications, such as the study of petroleum basins and mineral deposits. It is also closely related to geophysics if viewed as an attempt to obtain information about deep structures and mechanisms.

Mainly present or recent structures are discussed, viz., those belonging to the Alpine cycle *sensu lato*. These are, truth to tell, the best preserved and the most easily unravelled structures. Similar structures existed in older cycles and they are sometimes mentioned.

Format

Introduction: Concepts of the earth's crust and lithosphere
Part one: Oceanic structures
Part two: Extensional structures of the continents
Part three: Compressional structures of the continents

PREFACE TO THE THIRD EDITION

Since the previous edition, significant progress has been achieved in various domains and several regions. Excellent books have appeared on different aspects of major geological structures. It hence became necessary to revise some sections and to add new ones. The oceanic region, continental margins and collision chains have more particularly benefited from these new concepts. Gravitational spreading (collapse) of orogens has become a specially developed subject. On the other hand, in order to avoid too much enhancement in the volume of the book, some sections less directly related to the description of major types of structures have been deleted (Caucasus, for example).

CONTENTS

PART TWO
EXTENSIONAL STRUCTURES OF THE CONTINENTS

INTRODUCTION

NOTIONS ON THE EARTH'S CRUST AND LITHOSPHERE

The study of seismic wave propagation, especially of the P (longitudinal) waves, has enabled us to distinguish two components in the shallower part of the earth:

— The crust (thickness varying from 10 to 70 km, average 30 km);

— The underlying mantle, separated from the crust by a surface of discontinuity at which the seismic-wave velocities change suddenly (the Mohorovičić discontinuity, generally shortened to Moho).

The crust is passive in behaviour. The driving force for tectonic phenomena is situated in the mantle, mainly in the deeper part. It is in fact observed, mostly by means of seismic waves, that the surficial part of the earth is rigid over a thickness of the order of 100 km, this part is known as the *lithosphere* (the crust and upper mantle).

The zone below the lithosphere is the *asthenosphere*; here the seismic-wave velocities decrease, at least in its upper part. This low-velocity zone corresponds to a readily deformable material. It is probably at this level that isostatic compensations occur, as also the upper cells of the convective motions considered to be the causative force for most of the orogenic phenomena.

The surface of lithosphere consists of two types of crust, continental and oceanic.

CONTINENTAL CRUST

In the case of stable continental zones (viz., the major shields and platforms, such as those of Africa or Russia, which have not undergone deformations for several hundred million years), two parts of the crust can be distinguished:

— Upper crust (10 to 15 km), $d = 2.7$, $V_P = 6$ km/s;

— Lower crust (10 to 15 km), $d = 2.8$, $V_P = 7$ km/s.

Sometimes an intermediate layer exists between these two in which the velocities of P-waves vary from 6 to 5.5 km/s. This low-velocity layer (Conrad discontinuity) is a possible zone of creep and slippage whose origin is still debatable. The initial melting has been suggested as a possible cause but at a depth of 15 km, the temperature is only 400 to 500°C, which is much lower than the melting point of normal crustal rocks.

The upper crust is readily interpreted: below a variable thickness of sediments, it consists of more or less granitised gneisses and hence the name *granito-gneissic* (or sialic) layer. The upper crust outcrops mainly in major shields; it is found in ancient and recent mountain chains, where it is often visible throughout its entire thickness.

The lower crust is more difficult to interpret because of its inaccessibility in major shields. A small increase in P-wave velocity with an increase in pressure due to depth does not suffice to explain the observed increase in velocity. A change in lithologic composition has to be introduced. The density of the lower crust (2.8 to 2.9) corresponds to that of basalt and hence this zone is sometimes called the basaltic layer. One should search for answers to this problem in the orogenic regions where outcrops of the lower crust can be expected.

However, such outcrops are very rare, which suggests that at the time of folding of these belts cleavage could have taken place between these two crusts (at the level of low-velocity layer?) and only the upper crust, cleft and sliced, forms outcrops in general. Nevertheless, some sections exist which show the following parts:

— An upper group in which diverse metasedimentary rocks (gneisses, marbles, quartzites) and metavolcanic rock sills (amphibolitised) alternate, all metamorphosed into a facies of amphibolite to granulite or eclogite. One can cite examples of the 'kinzigitic' complex of the Ivrea zone of the Alps or the 'leptyno-amphibolitic' complex of the French Massif Central.

— A lower group, more massive, consisting of basic rocks (which are the source for the sills mentioned earlier) in which gabbros and peridotites with cumulate texture predominate. The mineralogical equilibria of this basal complex indicate a pressure of 7 to 9 kbar and a temperature of about 1150°C.

To summarise, the lower crust consists of intrusions of mantle material, basic to ultrabasic, into a sedimentary sequence, metamorphosed, as already mentioned, into a facies of eclogite to amphibolite.

Below the lower crust *layered peridotites, with a texture of tectonites* (porphyroblastic and mylonitic textures), are expected, which represent the upper mantle. This correlates well with geophysical observations (V_P = 8 km/s, d = 3.3).

The ancient Moho is never observed in the field because this zone does not outcrop except in folded regions, where it corresponds to a decollement or slip plane marked by mylonites and peridotitic breccia.

It may be noted that the ECORS seismic profiles often indicate that the lower crust often exhibits a layered structure in contrast to the relative uniformity of the upper crust (Fig. 1). This denotes the existence of several subparallel reflectors whose origin is still debated. The general tendency is

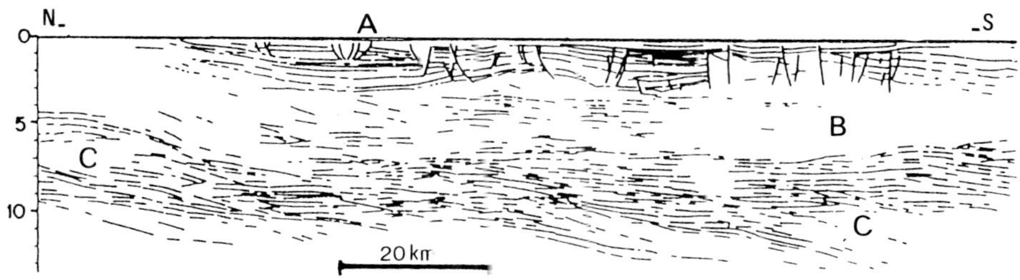

Fig. 1. ECORS SWAT profile No. 8 in Western Manche, between Plymouth and the Ile d'Ouessant (ECORS inform. No. 3. 1985).
Below a thick sedimentary cover (A) occurs the upper crust devoid of reflectors (B) and the layered lower crust (C).

to interpret them as cleavage planes satellitic to the Moho, which might constitute the principal one.

In the orogenic zones the thickness of the continental crust increases (Moho sinks), resulting in the formation of a 'root', which may be double the thickness of the crust (60 to 70 km).

Rheological Properties of the Crust and Upper Mantle

Seismic studies have shown that from the rheological point of view the crust is made up of two parts: a *rigid* or *fragile* upper crust and a lower crust in which the deformations are mainly produced by ductile stretching (*ductile crust*).

The different behaviour of these two layers may lead to their separation. We shall see an example of this in the Galicia bank in the Atlantic, offshore of NW Spain (p. 138): here, the ductile crust is stretched to its disappearance, such that the upper mantle has directly come into contact with the upper crust.

As for the upper mantle, one may think that the ductile regime is strengthened here due to prevailing pressure and temperature conditions. However, such is not the case. At present, a rigid upper zone and a ductile lower zone are distinguishable in the upper mantle, as in the case of the crust, such that in the zones of stretching the first may itself be fragmented or form a boudinage structure.

OCEANIC CRUST

The oceanic crust forms the bottom of the major oceans and differs from the continental crust essentially by its smaller thickness and the absence of a granito-gneissic layer.

a) Stable oceanic zones (= abyssal plains)

Below a variable thickness of sediments there exist:

Upper oceanic crust: Only this layer has been reached and partly intersected by boreholes (the deepest borehole has penetrated this layer over a section of about 2 km, near the Galapagos islands, below 275 m of Pliocene sediments). Its thickness is about 2 km, d = 2.5 to 2.7, V_P = 5 km/s. It consists of basaltic flows containing some consolidated sedimentary horizons.

Lower oceanic crust: Thickness 5 km, d = 2.8 to 2.9, V_P = 7 km/s. Its nature is uncertain since boreholes have not reached it. Dredge samples and subsurface observations (Goringe bank, SW of Portugal, Fig. 99) on oceanic open fault cliffs have indicated basalts, metamorphosed gabbros, amphibolites and serpentinised peridotites.

It was long thought that this layer is located in the extension of the subcontinental 'basaltic' layer. In fact, the situation is quite different, since the *ophiolites of the orogenic zones*, which represent fragments of the ancient oceanic crust broken by tectonics, have facilitated observation of a complete section of the oceanic crust. One can find both the *upper oceanic crust* (alternation of sediments and basaltic flows) and the *lower oceanic crust*. In other words, the section shows, from top to bottom (Fig. 2), the following sequence:

- Massive dolerite layer made up of a dense swarm of basaltic dykes, manifestly the source for overlying volcanism;
- *Gabbroic suite*, with cumulate texture[1];
- *Layered ultrabasic cumulates* (partly serpentinised peridotites) which form the base of oceanic crust; they are underlain by peridotites of the upper mantle, which are distinctly different from the former due to their texture of tectonites.

The *palaeo-Moho*, as in the continents, is difficult to observe since all these peridotites are strongly serpentinised and correspond to a zone of differential sliding between the crust and the mantle.

Moreover, all the ophiolites of the mountain chains of the Alpine cycle do not exhibit the ideal succession mentioned above. In the Alps, for example (Fig. 2), these are more often represented by serpentinised peridotites intersected by gabbros and diabase veins in an irregular manner. The upper volcanic flows may be completely missing. Oceanographic studies have indicated that this disposition is frequent on the slow ridges (such as Atlantic) in which serpentinised peridotites appear as outcrops.

---

[1]Rocks resulting from the gravitational accumulation of crystals in the midst of the magma. They are often layered.

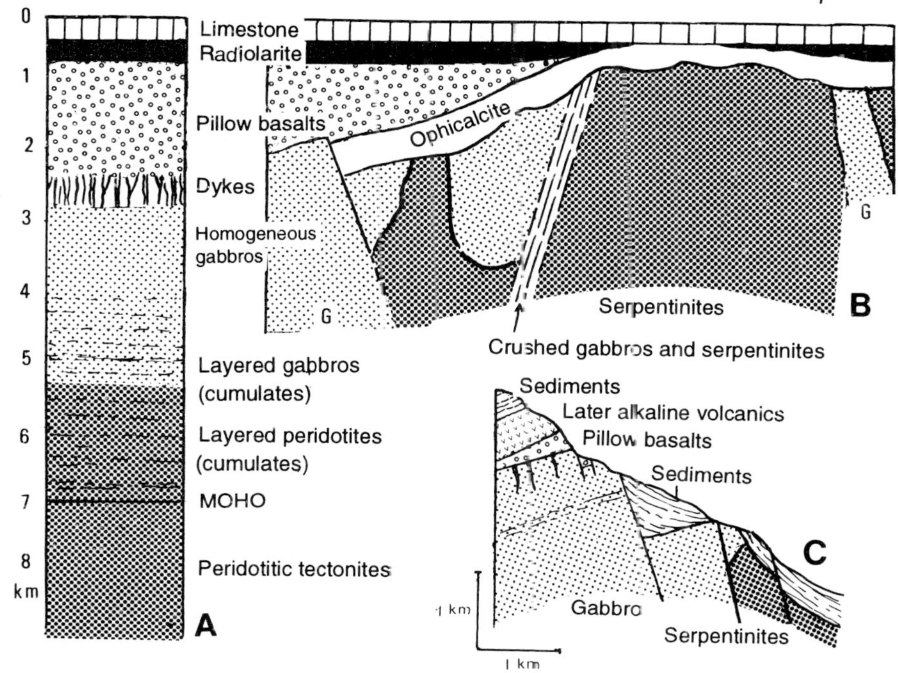

Fig. 2. Oceanic crust and ophiolites.
A—Cross-section of a classic ophiolite sequence. Oceanic crust is generally compared with such a sequence. The thicknesses given are only approximate and may be variable. When the thicknesses are large, this probably represents a sequence emplaced around hot spots (see pp. 40–44).
B—Abnormal ophiolitic suite observed in some sectors of the Franco-Italian Alps. It is characterised by extreme reduction of pillow basalts and gabbros (serpentinised peridotites forming the sea bottom) and the presence of serpentinous breccia with white calcite cementation (ophicalcites) whose origin (sedimentary or hydrothermal) is uncertain. Thickness of the section, about 1 km.
C—The Goringe bank, offshore of Portugal (see Fig. 99). It represents an oceanic lithospheric slab, tilted by about 20°, situated in the surroundings of the Azores-Gibraltar fracture. Direct underwater observations have facilitated reconstruction of a section of the Atlantic oceanic crust. The alkaline volcanic rocks at the top of the section were subsequently emplaced and have no relation with the oceanic crust per se.

b) Oceanic ridges

These are suboceanic volcanic reliefs in the form of long rounded crests running all along the major oceans. Neither sediments nor lower oceanic crust occurs in this zone (Fig. 3). On the contrary, a sort of bulge has developed below the upper crust, in which the seismic velocity reaches 7.5 km/s. This can be interpreted as a magma reservoir that supplies the ever-intensive volcanism at the top of the ridge. It might thus represent an abnormal upper mantle partly invaded by the products of partial melting, which is the cause

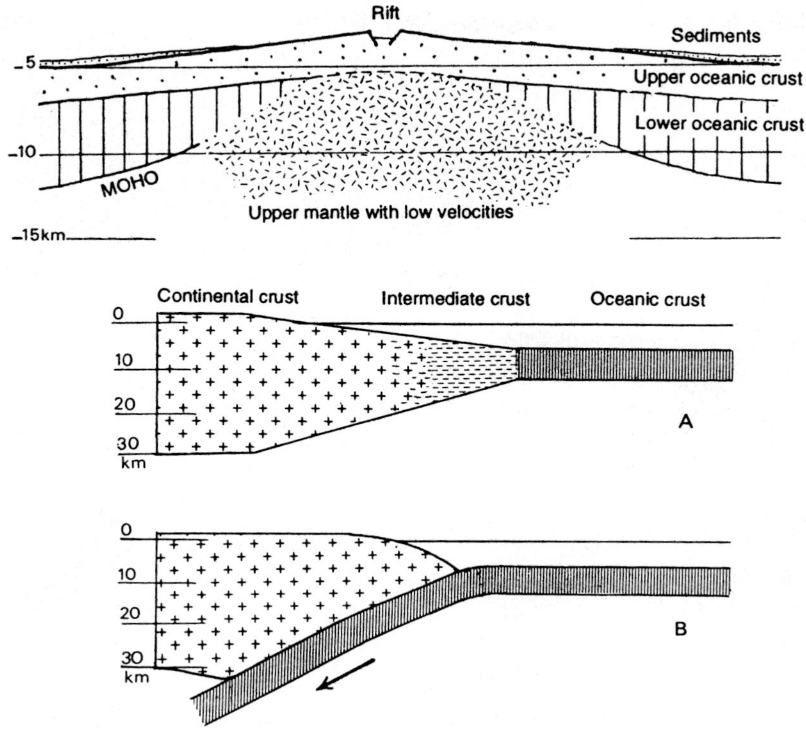

Fig. 3. Oceanic crust.
Top: structure across a ridge. *Bottom*: Its relations with the continental margins.
A. Juxtaposition: Passive margin (geophysical section: Compare with Fig. 98 in which data on the geological structure of these margins is introduced).
B. Subduction: Active margin.

of *low seismic velocities* and high thermal flux (contrasting with the low flux of the abyssal plains).

TRANSITION FROM CONTINENTAL CRUST TO OCEANIC CRUST

Two cases are possible (Fig. 3):

A simple juxtaposition: This type of transition is observed in 'atlantic' type margins (because frequent in the boundaries of this ocean). They are inactive (aseismic and non-volcanic). The continental crust gradually thins out towards the ocean (Fig. 3A), acquiring seismological characteristics intermediate between these two zones ($V_P = 6.5$ km/s). Hence it is referred to as an *intermediate crust*. Its origin is still not clearly known. However, it appears that due to the stretching induced by thinning out, the lower crust disappears and the granito-gneissic crust may be invaded by 'sills' of the local basic products of mantle origin.

A specific case of such juxtaposed margins are transcurrent fault margins (the contact is a strike-slip fault). In this case the thickness and characteristics of the continental crust are not affected and a sudden transition to oceanic crust occurs (p. 142).

Subduction surface (formerly known as the 'Benioff plane' or the 'Wadati-Benioff plane'): These are 'pacific' type margins as they are frequent in the boundaries of this ocean. The oceanic crust plunges below the continental margin along a contact marked by seismic activity from 20 km up to about 700 km (Fig. 3B). It was long thought that the seismic sources marked the surface of friction. In fact, they are situated in the middle of the plunging lithosphere and represent the internal tensions in the latter between 60 and 300 km. Beyond this zone, contrarily, the compressive regime predominates.

LITHOSPHERE AND ISOSTASY

Concomitant with the establishment by geophysicists of the existence of the crust and the mantle, it was proven that, excluding some points of the earth in rapid evolution by subsidence or folding, the field of gravity is roughly the same over the continents and the oceans and, consequently, a mechanism exists that regulates the irregularities of mass distribution or the nature of the crust above the mantle.

Based on the theory by the British geophysicist George Airy, it was established that this mechanism is hydrostatic. According to this theory, the crust 'floats' on the mantle and hence any high relief of the continental crust is compensated by a 'root' of the same crust buried in the mantle, exactly like an iceberg in the sea. This comparison is more justified when the ratio of densities of the two media is of the same order of magnitude in both cases. This hydrostatic equilibrium is known as *isostasy*.

Seismological studies have confirmed the existence of these roots, which may attain a thickness of 70 km and thus be double the thickness of normal crust below mountain belts. The oceanic crust is not more than 7 km thick and compensates the low density of overlying water.

Such a compensating mechanism implies that there could be fluid medium at depth, even if this fluid is extremely viscous. This medium was long considered to be the upper mantle. Now it is known that it constitutes the *asthenosphere*.

The asthenosphere is of the same chemical composition as the base of the lithosphere and differs from the latter only in physical properties. The corresponding boundary is at about the level of 1300°C isotherm. This boundary obviously is gradual and hazy, and, in fact, indicates very small partial melting of the mantle material.

Thus isostatic phenomena ought to be examined at the lithospheric scale and not solely at the crustal. From this point of view two types of isostatic compensations can be distinguished:

If the lithosphere is thick and rigid, when overloaded with a sufficiently voluminous object, it will react elastically by undergoing deformation over a vast surface area greatly exceeding that of the overloaded area around which a zone of depression appears. This compensation is known to be *regional.* A good example is provided by the vast volcanic complex of the Hawaiian islands. Here we find large accumulations of basaltic flows emerging from an ocean bottom of – 5000 m and culminating to + 4000 m, thus presenting a volcanic pile more than 9000 m high. Its weight has bent the oceanic crust up to a distance of 1000 km from the centre of eruption. Moreover, a ridge juts out around the depressed zone, indicating the elasticity of the curved lithosphere. To comprehend and visualise this phenomenon, a plexiglass ruler may be folded over the edge of a table: the ruler slides off the table and shows a slight bulge at the place of its bending.

The Himalaya, formed by large slabs of gneissic crust, piled up on the thick and rigid Indian lithosphere, is another example of this phenomenon. As a result, a depression 4 to 5 km deep exists at the India-Himalaya boundary. This depression is filled by sediments forming the Siwalik basin with a width of 200–300 km. As the compensation extends over a vast area, the root of such chains may be relatively small (only 50 km in this particular case). This definitely applies to all the high mountain chains associated with foreland basins.

If the lithosphere is less rigid, the regional compensation will be much smaller due to petrographic, structural or thermal reasons. A montane relief, for example, will be compensated only by its root but the latter will be much thicker than for a belt folded over a thick and rigid lithosphere. In other words, the weight of the mountain is counterbalanced only by the Archimedes force acting upon its root according to the Airy model. Moreover, this equilibrium is precarious and lasts only as long as the conditions that cause it, such as compression, persist. When the compression ceases, the elevated relief stretches itself very fast by gravity and the root is rapidly obliterated[2].

The Tibetan plateau is a good example of this second type of compensation. Its altitude and its root of 70 km are only due to the pressure exerted by the Indian lithosphere against the Asian plate (see p. 279).

[2]For further discussion of this subject, see pp. 82 and 283.

PART ONE

OCEANIC STRUCTURES

A major part of the ocean bottom consists of *abyssal plains* at –5000 m on average. These are nearly planar vast surfaces. Geophysical data shows that they represent the top of the undisturbed oceanic crust, covered by a generally thin layer of pelagic sediments. The uneven reliefs observed in these plains are most often of volcanic origin and can be divided into two groups according to whether they are seismically active or not.

CHAPTER 1

SEISMICALLY ACTIVE OCEANIC RELIEFS

RIDGES

Morphology

The most renowned oceanic ridge is the Mid-Atlantic Ridge (Figs. 4, 5, 7, 9), with a width of 1000 to 2000 km and an average elevation of 4000 to 2500 m from the bottom. Some parts of the ridge emerge on the surface, thus giving rise to the volcanic islands consisting of tholeiitic basalts (Jan Mayen, Iceland, Azores, Ascension).[1]

The disposition of the ridge in the middle of the ocean is highly remarkable but cannot be generalised for all ridges. The Indian Ocean Ridge, for example, is in the mid-point of the ocean only in its southern part, though its morphology is similar to the Mid-Atlantic Ridge. In the northern part it penetrates into the Gulf of Aden and continues through the Red Sea (see p. 21).

In both cases the top of the ridges is marked by a continuous trench, the 'rift', with an average depth of 1000 m and a width of 10 to 50 km. It is surrounded by seismically active tension faults marked by volcanic flows. These faults intersect its walls in a quite complicated mosaic. Moreover, the rift is cut into segments, like the ridge which holds it, by a number of transverse fractures which offset it into as many segments. It is these fractures which are also the foci of earthquakes (see pp. 20–24).

The South Pacific Ridge extends, as its name indicates, in the southern part of the Pacific Ocean, trending NE and is marked by Easter Island (Figs. 5 and 11). It thus approaches the American continent near California. Broken by a major strike-slip fault (the San Francisco or San Andreas fault), it is pushed along the Canadian coast (the Gorda and Juan de Fuca ridges) and disappears (for the modes of this disappearance, see Figs. 13 and 197).

---

[1] However, these structures correspond in fact to 'hot spots' coinciding with the ridge (see p. 40 and Fig. 25).

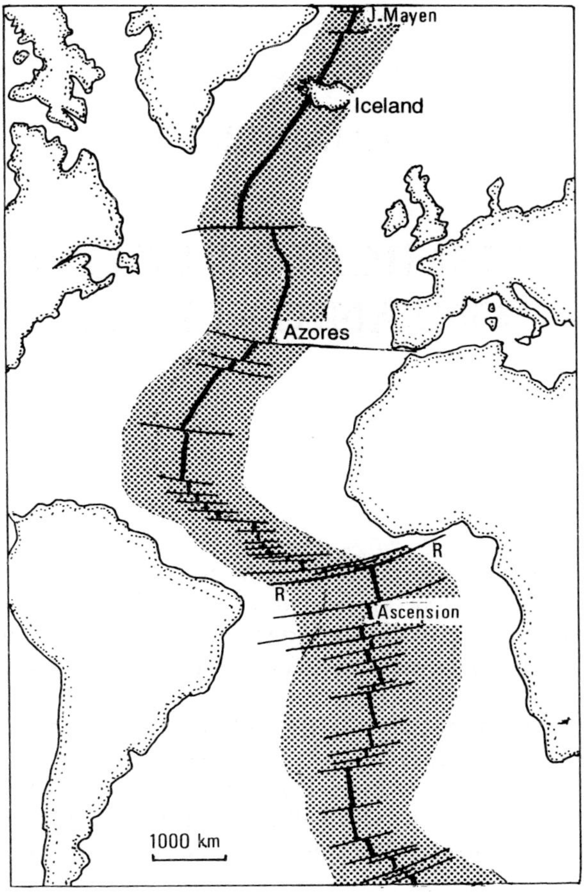

Fig. 4. The Atlantic ridge (rift: thick line). RR—Romanche fault.

This Pacific ridge differs from the Atlantic and Indian ocean ridges by the absence of the rift. It is separated into two other branches, one towards Chile (the Chilean Ridge) and the other towards Central America by the Galapagos and the Coco islands (Galapagos Ridge, with a rift).

The *volcanic structures* on the top of the ridges rarely show cones and in no case classical craters, so long as the volcanism remains submarine. Moreover, the axis of the rift in some ridges is indicated by isolated monticules hardly exceeding 300 m in height. Eruptions are essentially fissural, giving rise to digited mounds or pillows (bulges of lava piled one above the other). Neither explosions nor projections occur, as the gaseous phase remains dissolved in the magma due to the enormous pressure (200 to 300 times the atmospheric pressure) exerted on the lava.

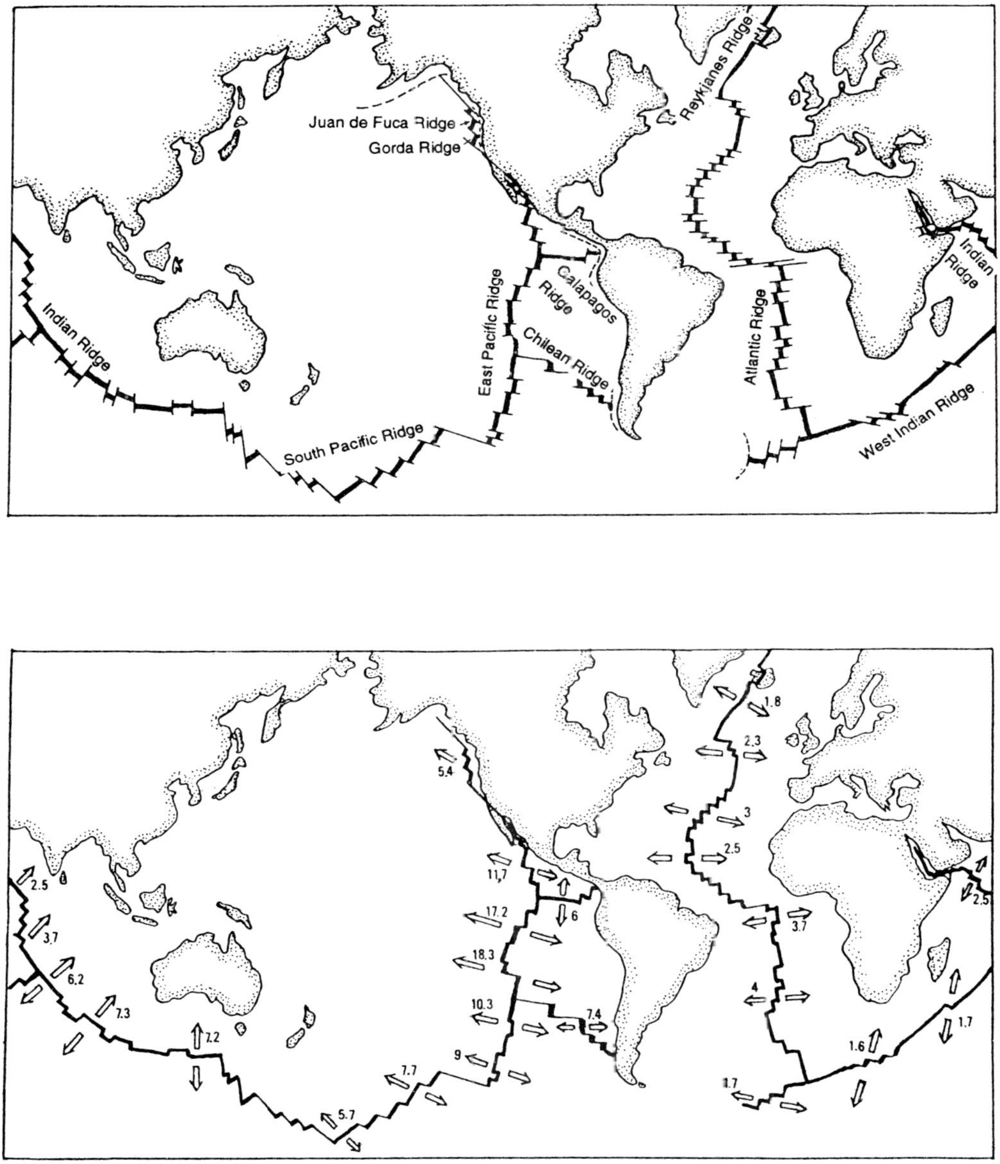

Fig. 5. Oceanic ridges and the corresponding accretion zones (direction and velocities in cm/yr).
Note the velocity differences between the Atlantic (slow motion) and the Pacific (rapid motion) and the gradual variation of velocities along some ridges (Indian ridge, for example, from west to east), which indicates the position of the measurement points relative to the pole of rotation.

In some ridges, however, normal planar basaltic flows may be observed, covering large areas and filling the topographic hollows.

The *hydrothermal activity* of some ridges was one of the major discoveries of the 1980s. The bottom of the rift in such cases shows real geysers of hot water (350°C in the Galapagos rift) concentrated with diverse metallic sulphides (Fe, Cu, Zn). It may also be built up of columns or irregular mounds, several metres in height, decked with crystals around which varied fauna often settle down.

The source of these hydrothermal springs is the sea water that has infiltrated into the fractures of the young oceanic crust and descended up to the vicinity of magma chambers, where it becomes heated and its pressure builds up. The hot water rapidly ascends through other fractures, dissolving in its passage the metals and sulphides contained in trace amounts in the already consolidated basalts.

Volcanism and hydrothermal activity obviously produce large heat flux, which is dissipated at the ridges, whereas the abyssal plains have very low heat flows.

The profile along a ridge does not show a constant depth. The ridges are in fact made up of upwelled sections, 50 to 100 km long, which are connected by transverse faults (transform faults, p. 23) or by juxtaposition of spreading centres (overlapping spreading centres: OSC).

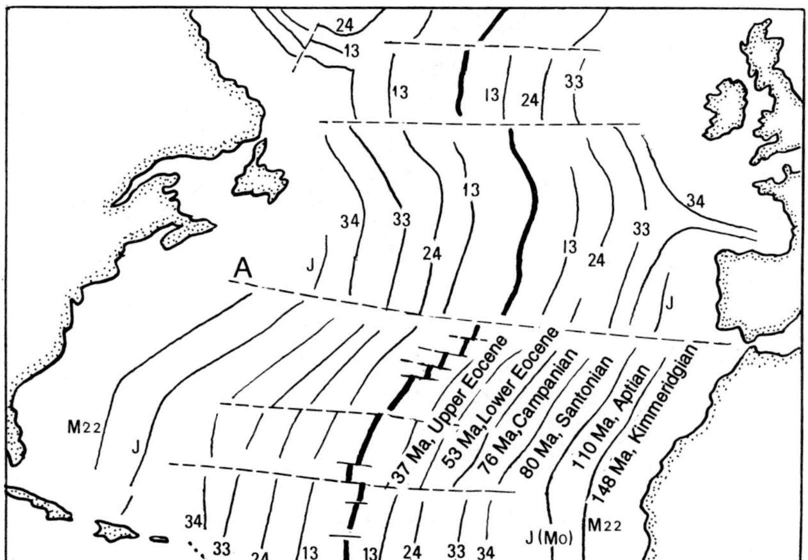

Fig. 6. Magnetic anomalies of the Central Atlantic. A—Azores-Gibraltar transform fault.

Significance of Ridges and Oceanic Accretion

It is now well known that the lateral slopes of the ridges consist of parallel basaltic bands, readily identifiable by their normal or reverse remanent magnetisation. These bands are distributed more or less symmetrically relative to the rift. Their age increases directly proportional to the distance from the rift, as indicated by palaeomagnetic, radiometric and conventional stratigraphic (age of associated sediments) methods of dating (Fig. 6).

A continuous deposition of basalt thus takes place in the rift axis over a relatively narrow belt of the order of 1 to 2 km, which represents the existence of extension fractures in the process of opening up. The deposition of basalt might compensate the rifting. But the latter being incessant, the basaltic band emplaced is in turn split longitudinally and its two halves diverge with a velocity which may vary from 1 to 20 cm/yr. This is the phenomenon of *oceanic accretion*, due to which the oceanic crust is constantly being formed. In this way the ridges constitute one of the three basic boundaries of the plates (Fig. 7), the other two being subduction zones and transform faults, which will be described later.

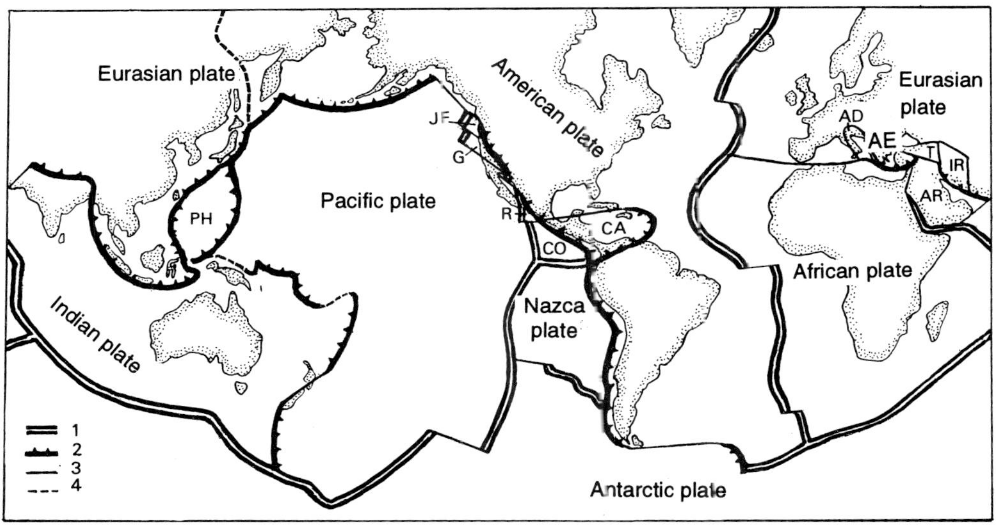

Fig. 7. Oceanic ridges (1), subduction zones (2) and transform faults (3) as plate margins. (4) Indistinct or uncertain margins.
AD: Adriatic plate, AE: Aegean plate, AR: Arabian plate, CA: Caribbean plate, CO: Cocos plate, G: Gorda plate, IR: Iranian plate, JF: Juan de Fuca plate, PH: Philippine plate, R: Rivera plate, T: Turkish plate.

Multiplying the mean rate of expansion by the length of the present-day ridge network, we obtain a figure of the order of 2 to 3 km^2 of oceanic surface created annually. As the total area of oceans is about 310 million km^2, it can be inferred that this ridge network has evolved over a period of one hundred million years (i.e., roughly, since the middle Cretaceous).

The rate of accretion along a ridge is not constant, at least when the phenomenon begins, with the result that it is fragmented during its formation into a number of more or less long segments separated by transverse fractures, as already mentioned above. Because of the accretionary phenomenon taking place on both their flanks, these fractures are seismically active, especially between the two offset segments of the ridge (see Fig. 12).

However, if these differences of behaviour are isolated as they rapidly become uniform, we can define a mean accretionary rate for each ridge. An accretion below 4 cm/yr is considered *slow* and in this case it has been proven that the ridge shows a rift (Atlantic, Indian Ocean ridges) whose axis is marked by small isolated volcanic cones. An accretion greater than 4 cm/yr is *rapid*, in which case the ridge shows no sign of rifting (South Pacific Ridge).

Slow ridges often yield basalts with porphyritic structure, i.e., with clearly visible phenocrysts. This indicates that the former had sufficient time to become isolated in the magma reservoir below the ridge. The basalts of rapid ridges are impoverished in phenocrysts (the fluid lavas of the lava pools are also completely impoverished in them). This attests to a very brief, close to zero, time of transition in the magma chamber.

Similarly, hydrothermal activity is feeble in the rifts of slow ridges and high in ridges in which the accretionary rate exceeds 6 cm/yr.

Ridge Rocks

The rocks of the ridges are essentially tholeiitic basalts of the type MORB (Mid-Oceanic Ridge Basalts) with 50% silica and low potassium content. These lavas are generated from the upper mantle by rapid decompression and partial melting associated with its rapid ascent.

Different types of lavas may also exist when a sufficiently vast magma chamber facilitates an intensive differentiation (rhyolites of Iceland and Tristan da Cunha, for example).[2]

The problem here is to find out what exists below this tholeiitic layer, 1 to 2 km thick according to seismic data, since boreholes have hardly exceeded 2100 m (2111 m in the Galapagos rift in 1994). Dredge samples of gabbros, serpentinites and amphibolites have certainly been obtained but their relations with the ridges cannot be observed. It is necessary to move

[2]The presence of such vast chambers generally indicates the existence of a 'hot spot' (see p. 40). However, the latter may also induce partial melting of the oceanic crust, constituting another possible origin of observed acidic products (Sigmarson et al., 1991, *Geology*, 19: 621–624).

away from them to find some dispositions which might provide an answer. This is notably the case of the Verma fracture zone (Fig. 4) or the Goringe bank, in the Atlantic, offshore of Portugal (Fig. 2). This section of the oceanic crust was recently uplifted and tilted eastwards. It is almost at water level (– 25 m) and its inclination facilitates the observation of several kilometres of its section despite a thick layer of sediments. Below the pillow basalts we find the following rock types:

— A suite of *vertical basaltic veins* mutually intersecting one another. These are probably the conduits of surficial volcanism.

— Very (4 km) thick *gabbros*, often 'flaserised' (i.e., of gneissic appearance, hence deformed) and metamorphosed to amphibolites.

— *Serpentinised peridotites* through which the ancient Moho probably passed.

This succession is also found in most of the 'ophiolites' of the orogenic zones, which can be interpreted as slabs of the sliced oceanic crust, uplifted and incorporated in a tectonic structure during its folding. They sometimes show thick sections of more than 10 km, attesting to the fact that not only oceanic crust (6–7 km), but also part of the mantle occurs there.

Most ophiolites belong to the Alpine cycle (their age varies from 180 to 60 Ma). However, older ones are also known, notably in the Appalachians, the Massif Central or the Alpine basement (Chamrousse, near Grenoble), where they are Lower Palaeozoic, as well as in Morocco where they are Precambrian. As the crust of the present-day oceans is not older than 150 Ma, these ancient ophiolites, evidence of the disappeared oceanic crust, are particularly valuable.

The ophiolites enable us to obtain an idea about the magmatic processes that create the oceanic crust below the ridges (Fig. 8).

The rise of the asthenosphere[3] leads to a reduction of pressure. The less refractory part of the peridotitic material (clinopyroxene and accessorily plagioclase) enter into melting. This liquid, corresponding to 5–20% of the total volume of the mantle rock, is uplifted and accumulates in extension zones in one or several magma chambers, while the remaining material is deformed and gives tectonites. Depending on the degree of partial melting, these are represented by dunites (high) or harzburgites (moderate).

Some rare seismic data suggests that such partial melting commences at a depth of about 60–80 km and that isolated magma chambers appear around –20 km, which directly feed the surficial reservoirs.

In these reservoirs the dense ferromagnesian minerals (olivine and pyroxenes) may partially undergo gravitational separation, giving rise, on

[3]The discovery of vast circular structures (10–15 km in diameter) in the large ophiolitic massifs led to the hypothesis that this ascent of the asthenosphere takes place in the form of juxtaposed magma diapirs. Some authors attribute the segmentation of the ridges to this process.

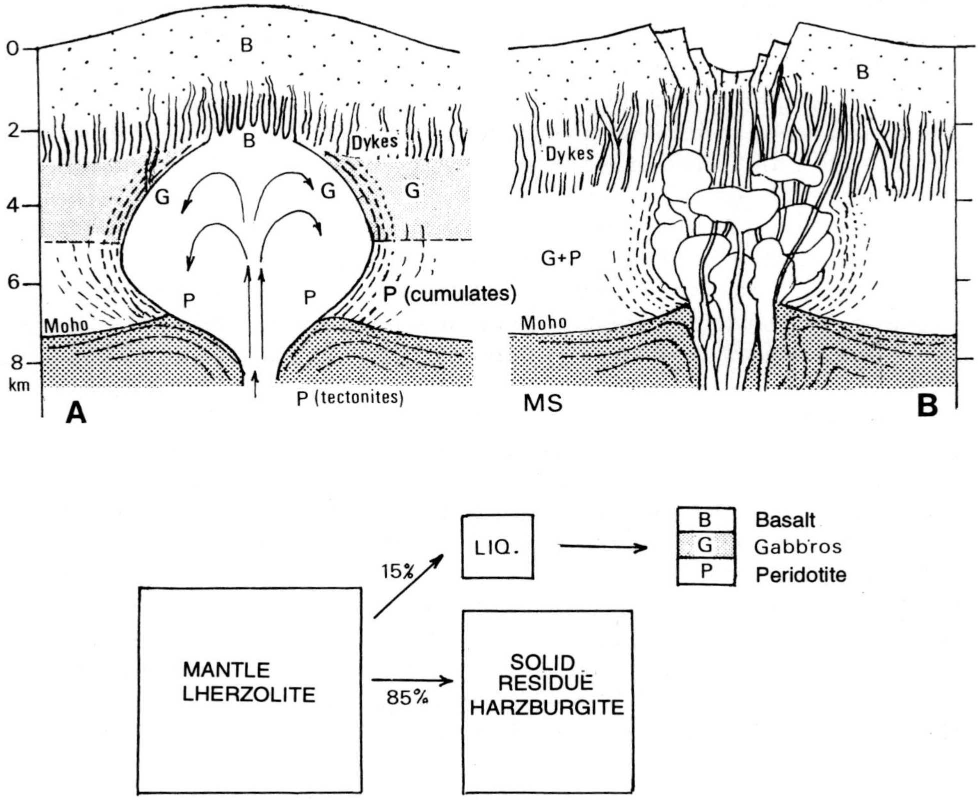

Fig. 8. Deep structural models of the ridges and corresponding magmatic differentiations. (A—rapid ridges, B—slow ridges). Horizontal scale arbitrary.
The content of the A-type magma chamber should not be considered as totally liquid. Recent data on seismic-wave velocities suggest that only the topmost bulge is the liquid fraction (position of the letter B). All the remaining chamber is only a 'sludge' loaded with numerous solid crystals in which the heavier fractions gradually undergo a gravitational stratification.

the bottom, to cumulate-type of peridotites rich in olivine and accessorily in clinopyroxene and plagioclase. Above this zone the magmatic liquid produces accumulative gabbros and basalts. The latter reach the surface through the numerous extension faults in the rift (dykes). Laterally, these different flows result, by cooling, in the rocks of the oceanic crust. However, their distribution varies depending on the type of ridge.

In the case of *slow ridges*, such as the Atlantic, it is found that the S-wave velocities are not reduced and, moreover, juxtaposed basalts, petrographically different from one another, are observed on the surface. Hence it is thought that small magma chambers exist (Fig. 8) which function intermittently, and that each is connected to the surface by a fracture. The oceanic crust that is built up laterally will thus show quite a thick zone of dykes and a certain irregularity in distribution of juxtaposed cumulates of gabbros and peridotites.

Rapid ridges, such as the Pacific, exhibit a reduction in S-wave velocities along their vertical section and, on the surface, a range of basalts that are more evolved but densely imbricated. Hence it is thought that in this case a single but larger magma chamber exists, which is constantly supplied by mantle material. Here crystal fractionation proceeds more readily with the result that the zones of peridotitic and gabbroic cumulates are thicker and more homogeneous, often with layered products.

The two types of oceanic crust thus produced are also found in ophiolitic suites.

Age of Ridges

The age of the oldest lava belts of the two flanks of the ridges under discussion does not exceed 80 to 100 Ma, which corresponds to the beginning of the Upper Cretaceous. This means that the present ridge network was emplaced during this epoch[4] (probably inducing one of the strongest known earth transgressions) and that since that time the ridges have enlarged, simultaneously widening the surface of oceans. This leads us to the phenomenon of sea-floor spreading.

Sea-floor Spreading

All the ocean floors, including the abyssal plains, are the result of accretion. Naturally, efforts have been made to verify this fact by attempting to find the characteristic basaltic bands. However, the sedimentary cover, which thickens towards the continents, and the absence of palaeomagnetic markers during the long period Albian to Santonian, make this research difficult.

Fortunately, the underwater drillings of the successive programmes of JOIDES (Joint Oceanographic Institutions Deep Earth Sounding), IPODE (International Programme of Oceanic Drilling Exploration) and ODP (Ocean Drilling Programme) have provided the ages of oceanic sections through those of overlying sediments. By and large the oldest sediments are found to occur in the vicinity of continental margins but they exist only here and there and not all along the oceans. Moreover, sediments older than the Jurassic are nowhere found (Figs. 9, 10, 11). All these facts lead to the following inferences:

1) The opening up of major oceans (or the renewal of their floor) is relatively recent and, in any case, not earlier than the Jurassic.

---

[4]Or still later, during the Eocene, as in the Indian Ocean (Fig.10), wherein a new ridge (East Indian ridge) has intersected, almost at a right angle, the southern part of the Cretaceous ridge which still remains active (West Indian ridge).

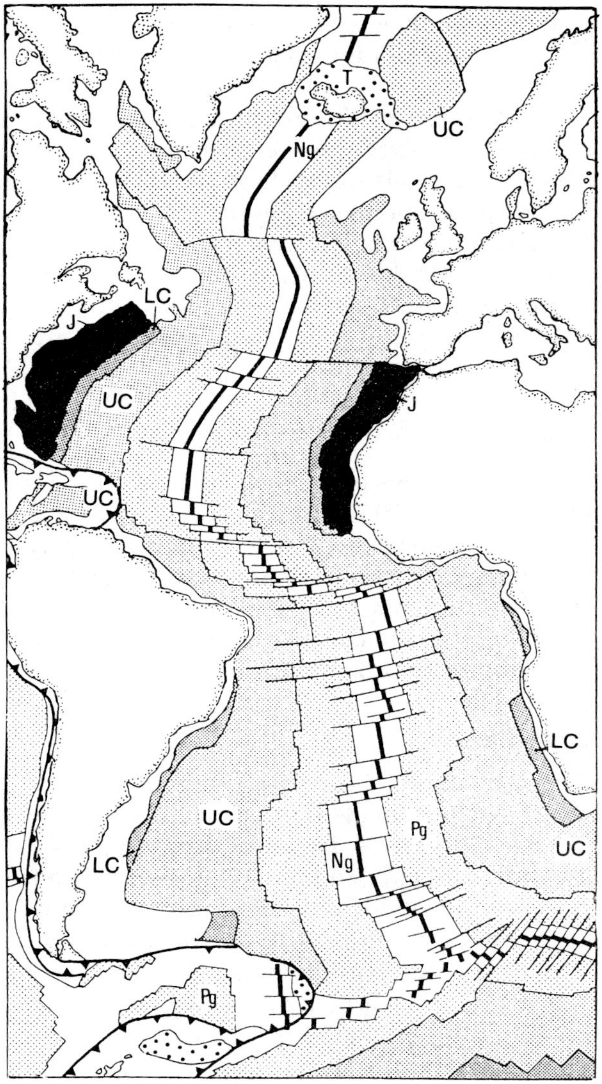

Fig. 9. Simplified geologic map of the Atlantic Ocean.
J: Jurassic; LC: Lower Cretaceous; Ng: Neogene; Pg: Palaeogene; T: undifferentiated Tertiary; UC: Upper Cretaceous;
The Jurassic is found only in the central Atlantic and marks the E-W trajectory of the ancient Tethys, which has disappeared everywhere else.

2) The oceans opened up through successive segments. Thus in the Atlantic (Fig. 9), Jurassic sediments exist only in the central part at Morocco and Mexico. This is explained by the fact that this segment is a fragment of the ancient Mesogea (or Tethys sea) which opened up from east to west. At

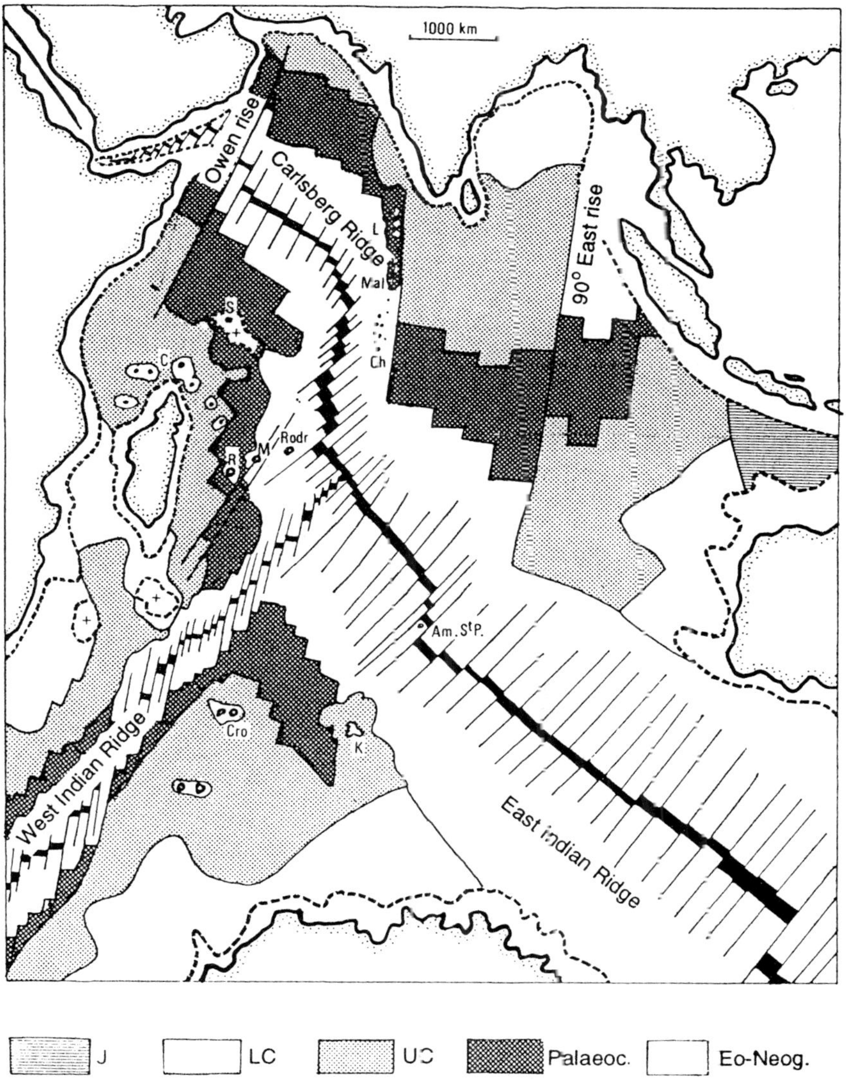

Fig. 10. Simplified geologic map of the Indian Ocean (from the Geological Atlas of the World at 29,000,000).

The relations between the ridges and the submarine crust varying in age at different points indicate that the West Indian Ridge has retained the same position since the beginning of the Cretaceous, and the Carlsberg Ridge since the Upper Cretaceous, whereas the East Indian Ridge (or the Indian Ocean Ridge as it is actually called) was emplaced in the Eocene, obliquely intersecting the earlier structures.

AmStP: Amsterdam-Saint Paul; C: Comoro Islands; Ch: Chagos; Cro: Crozet; K: Kerguelen; L: Laccadives; LC: Lower Cretaceous; M: Mauritius; Mal: Maldives; R: Reunion Island; Rodr: Rodrigues; S: Seychelles; UC: Upper Cretaceous.

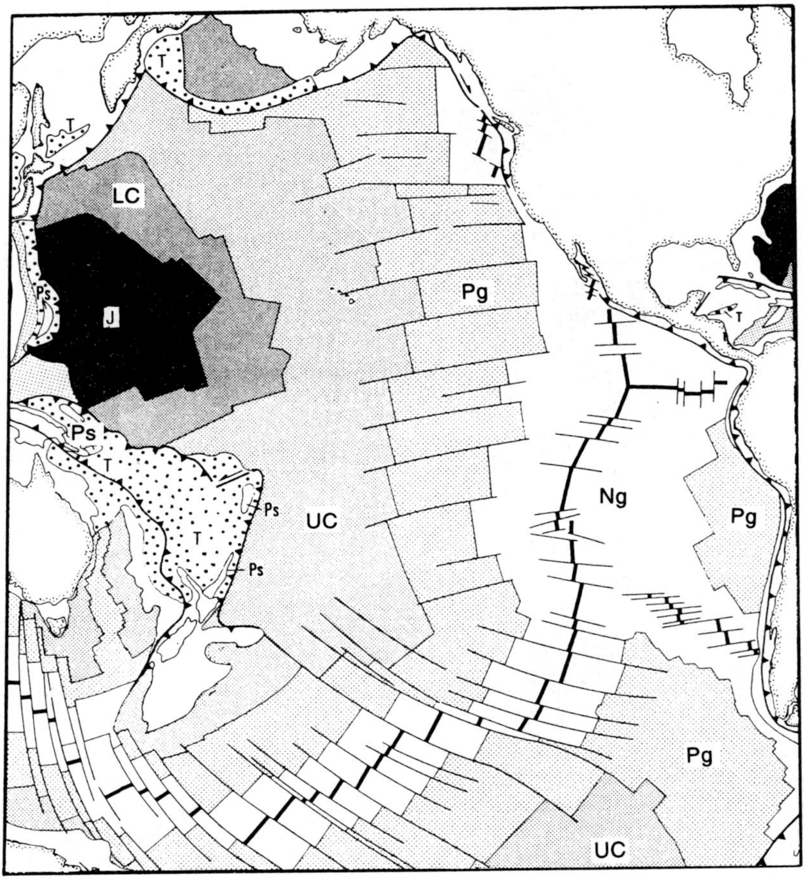

Fig. 11. Simplified geologic map of the Pacific Ocean.
J: Jurassic; LC: Lower Cretaceous; Ng: Neogene; Pg: Palaeogene; Ps: Pleistocene; T: undifferentiated Tertiary; UC: Upper Cretaceous.

the beginning of the Cretaceous, the South Atlantic and the North Atlantic opened up along a submeridional direction, perpendicular to the axis of the Tethys sea. This segment during its development up to the present pushed the Upper Jurassic and the Lower Cretaceous sediments to the coasts only of Morocco and Mexico.

3) Another consequence of ocean-floor spreading is the correlation between the depth of occurrence and the age of oceanic segments. In fact, as the oceanic crust spread away from the ridge, it concomitantly cooled, became denser and sank (*thermal subsidence*). A correlation exists between the depth of this crust and its age, and the same can be used as a dating tool.

Calculations based on the fact that the oceanic lithosphere loses heat by exchange with sea water without receiving it from elsewhere, have shown that the cooling and hence subsidence is proportional to the square root of time:

$$Z \text{ (depth in metres)} = 2500 + 350\sqrt{t} \ (t \text{ in million years}).$$

This relation pertains to an oceanic crust without sedimentary cover.
If overlying sediments are present, the relation becomes:

$$Z = 2500 + 350\sqrt{t} + 0.5\,H,$$

where H = thickness of sediments in metres[5].

This approximate calculation is valid only up to 70 Ma, as indicated by data on the North Atlantic and the South Pacific. Beyond 70 Ma the lithosphere being sufficiently cooled, a certain quantity of heat may be supplied by the underlying asthenosphere. The subsidence is therefore less and is fairly well expressed by the formula:

$$Z = 6400 - 3200\,\exp^{\left(\frac{-t}{62.8}\right)}$$

Transform Faults

The ridges are not continuous but broken by transverse faults which give rise to a number of small or large offsets along their axes. Thus the ridge axis is quite irregular. In fact it represents an image of the fracture from which the corresponding ocean has opened up. Its irregularity is related to the discontinuities of the original crust, especially to the pre-existing fracture network.

As oceanic accretion is produced on both sides of this fracture, its axis, with its successive offsets, remains constant (except when the rate of accretion is not the same for both segments, which is rare). Based on this characteristic, such faults are distinguished from the conventional strike-slip faults since the offset between two segments of the ridge does not vary (Fig. 12). The only movement observed in this case is that of oceanic accretion, i.e., the movement of the crustal rolling mat of the juxtaposed oceanic slabs.

It is only in the segment separating two parts of a ridge (segment AB of Fig. 12) that displacements are in the opposite directions and earthquake foci are observed. Beyond, i.e., on either side of A and B, displacements occur in the same direction so that seismic activity becomes feeble or is absent.

Transverse faults are morphologically indicated by an unevenness caused by the difference in subsidence of the two parts of the oceanic crust since they differ in age (and hence temperature and density), and also by narrow and deep troughs. For example, the Romanche fault[6] in the Central Atlantic is a 4-km deep trench which could have presented a natural cross-

---

[5] Z is the depth to the basement. The depth of the ocean bottom is $Z = 2500 + 350\sqrt{t} - 0.5H$, since obviously the thickness of the sediments heightens the sea bottom.

[6] Named after an oceanographic ship.

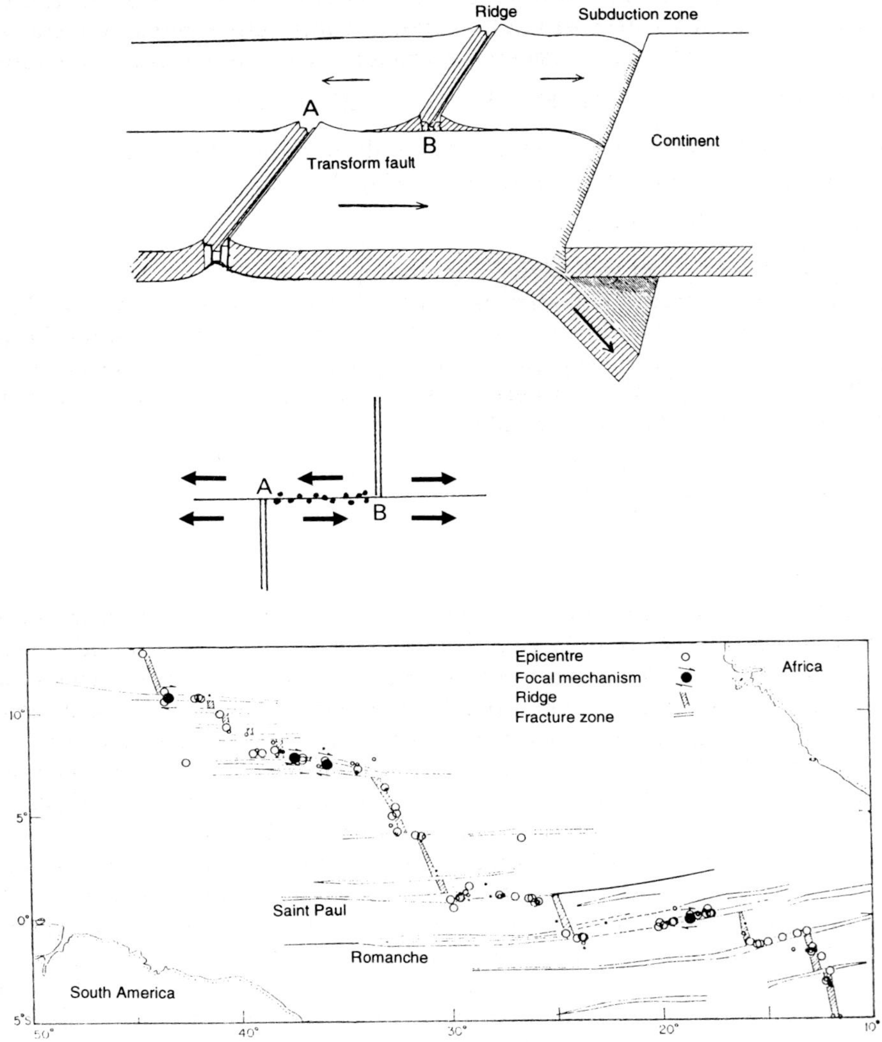

Fig. 12. Transform faults.
Along a ridge, between points A and B, sea-floor spreading occurs in opposite directions on either side of the fault, which is the cause of frequent earthquakes in this section. This is confirmed by the example chosen, the Atlantic Ocean Ridge, in the equatorial region (after Sykes, 1967).

section of the oceanic crust. Unfortunately, however, like all other faults of this type, the walls are matted with recent volcanic products or with serpentinite exuded from the altered peridotites of the lower oceanic crust.

These faults, totally different from the conventional strike-slip faults, are termed *transform faults* (term proposed by Wilson, 1965) since some of them terminate in a subduction zone. They thus form boundaries of an oceanic lithospheric slab which is generated at one end of their extremities and disappears at the other. In other words, following the faults, one can observe the *transformation* of an accretionary process into a subduction process.

However, the term has been extended to all faults with a horizontal component resulting from sea-floor spreading. Consequently, three groups of faults can be identified:

— Faults simply offsetting the rift of the ridge (*rift-to-rift faults*), such as the Romanche fault or the Saint Paul fault in the Central Atlantic (Figs. 12 and 104) or the Owen and Saint Paul-Amsterdam faults in the Indian Ocean (Fig. 10) or San Andreas fault (Fig. 13).

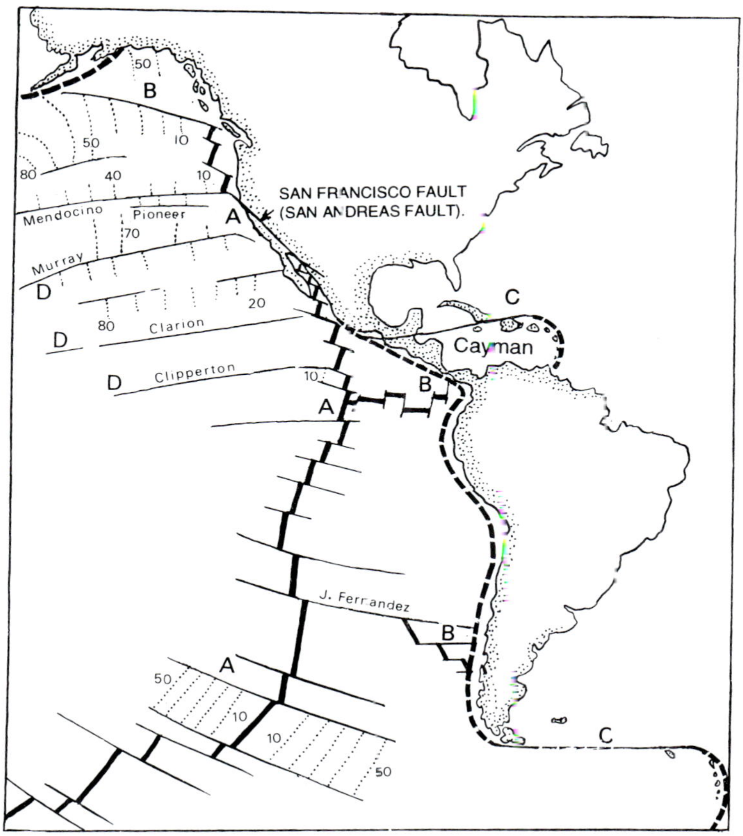

Fig. 13. Transform faults of the East Pacific Ocean.
A—rift-to-rift transform faults; B—rift-to-subduction trench transform faults; C—trench-to-trench faults; D—ancient transform faults, contemporary with a direction of spreading that was active up to about 10 Ma.

— *Rift-to-subduction trench faults,* which we have already mentioned as transform faults. Example: Juan Fernandez fault, forming the southern boundary of the Nazca plate, offshore of Chile (Fig. 13).

— *Trench-to-trench faults,* which are rare. The fault connecting Patagonia to the South Sandwich Islands as the south margin of the submarine continental plateau of the Falklands (Fig. 13) is a good example. One can also cite the North Caribbean transform fault zone (or Oriente fault) which runs along the southern coast of Cuba between the lower Caribbean and Central American trenches. This system presents a complex morphology. It is in fact marked by basins and ridges and shows a small oceanic region, the Cayman trench (Fig. 13), whose eastward continuation reveals extension faults or oblique en echelon folds in different areas (Fig. 14). All these faults have been inferred from the Seabeam and seismic reflection surveys. The suite of these structures has subsequently been compressed and overturned to the south. The structure is thus much more varied than transform faults which affect only ridges.

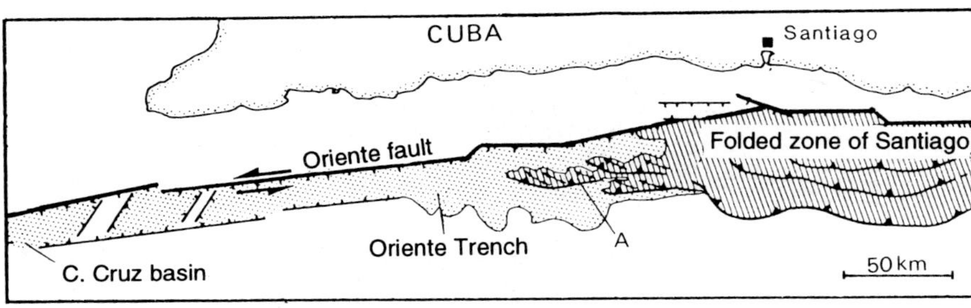

Fig. 14. Submarine trenches and diverse structures marking the North Caribbean transform fault (Oriente fault) on the southern coast of Cuba (after Calais et al., 1990, simplified).
From West to East, extension faults or en echelon folds (A) occur successively. South of Santiago, the material of the trenches also shows a folded structure with a southward vergence, representing a recent contraction.

ISLAND ARCS

Another type of seismically active relief is formed by the alignments of volcanic islands. It differs from ridges in two aspects: (i) volcanism, whose principal products are *tholeiitic* (known as arc tholeiites) to *calco-alkaline* and (ii) its association with a major submarine trench, parallel to the alignment of the islands, in which the depths may exceed 10,000 m. These major trenches represent the beginning of a subduction zone and hence a plate margin, its extension or compression regime depending on the location and the epoch.

A. Tonga-Kermadec or Mariana Type Arcs

Such arcs develop at any point of the oceanic domain, due probably to various reasons but mainly because of gravitational factors. Actually, on moving away from the ridge the oceanic lithosphere becomes denser than the underlying asthenosphere, which is of the same chemical composition but hotter. The equilibrium becomes unstable and a simple crack suffices to tilt one of the flanks and plunge it into the asthenosphere. It gradually assimilates into the latter.

Its subsidence into a hotter medium induces, as we shall see later (p. 156), partial melting in the overlying mantle at the surface of subduction, which results in the generation of more acidic and lighter magma. When these products ascend to the surface, they produce a volcanic island arc (arc tholeiites)[7], while granodioritic plutons are emplaced at depth.

Tonga-Kermadec archipelago, NW of New Zealand (Fig. 15), consists of tholeiitic and andesitic volcanoes, rarely rhyolitic, associated with volcanodetrital sediments and calcareous reefs of Eocene to Quaternary age. Recent extensional tectonics has tilted the Quaternary terraces.

Mariana Islands, between Japan and New Guinea (Figs. 15 and 16), are similar to the Tonga-Kermadec Islands. Their significance lies in the oceanic trench, largely open, which shows a number of extension faults. Sedimentation here is scarce and hence the volcanic rocks of the arc contiguous with the trench constitute the main outcrops in these islands, which represent a typical extensional system.

The New Hebrides (Vanuatu) and Fijian archipelagos present a slightly complicated disposition and history, consisting of the following successive stages (Fig. 15).

In *the Eocene* we have a single arc whose remnants are found in just the Fiji Islands, in the form of tholeiitic and andesitic lavas and volcanodetrital sediments deformed and metamorphosed to greenschist facies. This first group is also intersected by granodioritic and gabbroic plutons of Eocene age.

In the *Upper Miocene* a new period of activity occurred with the emplacement of more differentiated and more acidic lavas (andesites, dacites, rhyolites) unconformably lying upon the preceding group. This second phase has been interrupted by new compressive tectonics that has given rise to folded structures.

Extension phenomena recurred in the *Pliocene* and an intra-arc basin developed, separating the Fijian and Vanuatu arcs (Fig. 15, section A). All volcanic activity ceased in the former, which thus became a residual (or 'remanent') arc.

---

[7]With lower concentrations of incompatible elements (including Ti and P) and K_2O, by comparison with the MORB.

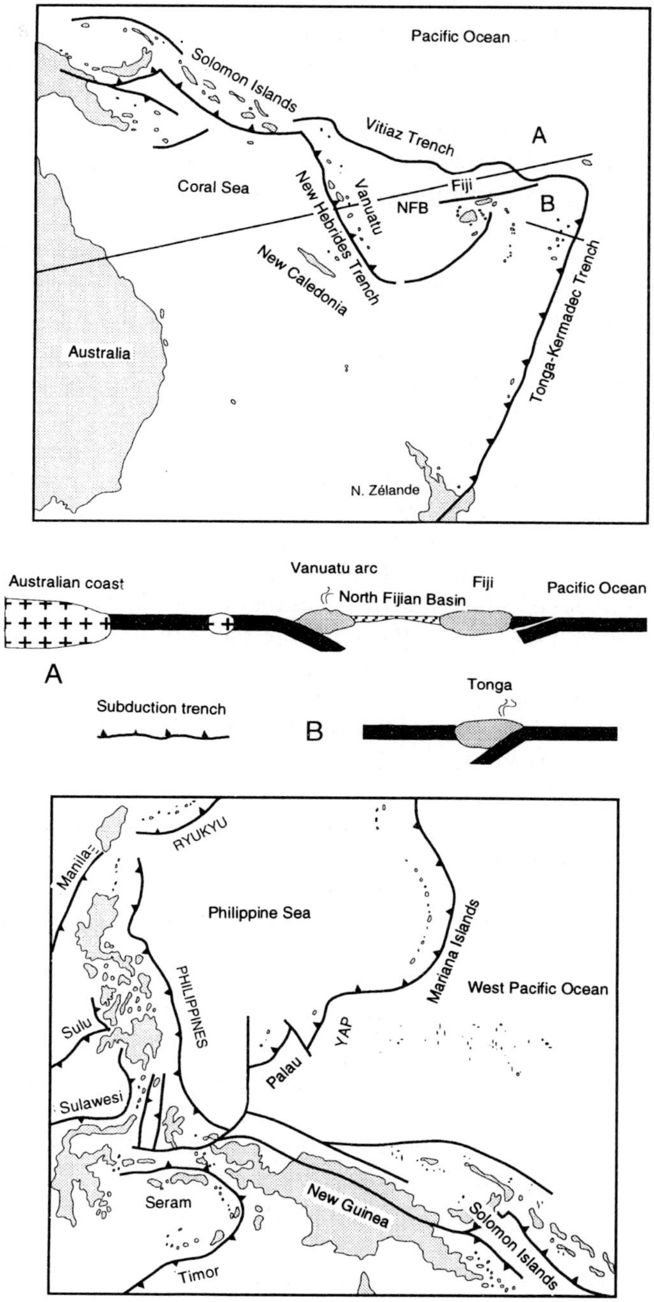

Fig. 15. Intraoceanic arcs of the West Pacific (cross-section after Lagabrielle, 1987).

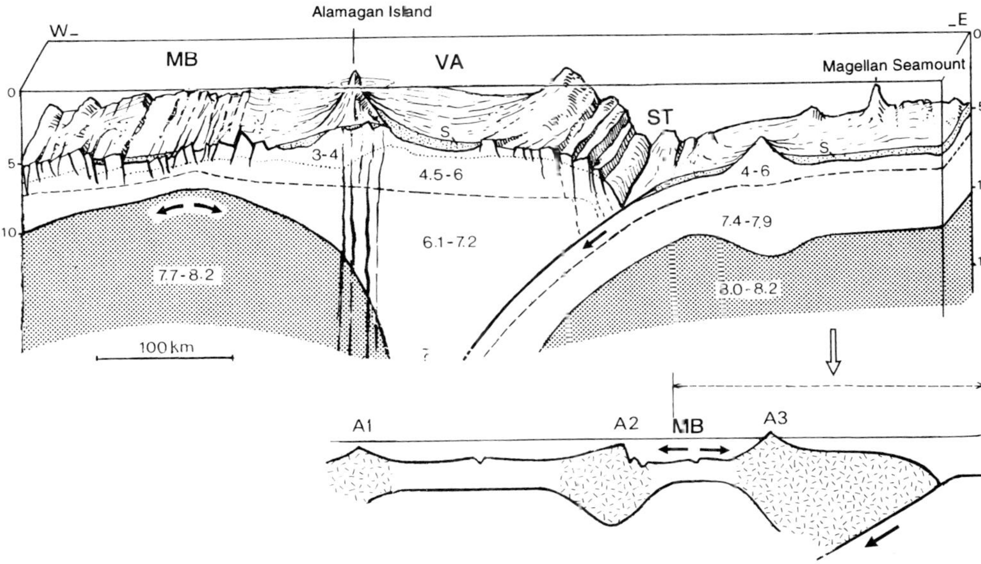

Fig. 16. Intraoceanic volcanic arc of the Mariana Islands (W. Pacific) (taken from Hussong and Uyeda, 1981, simplified).
ST: Subduction trench; VA: present volcanic arc (broken up into two ridges, one of them inactive, by a small intra-arc basin. (We have not detailed, for want of data, the deep structure of this arc, where granodioritic plutons ought to exist). MB: Mariana back-arc basin, probably in expansion, considering that the heat flow is somewhat higher than normal and that some inflows of fresh tholeiitic basalts occur.
Lower oceanic crust in light grey; upper mantle in dark grey. Numbers indicate seismic-wave velocities (in km/s).
Schematic section, given as annex, shows the position of the present-day arc (A3) in relation to two remanent (or residual) arcs: the West Mariana Ridge of the Neogene (A2) and the Palau Kyushu Ridge of the Neocretaceous-Palaeogene (A1). These arcs, at present inactive, are due to a process of comminution which often affects intraoceanic ridges by the development of 'intra-arc' basins.

This extension stage is significant and is found in other archipelagos, where it frequently gives rise to such residual arcs (see below). This indicates that the origin and evolution of these intraoceanic arcs take place in an *extension regime*.

During the *Quaternary* the Vitiaz Trench, north of the North Fijian Basin, was perhaps the priming of a new intracratonic subduction, with a vergence reverse to the preceding one. This subduction might explain the new and lighter basaltic volcanism which manifests in the Fiji Islands. A neotectonic activity has uplifted and tilted the erosion surfaces in the two archipelagos.

The **Lesser Antilles**, such as **Guadeloupe** and **Martinique** islands, represent a volcanic arc whose activity also started in the Eocene and gradually migrated westwards, thus moving away from the subduction trench, a process that often occurs above the subduction planes.[8] The substratum upon which volcanism has built up is not known, but it is probably the oceanic crust. In fact, at Désirade Island (eastern annex of Guadeloupe) the andesitic volcanism rests on a substratum of radiolarites (palaeontologically dated as Lower Cretaceous) and basic (tholeiitic) lavas, both metamorphosed to greenschist facies and intruded by acidic plutons (trondhjemites) of a radiometric age of Middle Cretaceous. Thus the basement of Désirade Island represents an intraoceanic Lower Cretaceous island arc subjected to tectonics and metamorphism before the Middle Cretaceous.

The importance of all these islands is that they indicate the *genesis of relatively acidic magmas from a basic oceanic crust.* Perhaps this represents the preparation for the build-up of a new continental crust. A question may arise as to whether the destination of the latter is to grow uninterruptedly at the expense of the mantle. This was certainly the case in the Precambrian. At present the balance appears to be rather static. In fact, isotopic tracer studies show that part of the continental crust joins the upper mantle, dragged along by the subduction process, whereas another, the deepest part, is gradually being 'mantlised' by digestion and intrusion of basic products.

Intraoceanic volcanic arcs are often associated with mainly submarine ridges parallel to them, which are either accumulations of sedimentary products, or residual arcs such as that of Fiji, already mentioned.

Sedimentary arcs are the products of accumulation, at the starting point of a subduction zone, of sediments continuously produced by the plunging oceanic lithosphere. Such an arc has been classically termed an 'accretionary prism'.

A good example of such a disposition can be observed in the case of the Lesser Antilles. The volcanic arc (Guadeloupe, Martinique, Saint Vincent, etc.) is doubled in the east by a thick accretionary prism that has been thoroughly investigated (by seismics, drilling, Seabeam, etc.). It consists of a series of piled up slices distributed in the form of an irregular fan (Fig. 17), as though the material brought about from the ocean were buried below the earlier material, thus gradually uplifting it.

Remanent or residual arcs: These are the inactive submarine ridges found at the rear of an active island arc from which they are separated by a small oceanic crustal basin. As the age and nature of their constituent materials are the same as that of the neighbouring active arc, it is considered that they represent strips of these arcs separated by rifting, i.e., by the appearance and development of an 'intra-arc' basin.

One can cite the example of the Aves Ridge west of the Lesser Antilles arc (Figs. 17A and 17B) or the residual arcs of the West Pacific (Fig. 18), which are more spectacular.

---

[8]The origin of corresponding volcanic products will be discussed with regard to island arcs derived from the evolution of an active continental margin (p. 154).

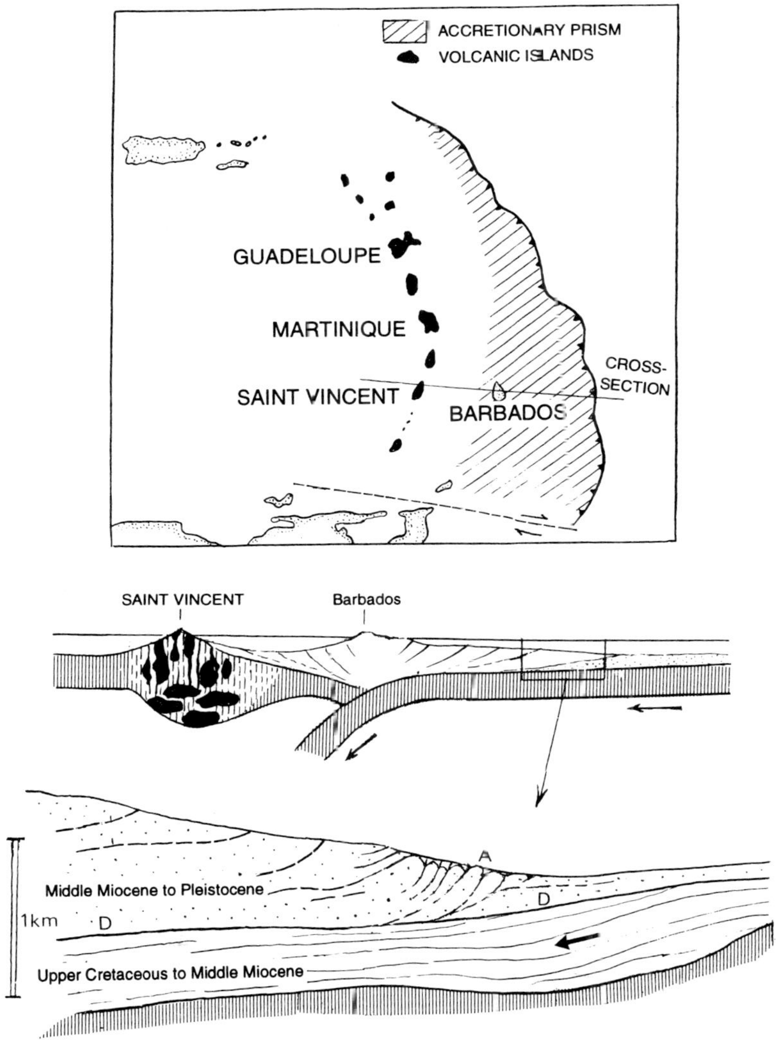

Fig. 17A. Intraoceanic arc of the Lesser Antilles.
Deformation is produced in the accretionary prism either along the nearly horizontal decollement surfaces (D) or by piling up of small anticlines separated by thrust faults (A).
A specificity of the Barbados prism is that its layers consist of only Neogene sediments whereas the sedimentary cover of the Atlantic basement varies from Quaternary to Upper Cretaceous. This peculiarity is due to cleavage of the said cover in the shales of the Middle Miocene. Only the overlying detached horizons enter into the prism. The underlying horizons (Lower Miocene to Campanian) remain at depth but it is not known whether they enter totally into subduction or undergo still further 'accretion'. Geophysical data and the recent discovery of Eocene-Oligocene sediments in the prism appear to favour the second interpretation as given in this Figure.
An additional remark regarding this cleavage is that an overpressure of fluids (25 kg/cm^2) is observed in the cleavage plane. This fact is significant since such overpressures have been used to explain the horizontal displacement of certain thrust sheets but are rarely detected in the thrust planes themselves.

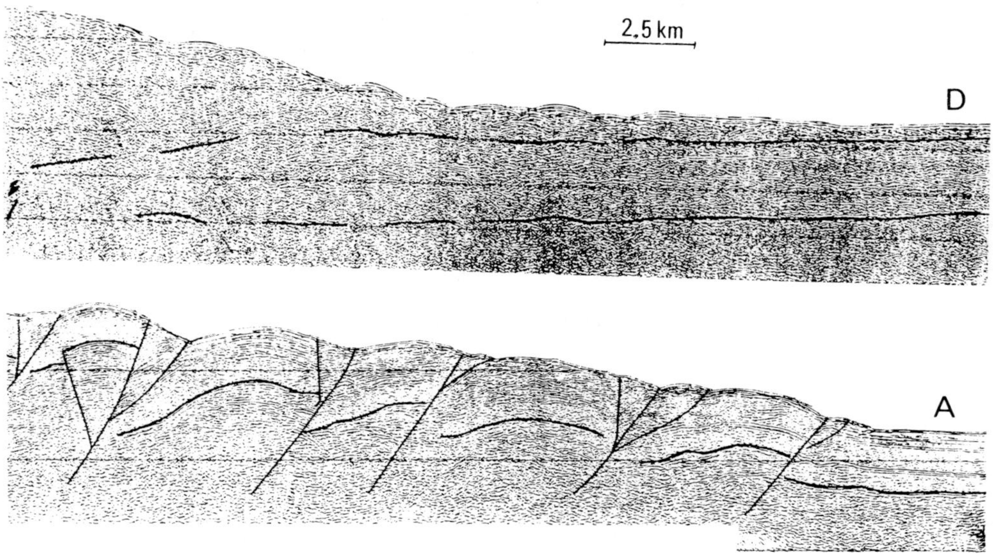

Fig. 17B. Seismic reflection profiles in the Barbados accretionary prism (after Biju-Duval et al., *Tectonophysics*, 1982).
These profiles illustrate the subhorizontal decollement tectonics (D) or the piling up of small faulted and overthrusted anticlines (A).

In the Philippine Sea (Fig. 18), from east to west, the West Mariana Ridge (1500 m depth, Neogene age according to DSDP drillings) and Palau Kyushu (–3000 m, Neocretaceous to Palaeogene) detached from the Mariana arc (Fig. 16).

Between the *Philippines and Borneo* the active arc extending from the Celebes to the Philippines (Sangihe arc) has given rise from S to N to the remanent arcs of Sulu and Palawan, respectively separated by the Celebes (Middle Eocene) and Sulu (Miocene) basins.

The *remanent arc of the Fiji islands*, as already mentioned, is separated from the New Hebrides by the North Fijian Basin and might combine with the Lau-Colville remanent arc behind the active Tonga-Kermadec arc.

On the contrary, the *West Melanesian residual arc* is isolated and represents an active arc on the verge of extinction.

The disposition of these arcs may be more complex locally. For example, *between New Guinea and the Solomon Islands* three arcs exist, with southward convexity (Figs. 18 and 19): New Britain, Woodlark and Pocklington, all three with the same Eocene basement. They clearly represent a dissociated arc whose frontal element, until a recent epoch, was the southernmost one, viz., the Pocklington arc, with its trench inactive however. The active trench of the system is now found at the foot of the New Britain Ridge in the prolongation of the Solomon and the Trobriand trenches. Thus this system is

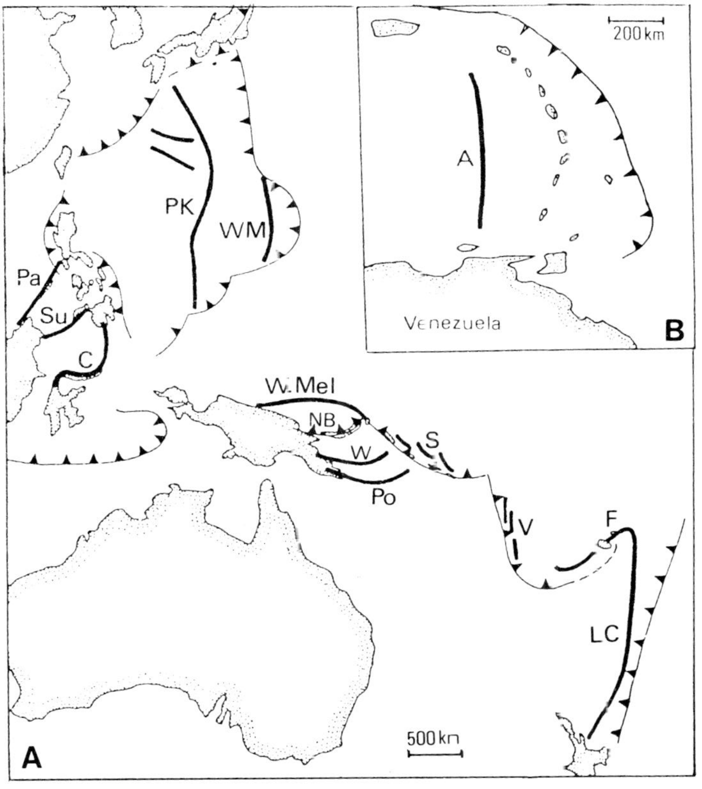

Fig. 18. Some examples of remanent arcs.
A—SW Pacific (after Karig, 1972, simplified).
Mariana system: WM: West Mariana, PK: Palau Kyushu.
Philippine system: Pa: Palawan; SU: Sulu; C: Celebes.
West Melanesian system (WMel): NB: New Britain; Po: Pocklington; S: Solomon; W: Woodlark.
Hebrides system: F: Fiji; V: Vanuatu.
Tonga-Kermadec system: LC: Lau-Colville.
B—Caribbean: Aves arc (A) west of the Lesser Antilles.

recently substituted, as an active ridge, to the West Melanesian arc, which is on the verge of extinction.

All these structural dispositions indicate a complex reorganisation of the regional stress field, probably associated with the modes of collision of the Pacific and Australian plates.

In any case the New Britain Trench and the Trobriand Trench will slowly engulf the small oceanic Solomon Basin, such that a collision between the New Britain arc and Woodlark arc will occur. We shall find such processes again in Part Three devoted to compressive phenomena.

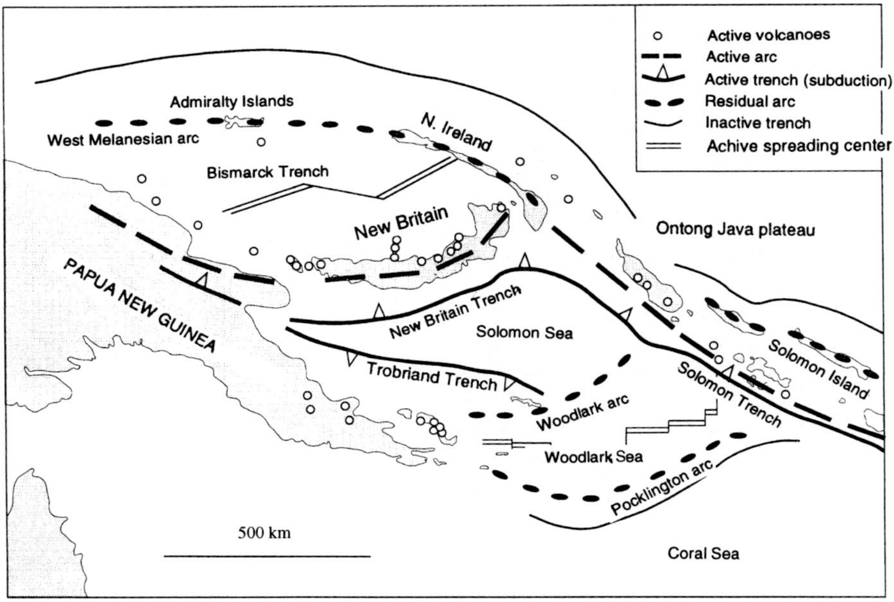

Fig. 19. Structural model of the West Melanesian archipelagos (after Karig, 1972, simplified).

B. Japan or Insulinde Type Arcs

These arcs, closer to the continents than the preceding ones, have a different and more complex origin. In fact, they show a granito-gneissic basement similar to that of the neighbouring continent because they have resulted from a specific evolution of the margin of the latter. Hence they will be discussed along with the active continental structures which produced them (see p. 145).

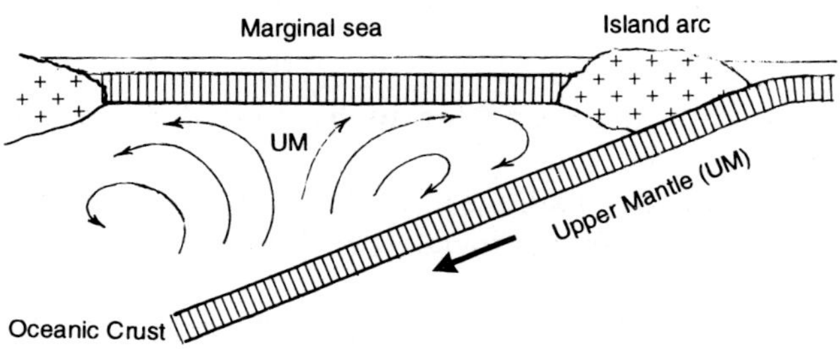

Fig. 20. Formation of a marginal sea by extensional opening of a continental margin under the effect of convection currents induced by the subduction of an oceanic lithosphere.

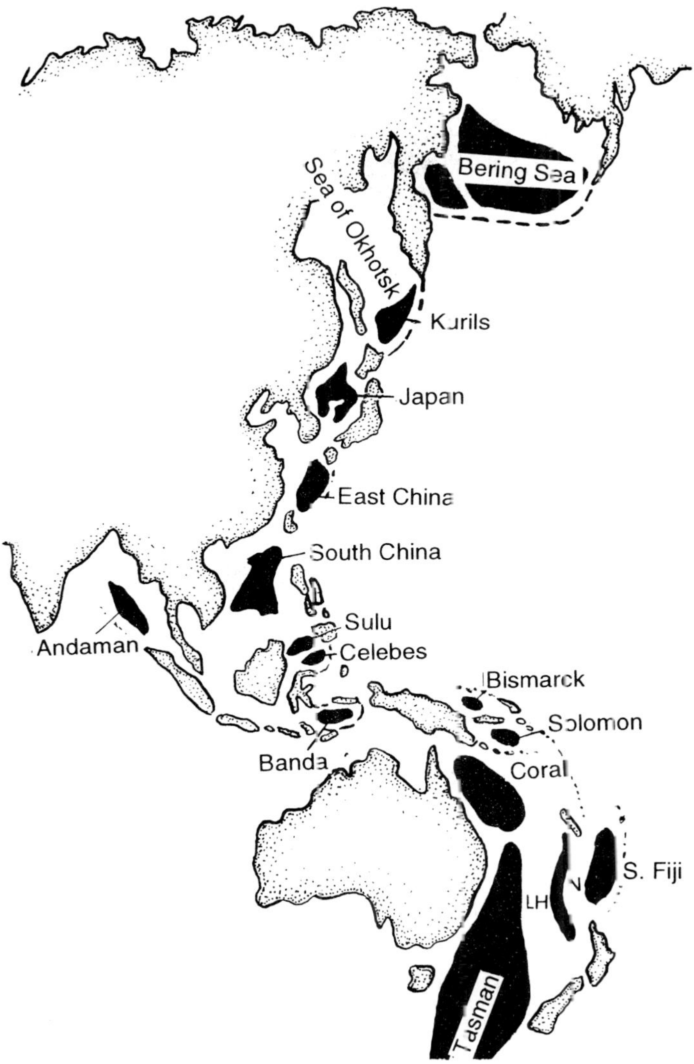

Fig. 21. Marginal basins of the West Pacific Ocean (LH—Lord Howe rise; N—Norfolk rise).

Fig. 22. Magnetic anomalies in the North Fijian Basin (after Malahoff, 1979, *in:* Auzende et al., 1988).

On the other hand, it is necessary to describe here the oceanic space which separates them from the continent and which is known as 'marginal sea'. Their origin is related to a very strong extensive evolution of the continental margin, resulting in its fragmentation and the appearance of a new oceanic type crust (Fig. 20). Such seas are frequent on the Asiatic margin of the Pacific (Fig. 21): Tasman Sea (between Australia and New Zealand), East China Sea (north of Taiwan), Sea of Japan, Sea of Okhotsk (Kuril basin), etc.

All these seas show a young oceanic type crust on a thin lithosphere, as in the vicinity of ridges. This fact is confirmed, among others, by gravity anomalies, high heat flows and low seismic velocities of the upper mantle. These characteristics indicate uplift of the upper mantle and the asthenosphere in these zones. The magnetic anomaly bands are often less distinct, rarely parallel to the near-by arc and those of the ocean beyond the arc (Fig. 22).

The marginal seas may contain fragments of continental crust, forming evidence of the original connection of the island arc with the near-by continent. We can cite the example of Yamato bank of the Sea of Japan (Fig. 23) or the Lord Howe and Norfolk rises between Australia and New Zealand (Fig. 21) as well as a part of the Mascarene plateau in the Indian Ocean. Their thickness is of the order of 10 to 15 km. Their margins also show normal step faults, like all passive margins.

Even though submarine volcanoes might disturb the morphology of basins, they show a flat sea floor, like that of abyssal plains but shallower. A direct correlation exists between their dimensions and their depth except when they occur in tectonic intersections, such as the Celebes Sea or Coral Sea (Fig. 24).

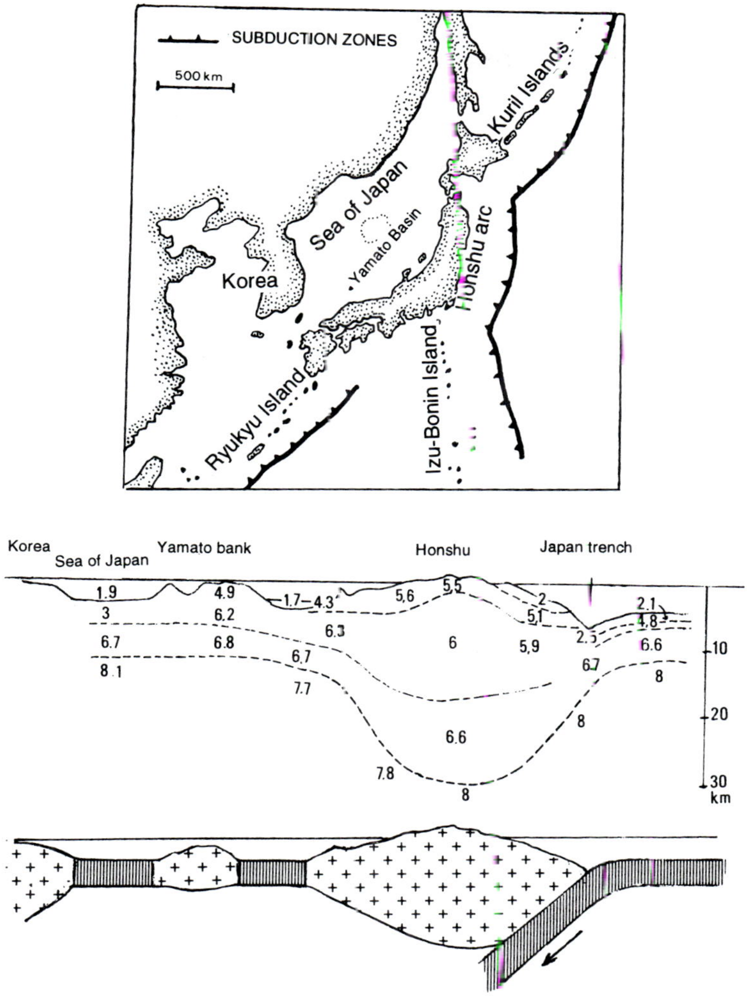

Fig. 23. Island arc of Japan.

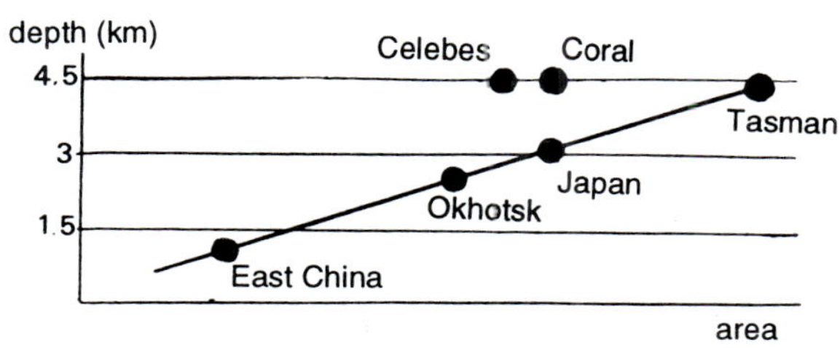

Fig. 24. Relation between the depth and area of the marginal seas.

The sediments of the marginal seas are generally very thick, which explains their flat sea floor. All the sediments are of Neogene age (except in the Tasman Sea where they are known to be Upper Cretaceous). This confirms the younger age of these basins, which could be the consequence of the folding of the Tethyan chains with reorganisation of the global stress field.

Seismic data shows that these marginal basins are *only situated directly above the subduction zones with steep slopes*. This is an obvious condition so that there exists a sufficient volume of upper mantle capable of giving convection currents thought to be the cause of these basins (Fig. 20). It is hence not surprising to find them in the West Pacific where the ancient (Jurassic and Cretaceous) thick and dense oceanic lithosphere subducts with steep slopes, contrary to the situation in the opposite margin of this ocean.

However, physical (analogic) modelling results (Chemenda, 1993) suggest that a coupling exists between the two lithospheres, the upper and the subducted, such that the subduction of the latter leads to fracturing of the upper lithosphere and opening of the back-arc basin in the region of a pre-existing weak zone.

CHAPTER 2

SEISMICALLY INACTIVE OCEANIC RELIEFS

VOLCANIC RELIEFS

• *Isolated oceanic volcanoes*: While some of the isolated oceanic volcanoes occur as emergent reliefs, such as the Bermuda Islands, those of Cape Verde, the Canary Islands and Madeira in the Atlantic (Fig. 25), Tahiti in the Pacific, Reunion or Mauritius in the Indian Ocean (Fig. 26) and Coco Island on the West Indian Coast the greater part is under water.

The origin of these volcanoes is variable.

Some were generated on a ridge and were later pushed towards the near-by continents by the sea-floor spreading process, with or without losing their activity according to the conditions of the fractures in the ridge. Examples are Saint Helena (inactive) and Tristan da Cunha (active), in the South Atlantic and the Bermudas (inactive) in the North Atlantic (Fig. 25).

Other oceanic volcanoes, on the contrary, have nothing in common with a ridge and mark the common fractures of the oceanic lithosphere. Moreover, their products are distinctly alkaline and non-tholeiitic (Canary Islands, Tahiti, Reunion and others).

The form of these underwater volcanoes is conical or truncated conical: in the latter case they are known as guyots or seamounts. Their flat top was formerly thought to be related to their emergence and subsequent erosion. However, it now appears that the cake-shaped form is of magmatic origin, the pressure crushing down the extrusion of a pasty lava. Accordingly, these forms are specifically submarine.

• *Aligned oceanic volcanoes*: These differ from ridges by in their seismic inactivity,[1] the presence of an underlying oceanic crust of normal thickness and the fact that only one extremity of the alignment has active volcanoes. Such is the case of the Hawaiian or Tuamotu-Pitcairn islands (Fig. 27). However, many of these alignments have now become extinct, such as the entirely submarine 90° East meridian ridge in the Indian Ocean (Fig. 26).

---

[1]Excluding, obviously, the small earthquakes associated with displacement of the magmatic material.

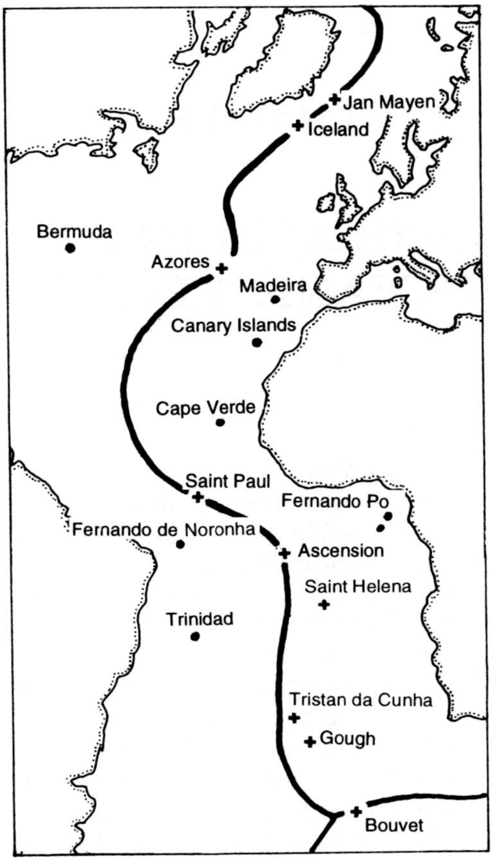

Fig. 25. Volcanic islands of the Atlantic Ocean.
All the islands are associated with hot spots but some are situated or generated on the ridge and hence have primarily tholeiitic volcanism (crosses), while others are located on the oceanic crust and their volcanism is alkaline (dots).

The origin of aligned volcanoes (as that of isolated volcanoes) is explained by the theory of 'hot spots'. They would indicate the existence of asthenospheric or still deeper hot spots that trigger the rise, in a vertical direction, of a column of molten products which ascend across the lithosphere towards the surface like a plume of steam. The alignments of volcanoes thus indicate the displacement of the lithosphere relative to the underlying hot spot and even give the direction of this displacement.

The *Hawaiian Islands* (Fig. 27) constitute a classical example of aligned volcanoes. While the volcanoes of these islands are still active at the SE extremity of the ridge, those situated more NW are extinct. The age of the lat-

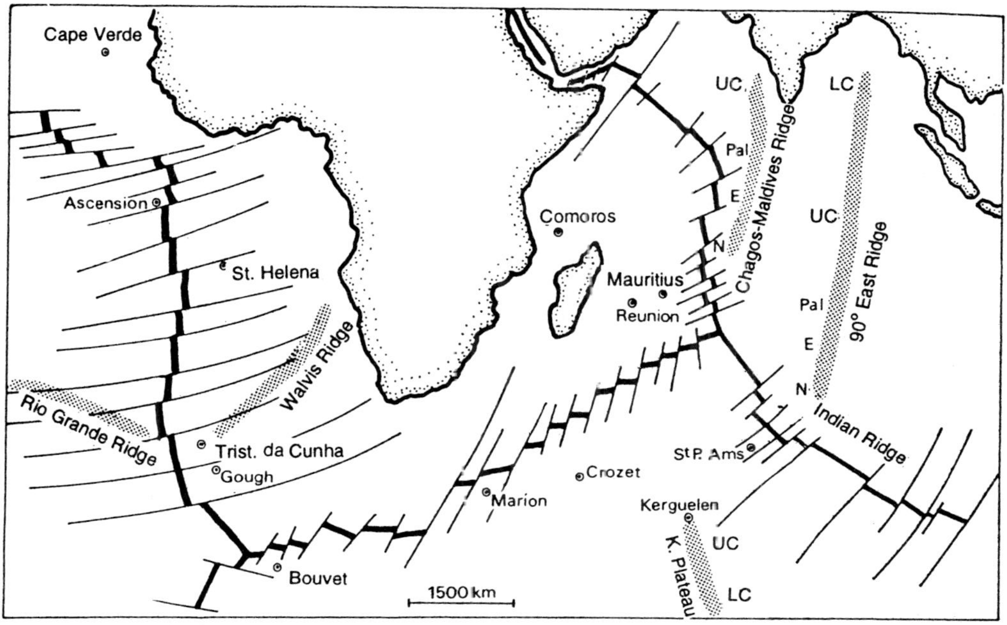

Fig. 26. Submarine ridges associated with hot spots in the South Atlantic and Indian oceans. As in the case of the Hawaiian Islands, these ridges indicate the direction of motion of the oceanic lithosphere that carries them. Most significant are the Chagos-Maldives and the 90° East meridian Ridges in the Indian Ocean. They indicate the northward migration of the Indian plate which ended in the Himalyan collision. These two ridges have certainly been produced by the hot spots of Reunion and Mauritius islands on the one hand, and the Saint Paul-Amsterdam on the other. But they have been separated by the later appearance of the Indian Ridge (see Fig. 10).
In the South Atlantic the nearly symmetrical ridges of Walvis and Rio Grande have emerged from a hot spot that coincides with the ridge. The Walvis Ridge highlights the northward ascent of the African plate which resulted in the collision of the Mediterranean mountain belts. The age of the oceanic crust that carries the submarine ridges has been indicated (E: Eocene, LC, UC: Lower and Upper Cretaceous, N: Neogene, Pal: Palaeogene).

ter is 40 Ma at the bend of the ridge and 70 Ma at the northern extremity, at the contact with the Aleutian trench where the ridge is subducted.

Such a bend is also found to be manifest in other archipelagos in the Pacific Ocean, such as the Line and Tuamotu islands or the Marshall, Gilbert and Austral islands (Fig. 27). Such a disposition represents a change in direction of displacement of the Pacific plate following a reorganisation of the ridge system since 40 Ma. It is thus a good confirmation of the fact that these volcanic alignments are related to a movement of the whole lithosphere since all these archipelagos belong to the same Pacific plate.

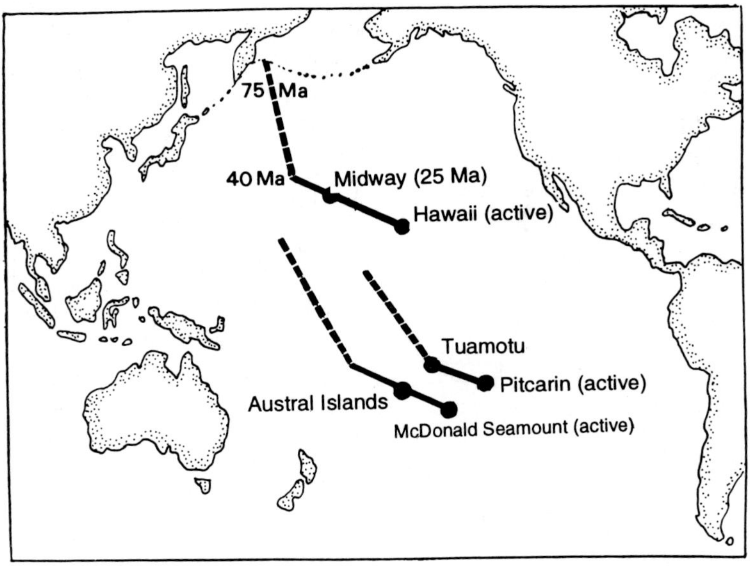

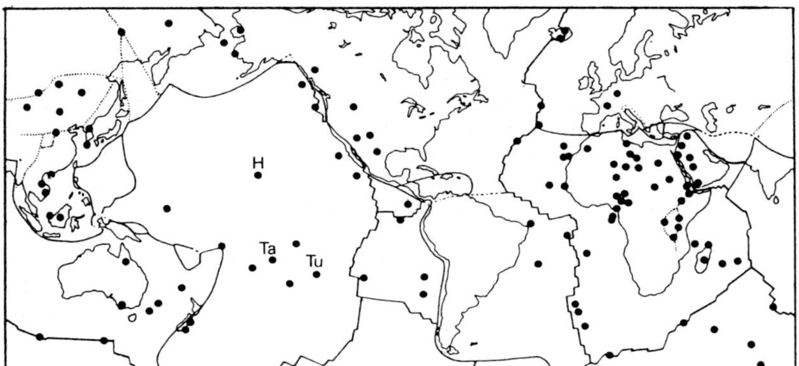

Fig. 27. Archipelago of the Hawaiian islands and the major hot spots in the world (H: Hawaii; Ta: Tahiti; Tu: Tuamotu).

If the hot spot is below an oceanic ridge, the magma plume there combines with the original volcanic mass of the ridge and increases the volume. This is the case with all the emergent structures of a ridge (Iceland, Azores, etc.) because they always show an association of tholeiitic and alkaline products. If the ridge is not displaced relative to the hot spot, the plume products will be thrown back from both sides of the ridge, thus giving rise to more or less symmetrically distributed masses of lava (the Walvis and Rio Grande ridges in the South Atlantic, Fig. 26).

The origin of the Chagos-Maldives and 90° East meridian fossil ridges (Fig. 26) is more complex. The former has certainly been associated with the hot spots of Reunion and Mauritius islands and the latter with those of Saint Paul and Amsterdam islands. But both have been separated from their point of origin by the late formation of the Indian Ocean Ridge. The parallelism of these two ridges indicates the northward motion of the Indian plate that resulted in the Himalayan collision.

The hot spot theory also explains the fact that the *first eruptions in the major intraoceanic structures were essentially tholeiitic and became alkaline only at a later stage.* The Hawaiian Islands are a classic example of this. The heat flow associated with the plume, while reaching the surface, might induce partial melting of the oceanic lithosphere, similar to the melting produced by the rise of the asthenosphere below the ridges due to simple extensional opening. The first products to erupt would thus be tholeiitic. The alkaline magma released from the deeper mantle would arrive after a certain time lag.

The number of hot spots assumed is uncertain and varies from 50 to 100 according to different authors. Some appear to have been active for several tens of million years (Reunion Islands, for example). They are found below all the plates, irrespective of whether the latter are continental or oceanic. Their distribution is quite irregular, however; they are less numerous below the Pacific and more abundant below the African plates (Fig. 27).

The effects of hot spots are essentially volcanic but sometimes also morphological due to the mass of hot and lighter material supplied by them at a depth of about 100 km. This has resulted in the formation of topographic upwarps with a diameter of about 1000 km and a height of 1 to 2 km around the islands of Hawaii, Cape Verde and Bermuda, and explains the origin of the Hoggar massif in the continental crust, which is linked with a superficial phonolitic activity.

These hot spots appear to be fixed relative to one another and might thus constitute a precise system of reference for estimating the absolute displacements of the lithosphere.

- **Oceanic plateaus** represent the purely oceanic expression of Large Igneous Provinces (LIP). The plateaus of Ontong Java (more than 1.8 million km^2) and Kerguelen (about 800,000 km^2) are the largest (Fig. 28). These regions possess an oceanic type crust, sometimes very thick (10 to 20 km) and hence show a smaller depth. They principally comprise magnesian basalts (OFB: Ocean Floor Basalts) emplaced during very rapid magmatic episodes. Below this crust exists a zone of dense material of high seismic velocities (7.2 to 7.5 km/s) interpreted as the arrival of superplumes that originate in a very deep mantle. One of these episodes may be situated in the middle of the Cretaceous, in correspondence with the major normal

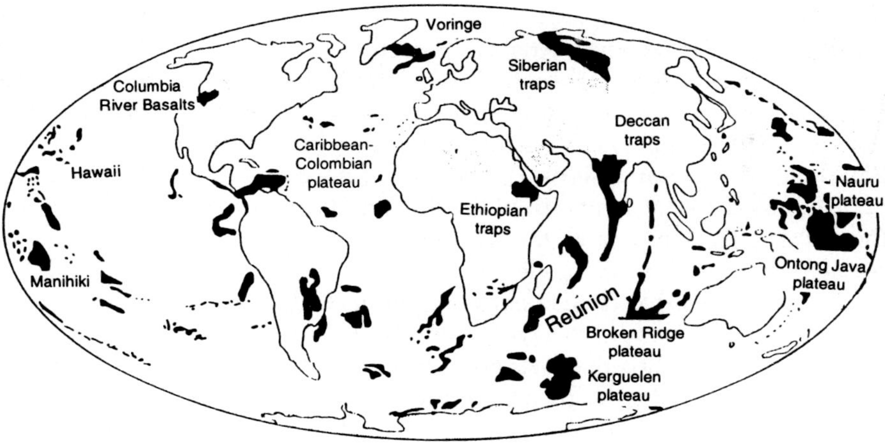

Fig. 28. Global distribution of Large Igneous Provinces (LIP) including the continental basalts (Traps), passive volcanic margins, oceanic plateaus, submarine ridges, oceanic-basin basalts and submarine volcanoes (after Coffin and Eldholm (1994) *Rev. Geophys.*, 32: 136).

magnetic period, and responsible for the emplacement of the major plateaus of Ontong Java, Nauru, Kerguelen and Caribbean.

NON-VOLCANIC RELIEFS

We limit ourselves to a description of one example, namely the *tectonic ridges on the Indian Ocean floor.*[2]

This intraplate deformation of the oceanic lithosphere is observed in the northern part of the Indian Ocean, south of the Bay of Bengal (south of Sri Lanka) and west of the 90° East meridian ridge. It is represented by large undulations of the E-W axis, with a spacing of 100 to 300 km and a vertical amplitude of 1 to 3 km, which makes the floor of the abyssal plain uneven.

Continuous seismic reflection profiles (Fig. 29) indicate the gradual deformation of the sediments and their trapping in the troughs separating the antiforms. Active thrust zones have also been characterised as due to compressive focal mechanisms, an important disposition since these may well have been the cause for the origin of intraoceanic arcs.

Gravity data confirm that the whole crust is involved and that the Moho itself is undulated.

---

[2] J.R. Cochran, D.A.V. Stow et al., (1988). Site 717. (*Distal Bengal Fan. Proc. ODP*, Init. Repts. 116). Texas A.M. University, pp. 42–48 and 197–210).

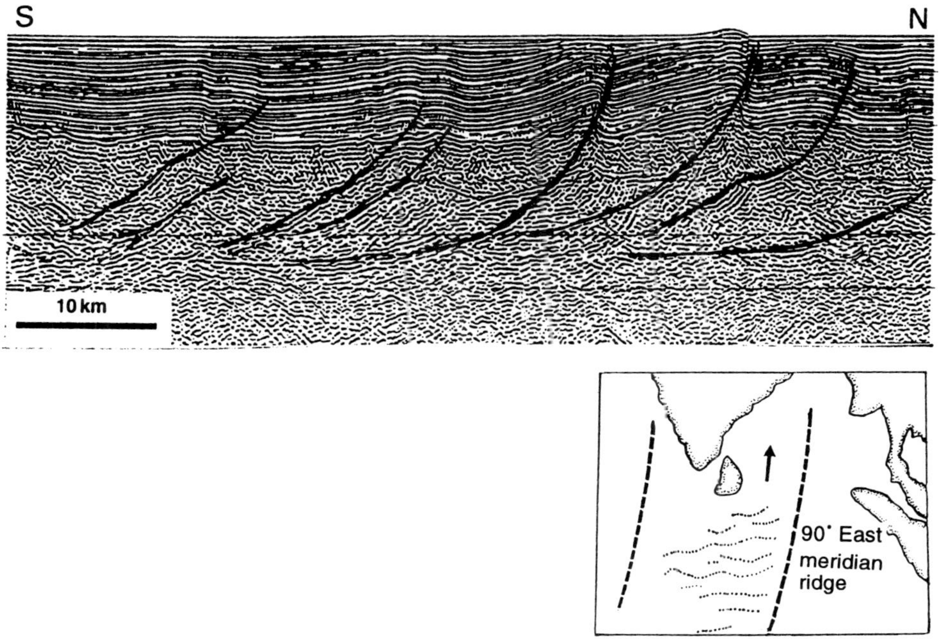

Fig. 29. Seismic reflection profile in the aseismic ridges in the northern part of the Indian Ocean (after Jestin (1994), Thesis, Paris University).
Inset: Arrow indicates the compressive stress field of the Indian plate.

The principal compressive component of the deformation is oriented almost N-S, i.e., perpendicular to the axis of undulations and the thrusts. This conforms to the general tectonic framework of the Himalayan collision (see p. 240).

The region also shows an abnormally high heat flow.

PART TWO

EXTENSIONAL STRUCTURES OF THE CONTINENTS

CHAPTER 1

SEDIMENTARY BASINS

Basins represent a surficial downwarping of the continental crust which is denoted by the general term *subsidence*.

Originally, this term was used to indicate sinkage of the base of coal basins. The sedimentation of these basins consists of a repetition, over thousands of metres, of coal beds and barren layers. It thus corresponds to an irregular and slow sinkage spread over long periods and accompanied by accumulation of thick sediments which include a shallow-water facies throughout the duration of the phenomenon.

This equilibrium appears surprising at first because, obviously, had sinkage not been gradual, the basin would have filled rapidly and, conversely, had the subsidence been too rapid, a deeper water facies would have been observed. It was hence thought that it is the weight of the sediments which determines the subsidence of the underlying crust and thus ensures the regularity of the phenomenon. In fact, a calculation involving the densities of the sediments present (Fig. 30) shows that a deposition of 1000 m sediments in an existing basin corresponds to an additional subsidence of only 400 to 600 m. It is thus necessary to invoke a process of deeper origin that triggers and maintains the continuous subsidence of the section under consideration.

At present, the term 'subsidence' is employed in a much broader sense, related neither to the rate nor to the rhythm of collapse, nor to the depth of deposits. Thus we used the term earlier to denote the gradual sinkage of the oceanic crust by cooling (p. 22).

Subsidence, in this broader sense, is produced by multiple causes, of which the principal ones are:

— an *overloading*, which may be due to a sedimentary sequence (its partial effect will be considered below), a major volcano, an ice sheet, a thrust block etc.;

— *cooling of the lithosphere*, which increases its density (described in the particular case of the oceanic lithosphere, p. 22) (thermal subsidence);

— *stretching of the lithosphere*, leading to its thinning and accordingly to a rise of the asthenospheric upper mantle, which is denser, thus inducing sinkage due to isostatic compensation (initial subsidence);

— *flexure of the lithosphere towards the base* and its forced sinkage in a compressive or a tensile tectonic regime (flexural subsidence).

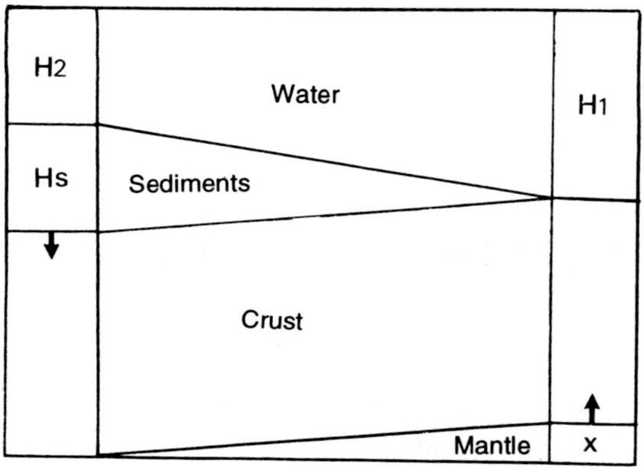

Fig. 30. Response of the lithosphere to local overloading on an Airy type crust (after Chorowicz, 1970).
H1 and H2: water layers (density rounded off to 1)
Hs: layer of sediments deposited (assumed density: 2.5)
x: displacement to be measured (density of the mantle: 3.3)
Geometrically, we can write:
$H2 + Hs = H1 + x$ (1)
In the case of isostatic equilibrium, we have,
$(1 \times H2) + 2.5\ Hs = (1 \times H1) + 3.3\ x$ (2)
Eq. 1 becomes: $H2 - H1 = x - Hs$
Eq. 2 becomes: $(H2 - H1) + 2.5\ Hs = 3.3\ x$
From these equations, we get:
$(x - Hs) + 2.5\ Hs = 3.3\ x$
$x = 1.5\ Hs/2.3 = 0.65\ Hs$,
i.e., 1000 m of sediments induce a subsidence of 650 m.

These different types of subsidence can be represented graphically as a function of time, as shown in Fig. 31.

Sedimentary basins can thus be divided into three major categories:

1. Basins whose evolution is mainly controlled by *lithospheric thinning (stretch basins).* These are essentially subsidence troughs, whatever be the mode of lithospheric stretching (rifts, pull-apart and back-arc basins), and sedimentary basins associated with passive margins.

2. Basins whose evolution is principally controlled by *cooling of the lithosphere,* which may be continental (intracontinental basins proper) or oceanic (abyssal plains).

3. Basins whose evolution is essentially controlled by the *geometry* or, more exactly, by the *flexure of the lithosphere (flexural basins),* which may be oceanic (subduction) or continental (foreland basins).

In reality, the evolution of a given basin most often proceeds with a combination or succession of these factors, for example, by stretching and cooling. In view of this complicating factor, we shall use a classification that is more

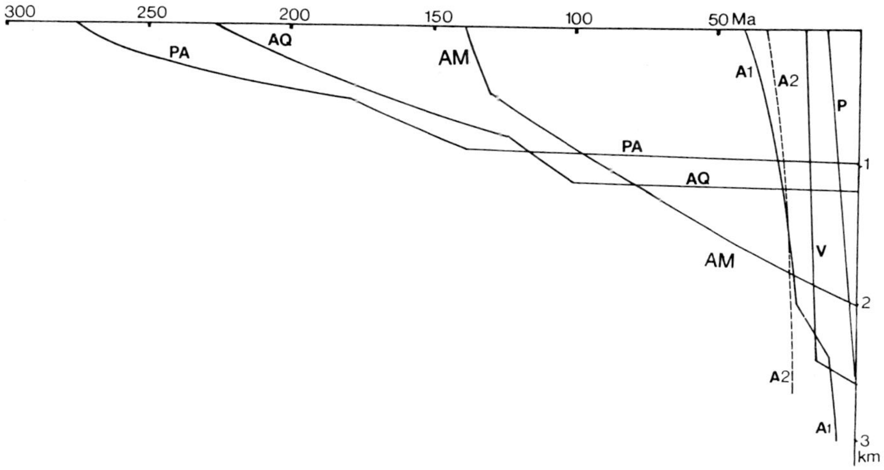

Fig. 31. Subsidence profiles.
PA: Paris basin (well ESSISES 1, after M.F. Brunet, simplified).
AQ: Aquitaine basin (well LACQ 301, after M.F. Brunet, simplified)
AM: Atlantic margin (borehole COST B2, after Steckler and Watts, modified)
A1: Peri-Alpine (swiss) molassic basin (well ANZIN 63, after P. Allen et al.)
A2: same as above (well SERVION 1, *id.*)
P: Central Pannonic basin (after Sclater et al., modified)
V: Vienna basin (after Sclater et al., modified)
These profiles show the variation of subsidence undergone by the substratum of the basin due to the effect of the weight of the sediments. They thus represent a real tectonic subsidence. Due to sedimentary loading, the substratum is in reality at a greater depth, almost double to that without loading.
Note that the Atlantic margin (AM) shows a typical profile with strong initial subsidence, followed by a feeble one (thermal subsidence). The Vienna basin (V) also shows a comparable profile but with a much greater initial subsidence. The intracratonic basins (PA, AQ) exhibit more moderate profiles, indicating a slow sinkage, at times accelerated during periods of extension. The foreland molassic basins (A1, A2) show a steeper profile, generally increasing, interrupted during tectonic movements of the basin. As for the Central Pannonic basin (P), its initial subsidence is not yet completed.

References

Steckler M.S. and Watts A.B. 1978. Subsidence of the Atlantic type continental margin off New York. *Earth. Plan. Sci. Let.*, 41: 1–13.
Brunet M.F. 1986. La subsidence de trois bassins sédimentaires français (Paris, Aquitaine, Sud-Est). *Bull. Elf. Aquit.*, 10: 23–32.
Allen P.A., Homewood P. and Williams G.D. 1986. *Foreland Basins: an Introduction*. Sp. Pub. Int. Ass. Sediment., 8: 3–12.
Sclater J.G. et al. 1980. The formation of the intra-Carpathian basins as determined by subsidence data. *Earth. Plan. Sci. Let.*, 51: 139–162.

descriptive than interpretative, and distinguish the following types of basins:

1. *Grabens (rift valleys),* in which subsidence occurs rapidly and with brittle faulting.

2. *Basins proper,* wherein subsidence is slow and gradual.

3. *Mixed basins,* in which rifting is associated or alternated with a more regular and continuous subsidence.

4. *Basins associated with mountain chains,* which are generally denoted by the name *molassic basins* and which are further classified as *foreland basins* (associated with lithospheric flexure) and *backland basins* (associated with a subduction process).

The sedimentary basins of continental margins, active and passive, will be discussed in the chapter devoted to these margins.

RIFT VALLEYS

These are narrow and elongated structures, bounded by conjugate normal faults with opposite throw directions (antithetic faults). Because of these fractures, rift valleys almost always exhibit alkaline type volcanism.

These structures are also known as *grabens.* They may be broken into secondary grabens by elevated segments known as *horsts,* in which case they are referred to as *germanotype* structures since they are typically observed in the lower Rhine valley. *Half-grabens* or *half-horsts* are bounded by faults with the same throw directions (synthetic faults), which results in the tilting of the elevated segments (tilted blocks).

Rift valleys may be surface structures of small size (dimensions of the order of a kilometre and maximum throws of about 100 metres).

When rift valleys are of continental scale, i.e., when they have a length of 100 to several hundred or thousand kilometres and a width of several tens of kilometres, it is established that modifications are produced in the crust and upper mantle below them. In this case they are known as *rifts* proper, which are obviously the largest subsidence structures.

Classical Rifts

Rhine graben

We shall consider the **Rhine graben** (Figs. 32 to 37) as a typical example of continental rifts.

This structure, with a length of 300 km and a width of 35 to 40 km, is situated along with the Bresse, the Limagne and other grabens on an extensional axis which intersected Europe, from Germany to the lower Rhone valley, during the Tertiary (Fig. 32). The deep structure of this graben is known

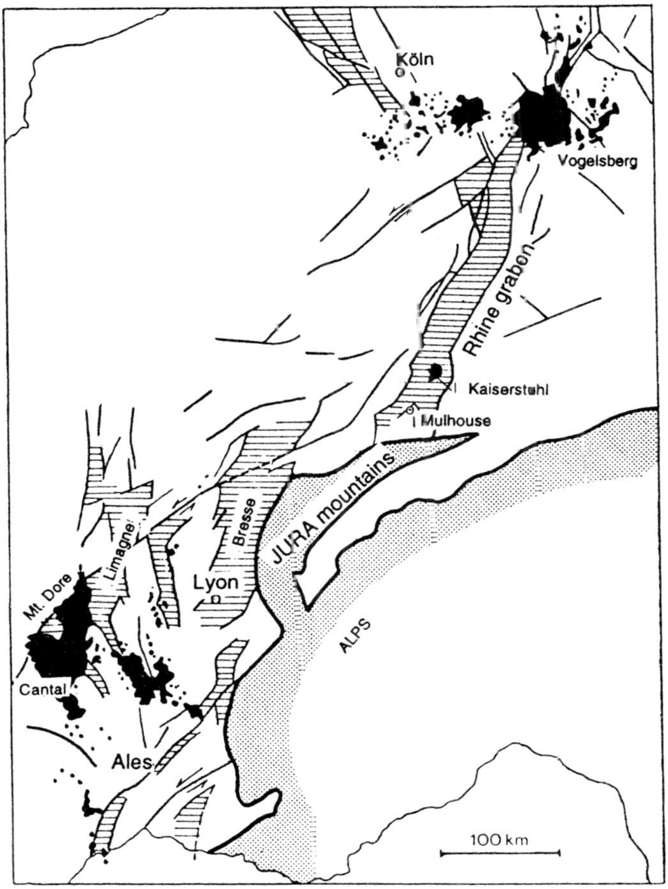

Fig. 32. Palaeogene rift valleys of western Europe. (in black: Neogene volcanic flows)

from the Mulhouse potassium mines, petroleum wells (Pechelbronn petroleum) and geophysical prospecting.

Its history and its origin are manifestly related to those of the Alps, though the nature of this relation is still not clear.

Morphology (Fig. 33): In its southern part, the Rhine graben is intercalated between the two crystalline massifs of the Vosges in the west and the Black Forest in the east (each of an elevation of about 1500 m). It is separated from the latter two by major faults ('Vosgian' fault on the French side) that are of normal, synthetic type and in the shape of a bayonet.

At the foot of these major faults a zone of hillocks occurs in which intensively faulted Mesozoic formations are found to outcrop (Fig. 34). This zone of 'fault clusters' offers variable width, the maximum occurring in the

Fig. 33. Rough sketch of the Rhine graben (after H. Cloos, 1955).
Ancient massifs are shown in grey. In the Oligocene the rift extended towards the Hesse. Since the Neogene, it is associated with the Lower Rhine rift (or Rhineland basin).

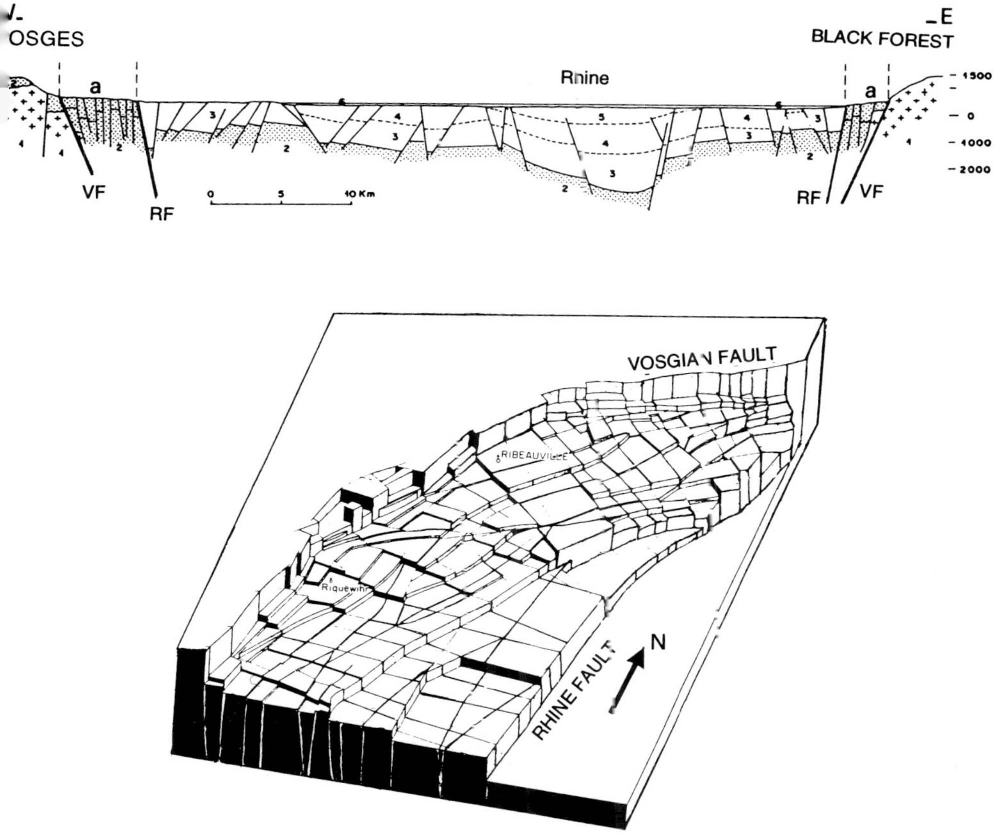

Fig. 34. Rhine rift (after Sittler, 1974).
Top: transverse section: a: Zone of fault clusters; RF: Rhine fault; VF: Vosgian fault. 1. basement, 2. Mesozoic, 3. Lower Oligocene, 4. Middle Oligocene, 5. Upper Oligocene, 6. Plio-Quaternary.
Bottom: Block diagram of the northern part of the Ribeauville fault cluster. A complex network of longitudinal faults, N 10° to N 45°, is developed between the Vosgian fault and the Rhine fault, intersected by readjustment transverse faults oriented N 90° to N 150°.

Saverne region. The hilly terrain is separated from the Rhine plains by new faults, known as 'Rhine faults', which are invariably of the synthetic type, with strong throws but least visible in the topography (Fig. 34).

The rift proper itself, filled with Tertiary sediments, shows a suite of small horsts and grabens, marked by geophysical anomalies. The basement is at a depth of 1200 to 3500 m (in the Mannheim 'trough').

The submeridian direction of the Rhine graben is oblique relative to the SW-NE Variscan structures (Fig. 32) and to the folds that affected the region at the end of the Cretaceous parallel to the preceding ones.

Evolution of the rift: In the Mesozoic, sedimentation is relatively uniform over the entire Rhine block and has a thickness of about 1500 m. There is no indication for the occurrence of a future rifting. The isopic lines are parallel to the Variscan structures.

At the end of the Cretaceous as a result of 'Laramian' or 'Eoalpine' folding, ridges are produced invariably in the Variscan direction, immediately attacked by erosion.

At the beginning of the Tertiary peneplanation is completed, accompanied by an alkaline (Lower Eocene) volcanic activity which may represent the beginning of extension. The latter becomes stronger from the end of the Eocene by the appearance of thick lacustrine sediments, and especially at the beginning of the Oligocene, when the southern part of the rift undergoes subsidence, invaded by the sea pursuing from the Alps. Thus marls are deposited, sometimes sandy, containing rock salt and potassium salts that are being mined at Mulhouse, as also the source rock of the Pechelbronn petroleum.

Since the *end of Oligocene* a general regression is produced, marked by the recurrence of lacustrine sediments. It represents an uplift probably related to that of the Alps because it attenuates gradually northwards. It prolongs throughout the Miocene with the result that, by the end of this period, the southern part of the basin is even eroded.

A second phase of extension takes place at the *end of Miocene,* contemporary to the new alpine tectonic crisis, in a different stress field. It results in a sinistral transverse faulting of the margins of the basin during which the latter undergoes subsidence once again but this time in its southern part where, as already mentioned, maximum subsidence is observed ('Mannheim trough'). The sedimentation remains essentially fluviatile. On the whole, there has been a tilting movement from S to N since subsidence commenced in the south in the Palaeogene and continued northwards in the Neogene.

A study of detrital products descended from the southern crystalline massifs showed that the uplift of the basin margins took place irregularly, mainly at the beginning of the Oligocene and then in the Plio-Quaternary, i.e., immediately after each of the major extension episodes.

Thus there was no subsidence of the top of an original dome, rather it resulted in the uplift of already formed margins.

Volcanism: The evolution of the Rhine graben is accompanied by fissural alkaline volcanism. The dominant lavas are essexites (tephrites, limburgites, theralites, nephelinites) and phonolites. This volcanism shows specific chemical characteristics; it is, in effect, rich in Na and $CaCO_3$ (to the extent that lavas made up of this carbonate or carbonatites exist).

It has been said that volcanism commenced in the Lower Eocene but it is essentially in the Upper Miocene that it developed and terminated. In French territory the volcanic structures are small and practically reduced to dykes (Riquewihr, Reishoffen, etc.). In Germany they are much larger: Kaiserstuhl in the south, renowned for its carbonatites, and Vogelsberg in the north (Fig. 33). It is in the former that the most intensive uplift of the upper mantle (24 km) is observed (Fig. 35).

Tectonics: The dislocations responsible for subsidence of the Rhine graben are normal faults of Rhenan (N-NE) and Variscan (NE) trends, the former corresponding to an ancient lineament revived by the axis of European plate extension during the Tertiary. On the boundary of the basin,

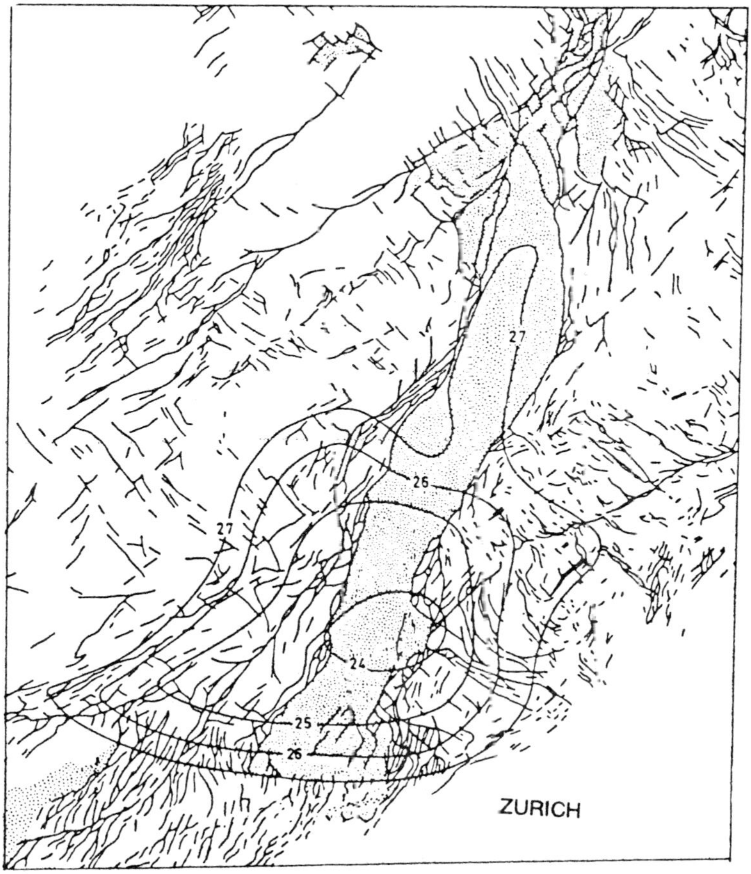

Fig. 35. Fractures in the margins of the Rhine graben (after Illies, 1974, simplified). They represent a revival of the ancient Variscan (NE-SW) faults intersecting the basin obliquely in a NNE-SSW direction. A number of NW-SE tension fractures are added to them due to the Neogene-to-Present stress field. The isobaths (in km) of the top of the upper mantle are superposed on the map. Its uplift is obvious in the southern part of the basin.

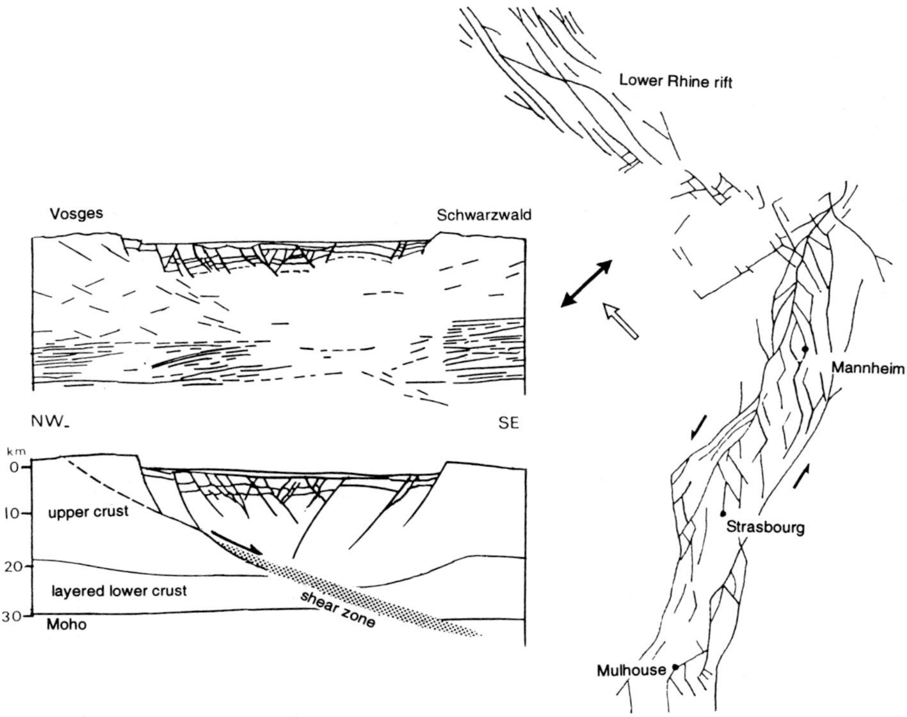

Fig. 36. Fractures in the Rhine graben.
In plan, the stress field appearing in the Neogene (white arrow) has produced sinistral transverse faulting of the basin margins, thus inducing the development of NW-SE riedels in an en echelon pattern.
The corresponding extension (black arrow) opened up the Lower Rhine rift along the NW-SE direction, which is the same as that of the stress field.
In section (according to an ECORS profile), an inclined shear zone is developed which appears to control the extensional opening (on this subject, see pp. 69, 72, 134, 146 (after Brun et al., 1992, simplified).

the Vosgian (or outer) faults are of synthetic type, with a dip of 40 to 60° E, which is generally smaller than that of Rhine (or inner) faults with 70 to 80°. The throw of the former (of the order of 800 m maximum) is, on the whole, smaller than that of the latter (1500 to 1800 m). Seismic reflection data indicate the termination of these faults at a depth of 7 km (transition to a ductile crust). In the centre of the rift, the faults are frequently NW to NNW (Fig. 36); these are generally interpreted as en echelon riedels of a sinistral faulting of the two margins of the basin.

The *deformation mechanism* is complex with two tectonic episodes:

1. *Eocene-Oligocene episode*: Structural analysis of deformations primarily indicates the effect of a north-south compression of Upper Eocene age,

roughly contemporary to alpine folding. It produces conjugate systems of transverse faults with NW-SE dextral and NE-SW sinistral members. The tension fractures also confirm this submeridian compression. But from the Oligocene onwards, σ_1 becomes vertical (vertical fault striae superposed on horizontal striae of the preceding event) and induces the generation of normal faults in an E-W extensional field. However, the structural heritage gives the basin a slightly inclined N 20° trend. Reactivation of the earlier transverse faults into normal faults and the obliquity of the rift explain the zigzag appearance of the boundaries.

In any case, the Oligocene subsidence of the basin is due to an E-W extensional movement and to the appearance of N-S antithetic faults forming the margins of the sinking blocks; σ_3 is hence slightly oblique to the rift whose direction, it may be recalled, is controlled by the ancient structures.

2. *Neogene episode*: In this period the stress field is modified: the compression becomes NW-SE. The basin then acts as a corridor of sinistral transverse faults (while, in the same epoch, the Lower Rhine rift is developed parallel to the new axis of extension, Fig. 36).

If the various subsident blocks are put together again, a space of 4 to 5 km still remains, which must represent the value of the Tertiary extension. The ratio of successive widths (before and after extension), denoted by β, is 6/5, or 1.2; in other words, there has been a crustal thinning of 15 to 20%.

It may be added that an extension of 5 km in 50 Ma gives an average rate of 0.1 mm per year. This is markedly removed from the rates of ocean-floor spreading, which make us examine the southern edge of the Rhine graben. In fact, the graben ends near the Jura mountains but is continued farther westwards by the Bresse graben (Figs. 32 and 35). Here a transform fault occurs between the two, exactly like the offsets of the oceanic rifts. On the surface, this zone exhibits an en echelon fault cluster, representing a slow rate of spreading.

Geophysical data: The important geophysical data are as follows (Fig. 37):

1. *Strong negative gravity anomaly* directly above the graben associated with sedimentary filling and the mass deficiency caused by partial melting of the asthenospheric 'swelling'.

2. *High heat flow*, long known from the heat flow values prevailing in the potash mines (the geothermal degree is 10 to 20 m instead of the normal 30 m), which provides a possibility for exploitation of geothermal energy (the Soultz-sous-Forêt well).

3. The *present rate of subsidence* of the rift, according to precise geodetic measurements, is of the order of 0.2 to 0.7 mm/yr.

4. This is a *high seismic region*, especially in its northern part, where the present activity appears to be concentrated. A seismic study primarily showed that besides vertical movements, there also exist sinistral horizontal displacements (5 cm/1000 yrs ?) in a NNW-SSE compressive stress field.

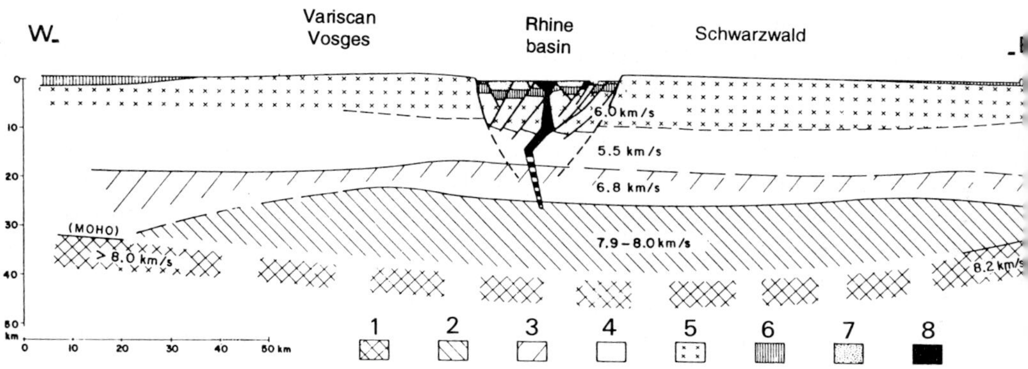

Fig. 37. Structure of the crust below the Rhine graben inferred from P-wave velocities (after Sittler, 1974).
1. normal mantle; 2. mantle with slightly reduced velocities; 3. deep continental crust (note its thinning out); 4. middle continental crust (low seismic velocity zone); 5. upper continental crust; granito-gneissic; 6. Triassic and Jurassic sediments; 7. Tertiary and Quaternary sediments of the basin; 8. basaltic volcanism.

Moreover, the seismic wave velocities indicate that the deep structure of the crust is abnormal below the basin. Throughout its width and that of the bordering crystalline massifs, there is an upwarp of the upper mantle in which the seismic velocities are slightly lower than in the normal mantle, with the Moho being indistinct. Here it represents a common reaction of the upper mantle in decompression zones, due perhaps to partial melting phenomena, but it does not constitute the source for the overlying volcanism whose origin is much more deep seated.

One may also observe, below the basin, a thinning of the deeper crust and the existence of a low-velocity zone (5.5 km/s, instead of 6.5 km/s) between 10 and 20 km depth.

Mechanism of deformation of the graben: Under the effect of extension, the crust breaks on the surface along a number of conjugate normal faults, whereas the ductile crust at depth thins out by stretching. The normal faults of the surface tend to be gradually recumbent in the depth, thus becoming 'listric' faults before being blurred in the transition zone from the rigid or fragile crust to the 'ductile' crust.

Because of this stretching of the ductile crust, the upper mantle is uplifted and is hence subjected to expansion, perhaps accompanied by partial melting. The latter makes the crust lighter and promotes its ascent. This is followed by a high heat flow which is readily dissipated at the level of the basin itself (certainly, for a part, because of groundwater circulating in the faults of the basin margins). However, laterally the thermal flux heats up the surrounding formations, decreasing their density and increasing their volume. These formations undergo uplift: thus *thermal intumescence* develops slightly later than the beginning of the phenomenon.

It was long thought that a converse situation occurs, viz., that the rifting phenomenon is initiated by an uplift due to growth of an asthenospheric swelling. The dome is fractured by tension and sinks during asthenospheric debulging (more appropriately, 'doming' precedes 'rifting'). However, the facts invalidate this notion except in a few very rare cases, for example the Oligocene rifting of the Ligurian Gulf across the uplifting Corsica-Sardinian massif. These cases invariably involve mechanisms other than asthenospheric upwarping, for instance folding in this particular case.

The irregularity of uplift of the margins (for example the reliefs in the boundaries of the northern part of the Rhine basin are much smaller than in the southern parts) is probably related to the heterogeneity of material in these parts, mainly with regard to thermal conductivity.

OTHER EXAMPLES

Gulf of Suez (Figs. 38 to 40)

With a NW-SE orientation, this gulf opened up during the Miocene in the midst of the Arabo-African craton, within the framework of anticlockwise rotation of the Arabian plate relative to the African plate. It is still under sea water but its depth does not exceed 80 m.

Stratigraphy of its sediments: Four groups of sediments that lie unconformably one above the other can be distinguished here:

Group A: Lower Miocene: Lacustrine or continental detrital deposits, at times evaporitic, overflowing the sides of the rift. Hence the latter was not yet formed. Numerous dykes and basaltic flows occur.

Group B: Middle Miocene: Pelagic marls and reef limestones (the rift is formed and is invaded by the sea).

Group C: Upper Miocene: Evaporite sequences, indicating that the milieu has become confined.

Group D: Plio-Pleistocene: Algal or reef limestones (the sea reinvades the rift) intersected on the boundaries by continental deposits (uplift of these boundaries).

Tectonic structure: The graben is demarcated by NW-SE normal faults, forming the boundaries of the tilted blocks. But the structure is always dissymmetric, one of the margins being generally controlled by a major fault, whereas the substratum rises with a mild slope on the opposite border (Fig. 39).

The network of these major faults is complicated by NNW-SSW transverse faulting (parallel to the Gulf of Aqaba and to the major fault which forms its boundary) and the NNW-SSE fault (known as the Duwi trend). The origin of these faults follows an evolutionary pattern:

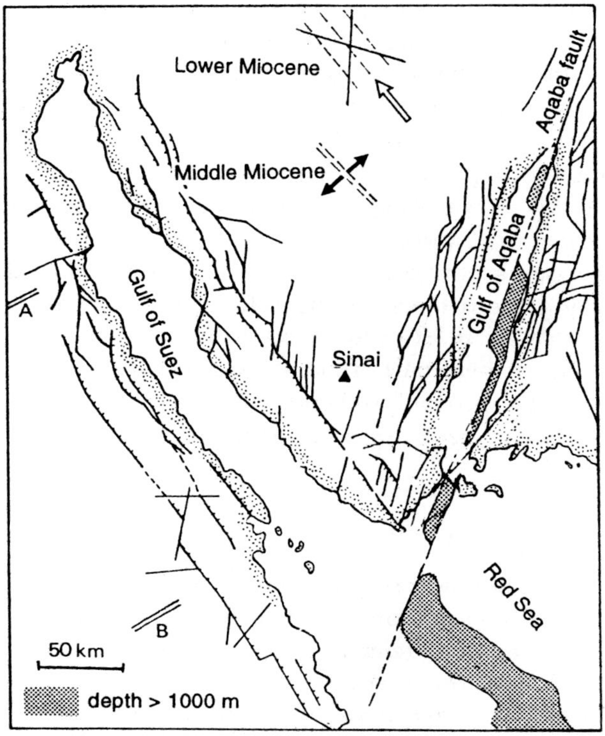

Fig. 38. Structural map of the Suez and Aqaba gulfs (after Ott d'Estévou et al., 1987; Chorowicz et al., 1987).
Grey zones (depths greater than 1000 m) represent segments of the thinned-out continental crust. Also see Figs. 46 and 47.

—The Aqaba and Duwi type faults are conjugate strike-slip faults corresponding to a NW-SE compression (horizontal σ_1). Tension cracks also develop parallel to σ_1. All these fractures are older because sealed by terminal deposits of group A.

—The NW-SE faults are classical normal faults which facilitate opening of the rift in a NE-SW extension field (vertical σ_1) that is initially simple (tilted blocks) and later becomes more diffusive ($\sigma_2 = \sigma_3$), resulting in fragmentation of the blocks into a horst-graben sequence.

History of the rift: Based on the preceding descriptions, the evolution of the graben can be inferred as follows (Fig. 40):

1. At the *beginning of the Miocene* and in a continental regime, the NW-oriented compressive movements (associated with Mesogean chain folding) led to the formation of a conjugate strike-slip fault network (Aqaba and

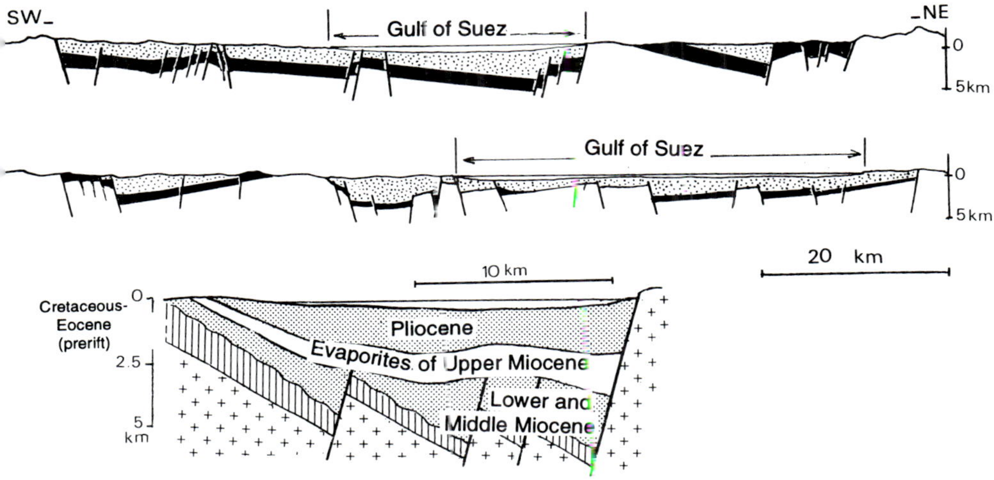

Fig. 39. Transverse section of the Suez rift.
Top: Tectonic style (after Ott d'Estévou et al., 1987):
in white: ancient basement; black: Mesozoic formations;
grey: Tertiary formations
Bottom: Relation to sedimentation (after Perrodon, 1983).

Duwi faults) and the NW-SE tension faults, providing a passage to alkaline basalts. The whole system is slightly subsident. No 'doming' is perceptible.

2. During the *Middle Miocene* a clear extension regime is developed and induces subsidence of the rift, which is invaded by the sea. Structuring of the tilted blocks follows the NW-SE normal faults, often offset by the ancient strike-slip faults of Aqaba and Duwi, which are reactivated, resulting in the zigzag structure. Changes in the direction of tilting of the blocks occur from both of the transverse discontinuities resulting in a highly dissymmetric basin. Later, tilting of the blocks ceases, giving the impression that the major phase of crustal stretching is completed. These movements are replaced by major vertical motions, forming the horst-graben structures, accompanied by uplift of the margins[1] . The width of the zone involved reaches 400 km.

All these structures are sealed by the marine sedimentation of group B.

3. In the *Upper Miocene* faulting activity is reduced in a regressive regime wherein the evaporites dominate.

4. During the *Plio-Pleistocene* tectonic activity is rejuvenated. Uplift of the margins accelerates, a part of the intermediate beds emerges and subsidence of the central basin, invaded by sea, increases. The Aqaba fault then functions as a major strike-slip fault (which is only the southern end of the

[1]This is confirmed by the study of fluid inclusions in apatites of the Sinai granites, indicating an uplift of 3000 m in the last 9 Ma.

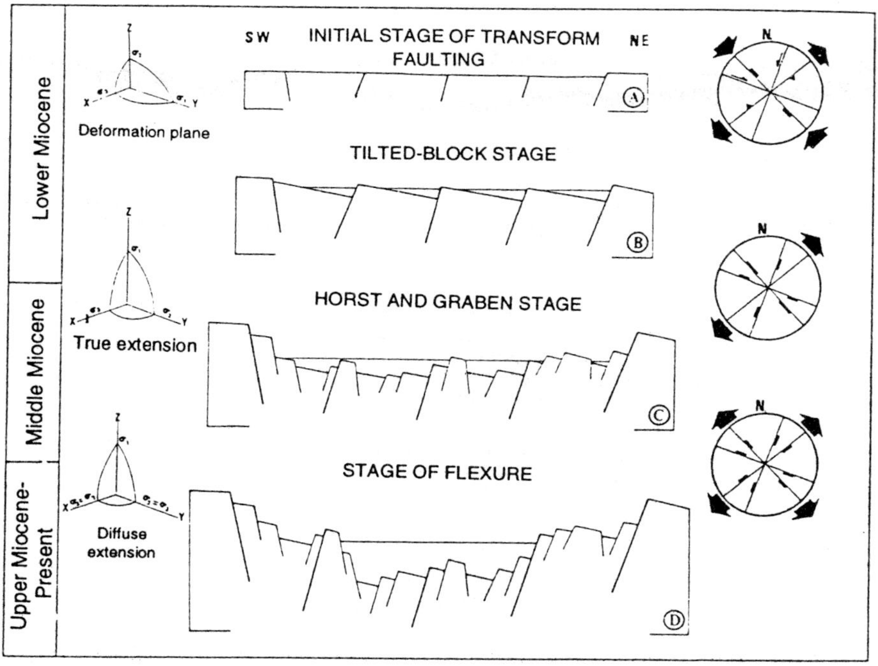

Fig. 40. Evolution of the rift in the Gulf of Suez (after Ott d'Estévou, 1987).

Jordan fault, see p. 72), 'decoupling' the evolution of the Gulf of Suez (which remains an intracontinental basin) from that of the Red Sea, which continues to develop as a crustal fracture (see p. 112).

Uplift of the boundaries results in topographic reliefs reaching 1750 m in the west and more than 2500 m in the east (the Sinai massif). Total throw is of the order of 5000 to 6000 m.

The present width of the rift (80 km), compared to its historic width that can be reconstructed from the blocks (about 55 km), gives an approximate extension of 50% ($\beta = 1.45$).

Rift of East African Great Lakes (Figs. 41–44)

The rift extends from Zambezi in the south to the Red Sea in the north over a length of 6000 km, with an average width of 40 to 60 km. It is divided into two branches, Eastern and Western, both connected by a lineament (intercontinental strike-slip fault?) which is clearly observed on space images and is known as the Assoua lineament. The western branch also presents a complex shape due to the Tanganyika-Malawi strike-slip corridor, which is a dextral transtension fault zone parallel to the Assoua lineament. All these lineaments are developed on Precambrian fractures.

The eastern branch terminates in the south in a curious pattern: it is not abruptly interrupted by the Assoua lineament but spreads out in the form of a 'duck's foot' by seismically active radial faults (Fig. 41), indicating that the extension is not totally stopped on the south-western side by the strike-slip fault lineament.

The rift opened up in the Miocene and its sedimentation has remained continental or lacustrine. Deposits may reach a thickness of up to 8000 m (Fig. 42) and the discontinuities observed there are of climatic or tectonic origin (tilting and faulting of the blocks).

Volcanism: The East African rift exhibits the volcanism common to this type of structure mainly in the eastern branch. This indicates that the extension here is greater, as would be expected because of its relation with the Afar depression. It is at the intersection of this eastern branch and the Assoua lineament that the largest volcanoes are situated (Kenya and Kilimanjaro).

The volcanism is of alkaline type with a certain enrichment in sodium as in the case of all the rifts. This abundance of sodium is the reason for the name Natron lake (eastern branch, south-western Nairobi, Fig. 41). Natron is a natural carbonate of sodium (CO_3 Na, 10 H_2O) which was used by the ancient Egyptians in the preparation of mummies.

Uplift of the margins: Uplift of the margins is also observed here with the lakes situated between two highly regular montane ridges whose mean elevation is 2000 to 2500 m except in the Ethiopian rift where it frequently exceeds 3000 m.

The Ruwenzori massif, a 120 km long and 40 km wide Precambrian block which is raised to more than 5000 m (5119 m), between Albert and Edward lakes (Fig. 43), is often attributed to such uplifts. In fact, the cause of this uplift is not very clear since the block appears to be well inside the rift and not on its margin. Moreover the seismic foci in this region occur abnormally deep for a rift zone (27 to 40 km, Fig. 44). This uplift might then have been the isostatic response to a local crustal thickening (compression or tilting of a block between the zones of asthenospheric upwarping ?).

Anyway, these uplifts are presently persisting in most cases and probably by jerks as indicated, for example, by a peculiar reversal of the drainage pattern of the Kafu river (north of Lake Victoria). The upper part of the drainage basin has become a lake with several branching parts (Lake Kyoga) (Fig. 43), while the block that separates it from Lake Albert is uplifted.

Seismicity: The epicentre map of 26 major earthquakes in eastern Africa from 1963 to 1970 (Fig. 44) confirms that the earthquakes are associated with the rifts but may also extend beyond their boundaries (note the high seismicity of the Tanzanian 'duck's foot'). In fact, the entire East African continent has been subjected to extension.

In the case of earthquakes associated with the rift a still unresolved problem is that of the difference in the seismic focal depths between the western branch (12 to 15 km) and the eastern branch (4 to 10 km). It has been noted that a certain relation exists between the depth of these foci and the

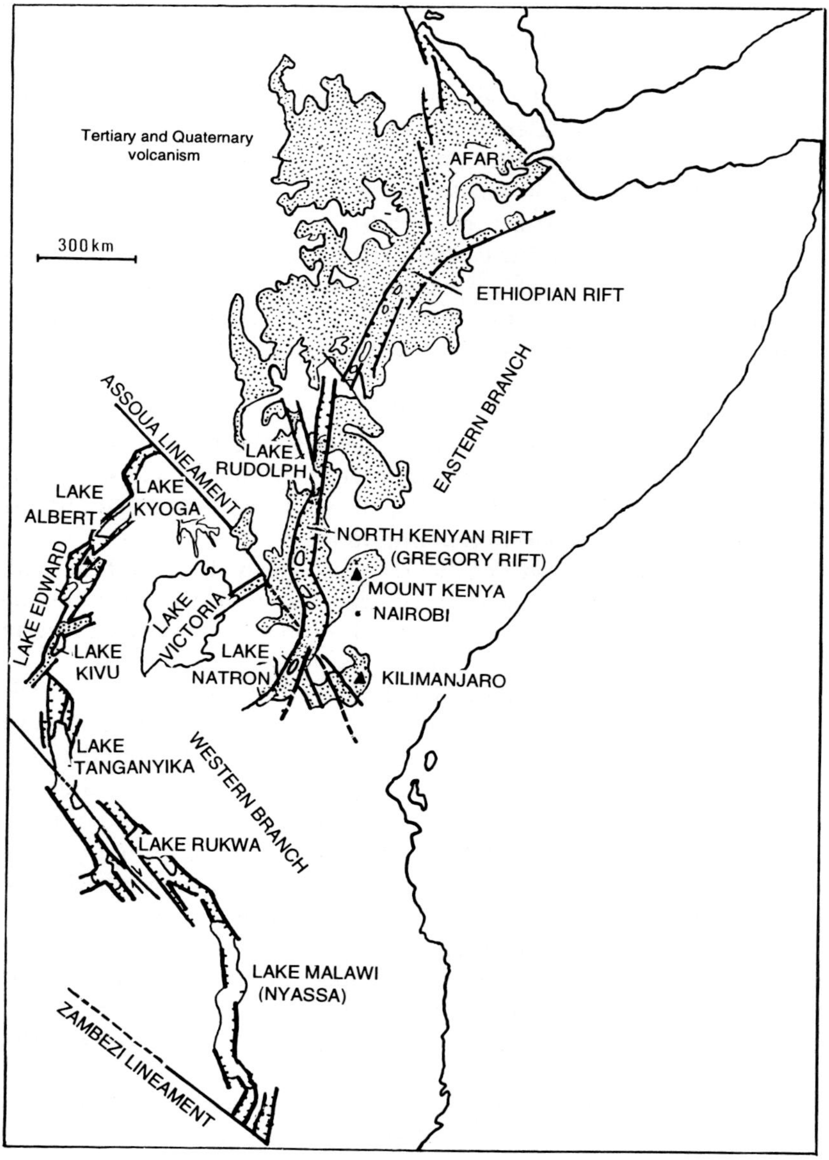

Fig. 41. Simplified structural map of the East African rift (after Chorowicz, 1983). Actually, this rift is formed from two branches which are connected by the Assoua lineament. In this zone of connection, Lake Victoria, which is shallow and has a low rate of sedimentation, separates the two branches and differs from the lakes that mark the latter. Its origin is related to the uplift of the margins of these rifts, both in the west and the east. These uplifts produce a natural basin in-between them.

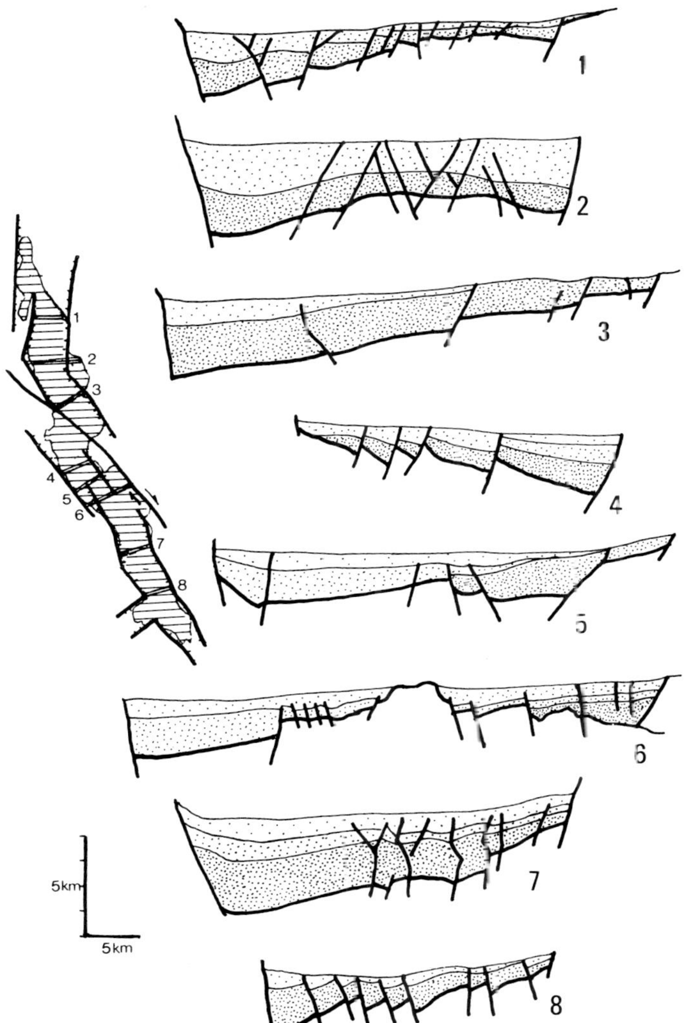

Fig. 42. Structure of the East African rift in the region of Lake Tanganyika (after Sander and Rosendahl, 1989).
Dotted zone: Mio-Pliocene sediments.

dips of the marginal faults; the smaller the dip, the shallower the focus. It follows from this relation that extension of the East African rifts might be due to a major listric fault, which is flat and dissymmetric below the rift (Fig. 44 A-B). We shall find a similar explanation for other structures of extension (pp. 72, 81, 134, 146).

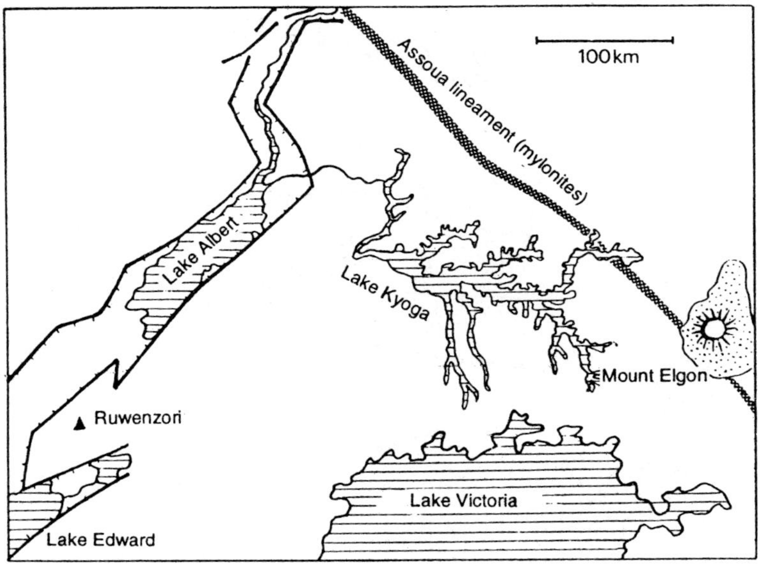

Fig. 43. Lake Kyoga and its relation to the East African rift.
The lake represents the ancient upper drainage region of Kafu River, flowing from Lake Albert and separated from the latter by the recent uplift of the eastern margin of the rift. This uplift has caused a reversal of direction of water flow.

Lake Baikal Rift (Fig. 45)

Unlike the preceding structures, the Lake Baikal Rift exhibits a peculiarity, namely an extension on a thick crust (45 to 50 km). Situated in the southern part of the Siberian platform and constituting a 2500 km long depression, its shape is inherited from earlier structures of Precambrian and Palaeozoic ages. The lake itself, with a length of 670 km, is 1200 m deep and is surrounded by reliefs of 1000 to 3000 m; the crystalline basement occurs 5000 m below the lake.

The cross-section of the rift is dissymmetric (greater throw in the NW) and the associated volcanism, alkaline in composition, situated only on its SE limb.

Sedimentation indicates two successive periods:

From the Oligocene to the Lower Pliocene (34 to 4 Ma) fine sands, silts and shales were deposited, which are essentially lacustrine. In other words, they represent fine-grained sediments, indicating topographies of very low feeding. Moreover, these deposits extend considerably beyond the margins of the present-day rift which was still not fully developed by that time. The thickness of the deposits varies from 2000 to 4000 m, which, corresponding to 30 Ma, gives a low subsidence rate (0.07 to 0.15 mm/yr).

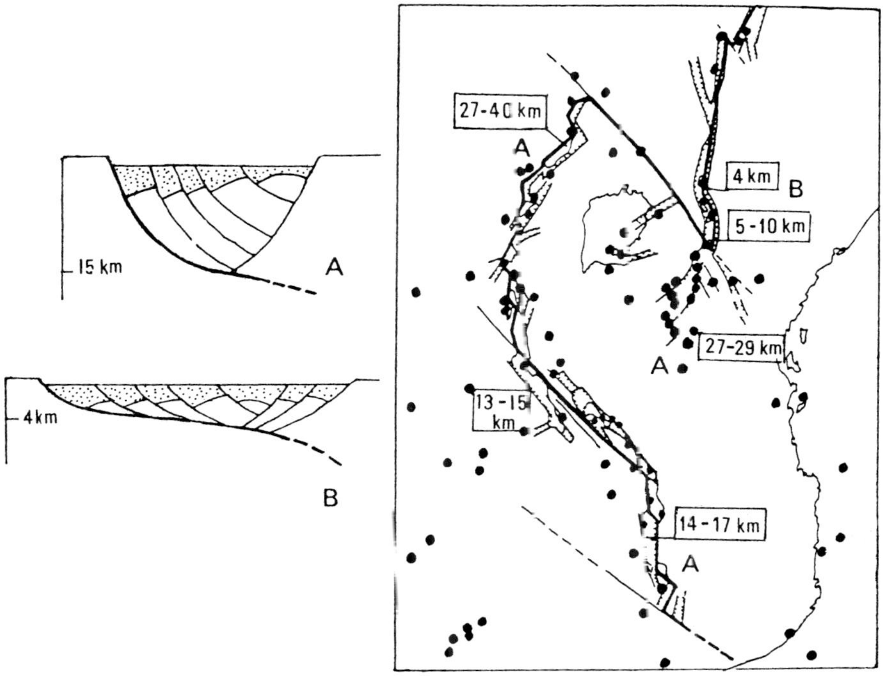

Fig. 44. Seismicity in the East African rift between 1963 and 1970 (taken from Chorowicz, 1983 and Morley, 1989).
Right: distribution and depth of seismic foci.
Left: interpretative cross-sections of rifts of deep foci (type A) and shallow foci (type B). They intersect a low-dip 'detachment fault'.

From the Upper Pliocene to the Present (3 Ma), the sediments, strictly limited to the rift, are more coarse-grained, torrential to fluviatile, and occur mainly on the rift margins where the local, pebbled conglomerates indicate an intensive uplift of the boundaries. The thickness of the sediments is of the order of 1000 m, giving a subsidence rate of 0.35 mm/yr. If the uplift of the margins is taken into account, we obtain a subsidence rate of 0.66 mm/yr, i.e., 4 to 9 times higher than that of the preceding stage.

These two stages represent, as in the case of the previous examples, the results of Neogene compressive phases, those of the Himalayas in the present case (see p. 237).

Geophysical data show that the Moho is at a depth of 30–40 km below the rift, as against 45–50 km below the neighbouring platforms (crustal over-thickening of orogenic origin) and that below the stretched zone there exists a dome of the abnormal upper mantle with lower seismic velocities (8 km/s, instead of 8.3 to 8.4 km/s). This asthenospheric dome approaches the Moho over a horizontal length of 200 to 300 km, without, however, causing thinning of the ductile crust except immediately below the rift.

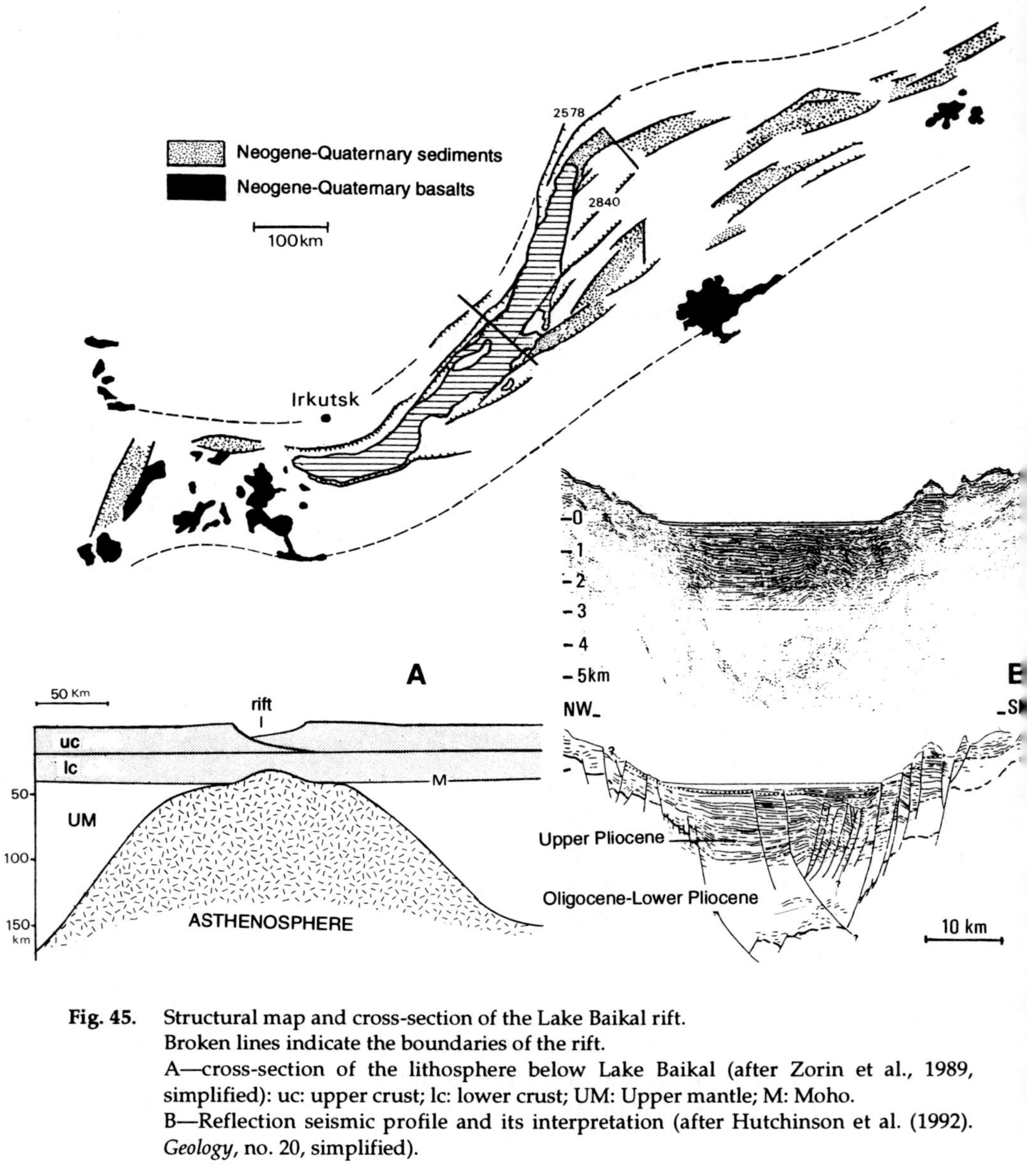

Fig. 45. Structural map and cross-section of the Lake Baikal rift.
Broken lines indicate the boundaries of the rift.
A—cross-section of the lithosphere below Lake Baikal (after Zorin et al., 1989, simplified): uc: upper crust; lc: lower crust; UM: Upper mantle; M: Moho.
B—Reflection seismic profile and its interpretation (after Hutchinson et al. (1992). *Geology*, no. 20, simplified).

The Baikal rift also exhibits high heat flow and negative gravity anomaly due to the large thickness of the sediments it contains. Seismicity is high and mainly concentrated in the marginal faults.

'Pull-apart' Rifts

Pull-apart rifts are also known as *rhombochasms* or *longitudinal extension basins*. These are rifts whose openings are due to the longitudinal sliding of the margins along a strike-slip fault in a broken line (see Fig. 46). Thus a diamond-shaped block is created which undergoes subsidence. Note that it no more represents a classical extension regime but only strike-slip faulting. In other words, these rifts may also be produced in a compression regime (σ_1 horizontal).

The most classical example of a pull-apart rift is that of the *Dead Sea* along the strike-slip fault of Jordan (Figs. 46, 47).

Even though this alignment of rifts appears to be preceded by an elongated depression in the Upper Cretaceous, it is only in the Oligocene-Upper Miocene period that these indicators of rifting manifest themselves. However, the first synrift sediments are of Miocene age and, starting from the Pliocene, the deposits of the basins distinctly differ from those of neighbouring platform areas. This observation is to be considered in parallel with the alkaline volcanism associated with these structures, which started around 10 Ma and reached maximum between 9 and 7 Ma.

Geophysical data indicate that the crust below the Dead Sea is continental but is thinner by 5 to 8 km relative to the normal crust. Transition to the upper mantle takes place by a 4 to 5 km thick intermediate type of crust (continental crust injected by basic products ?).

Geophysical data also show that the disposition of the two, N and S, extremities of the basin is not symmetric. There are some major listric faults in the south that are very flat compared to those of the north, whereas the northern slope is more regular and non-concave. It was hence thought that the former might be associated with a major cleavage surface ('detachment fault' of Anglo-Saxon authors) at depth extending under the whole of the depression (Fig. 46). We shall find a similar disposition in most of the other grabens.

Another example of pull-apart rifts is the *Salton Sea depression*, near Los Angeles, along the San Andreas fault (Fig. 48).

In France, such fossil grabens occur in the midst of Variscan structures but they have been partially closed by the later contraction phases of this orogeny.

In the *Armorican Massif*, pull-apart rifts are represented by the South Armorican major strike-slip faults (Fig. 49), where they are opened up in a compression regime. These are the basins of Chateaulin, Laval, Ancenis, etc. of Lower Carboniferous (Visean) age. All of them show a two-phase evolution:

—*A period of opening up* in which sedimentation is unstable and disturbed by incessant synsedimentary movements due to the deposition of coarse detrital sediments ('debris flows') and even by veritable sedimentary

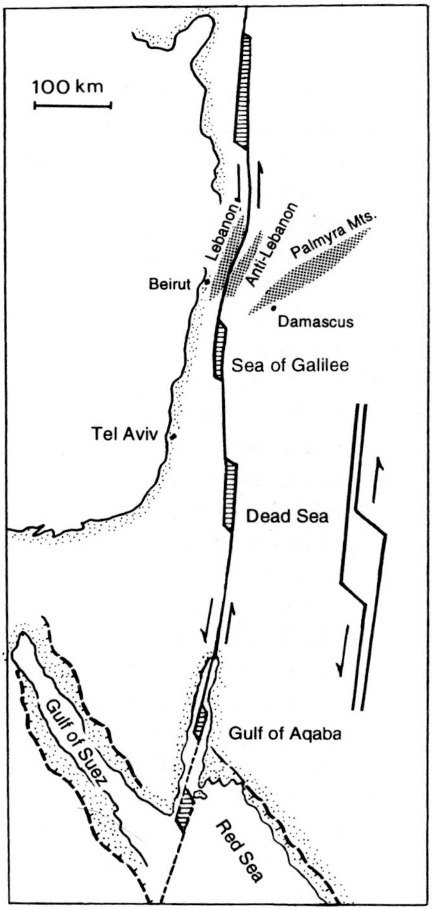

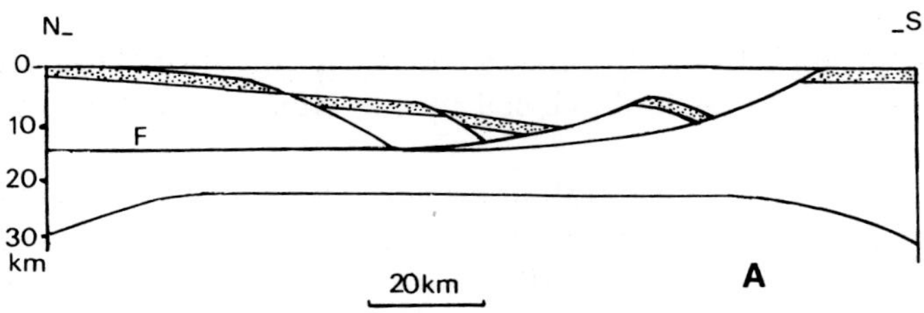

Fig. 46. Pull-apart basins along the Jordan fault.
For the Gulf of Aqaba and the Dead Sea, also see Figs. 38 and 120.
A—Cross-section of the Dead Sea showing the existence of a 'detachment fault' (F) along which extension takes place dissymmetrically. Dotted zones indicate sediments.

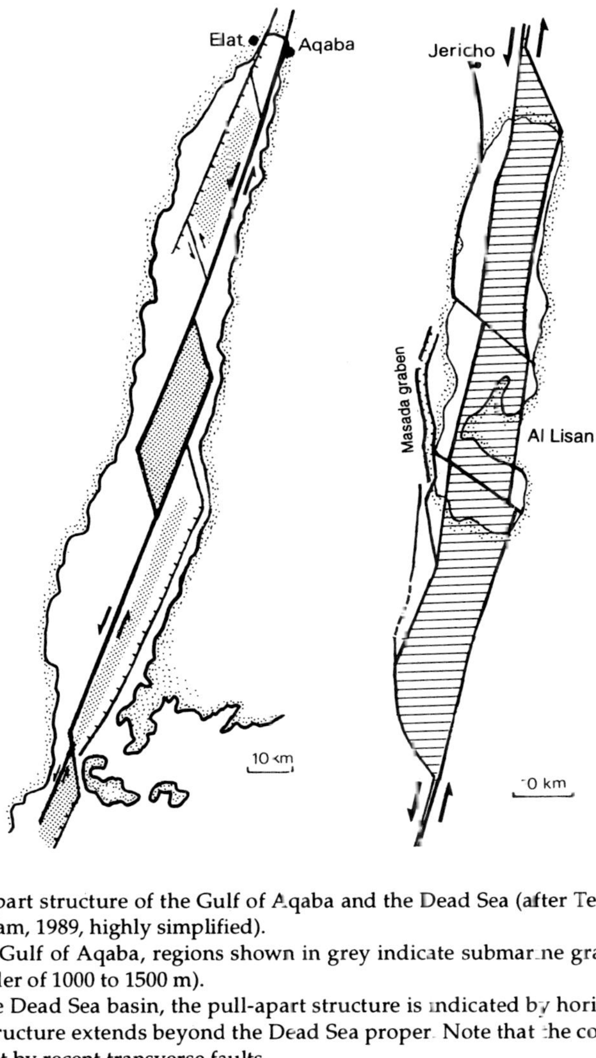

Fig. 47. Pull-apart structure of the Gulf of Aqaba and the Dead Sea (after Ten Brink and Ben Avraham, 1989, highly simplified).
In the Gulf of Aqaba, regions shown in grey indicate submarine grabens (depths of the order of 1000 to 1500 m).
For the Dead Sea basin, the pull-apart structure is indicated by horizontal hatching. The structure extends beyond the Dead Sea proper. Note that the corresponding rift is offset by recent transverse faults.

klippes descending from the dextral strike-slip fault margins. Synsedimetary volcanism is also observed, which is often acidic (quartz-keratophyres) but with some basaltic flows of tholeiitic affinity.

—*A period of sedimentation*, with rapid rate of sedimentation and progressive discordance of younger members over the older ones.

The basins were subsequently folded, stretched and granitised in the Namurian-Westphalian period (Sudetic orogeny).

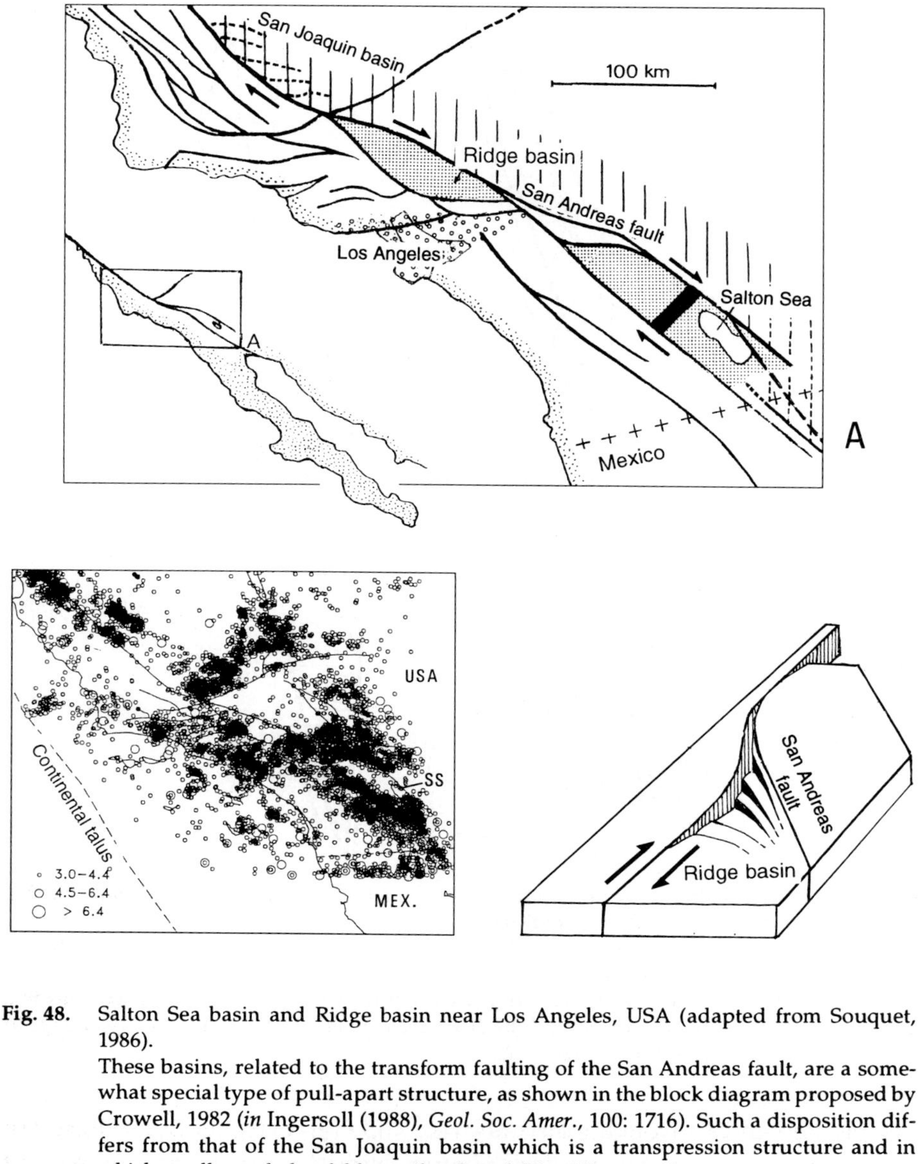

Fig. 48. Salton Sea basin and Ridge basin near Los Angeles, USA (adapted from Souquet, 1986).

These basins, related to the transform faulting of the San Andreas fault, are a somewhat special type of pull-apart structure, as shown in the block diagram proposed by Crowell, 1982 (*in* Ingersoll (1988), *Geol. Soc. Amer.*, 100: 1716). Such a disposition differs from that of the San Joaquin basin which is a transpression structure and in which small en echelon folds are developed (Fig. 69).

The main figure depicts the Colorado palaeodelta (Los Angeles Basin), offset by 300 km relative to the present mouth of this river, point A (after Howard (1996), *Geology*, 24: 783–786). A seismicity map of southern California is also included (after Chen et al., (1991), *Tectonics*, 10: pp. 577–586). Activity of the fault network is evident, especially around the Salton Sea (USA).

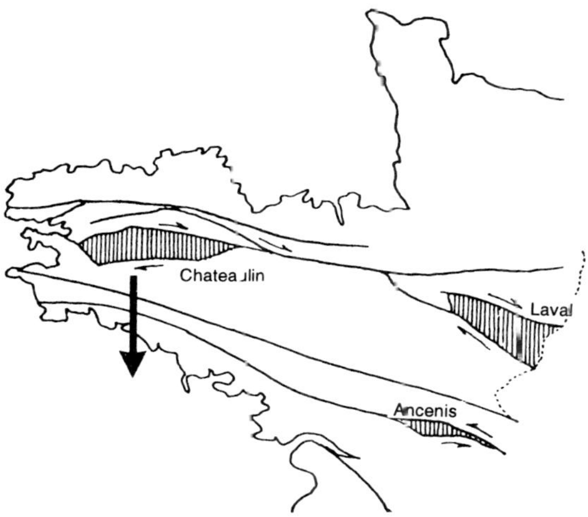

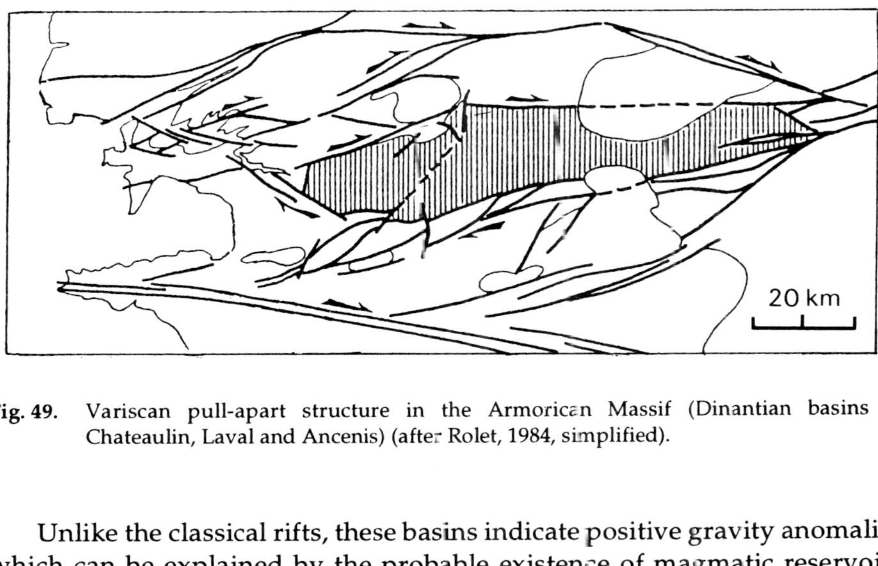

Fig. 49. Variscan pull-apart structure in the Armorican Massif (Dinantian basins of Chateaulin, Laval and Ancenis) (after Rolet, 1984, simplified).

Unlike the classical rifts, these basins indicate positive gravity anomalies which can be explained by the probable existence of magmatic reservoirs close to the surface.

In the *Massif Central* pull-apart basins mark corridors of strike-slip faulting. Important among such rifts are the 'Grand Sillon Houiller' of nearly N-S orientation (with, for example, the Commentry basin) or those of the Rodez region of E-W trend (with, for example, the Decazeville basin) (Fig. 50).

However, the disposition of these rifts is complicated by the fact that the direction of σ_1 has varied over the course of time in a horizontal plane, changing from N-S to E-W. Thus simple extension basins have subsequently evolved as pull-apart basins and vice versa, following their initial orientations.

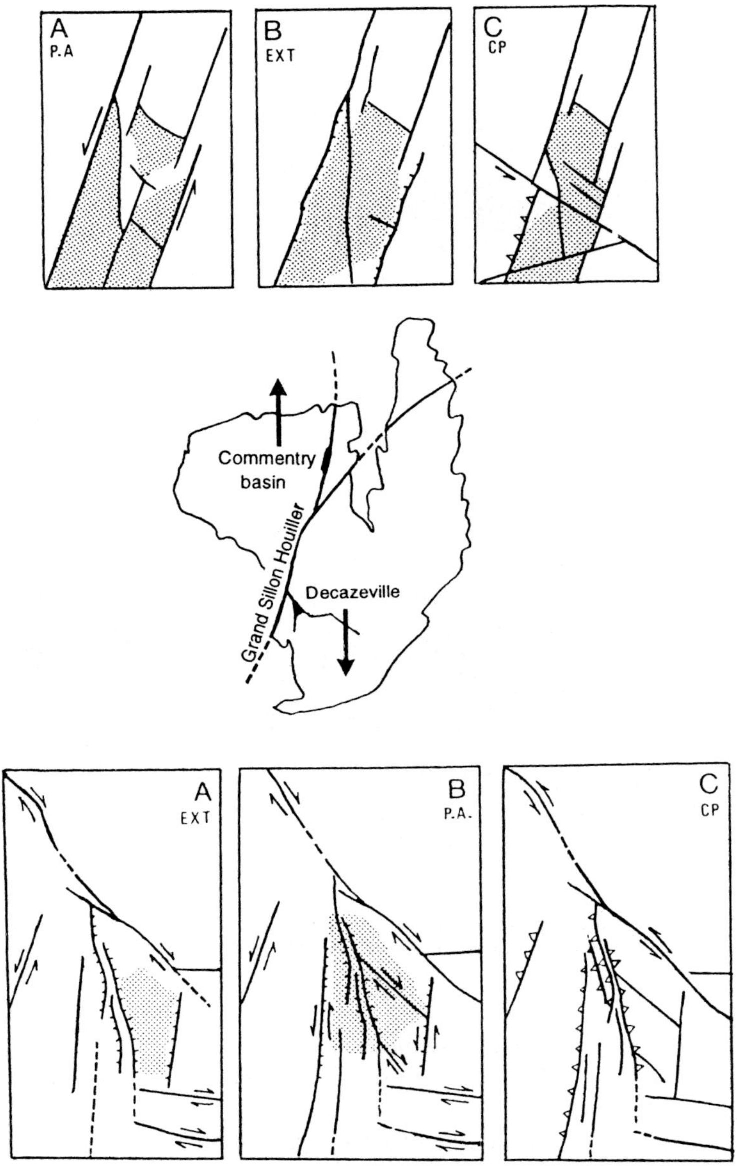

Fig. 50. Structural evolution of the Commentry and Decazeville coal basins (Massif Central) during Stephanian A, B, C (after Bonijoly and Castaing, 1984 simplified and modified).
P.A: pull-apart; EXT: Extension; CP: Compression

Confining ourselves to the two basins cited, viz., Commentry and Decazeville, the basin evolution can be summarised as follows:

Westphalian D. – Stephanian A.: σ_1 N-S

Decazeville = graben of E-W extension

Commentry = pull-apart basin

Stephanian B: σ_1 NW-SE

Decazeville becomes a pull-apart basin

Commentry becomes an extension basin

Stephanian C: W-E compression, resulting in the pinching out and folding of the two basins.

'Basin and Range' Type Extension

The Basin and Range province is situated in the western part of the USA, between Sierra Nevada in the west, the Wasatch mountains and Colorado plateau in the east and the large volcanic plateau of Snake River in the north. In the south the region extends up to Mexico on both sides of western Sierra Madre (Fig. 51). It represents a thick crustal region (ancient Laramian orogeny) as indicated by the high mean altitude of the region.

A regular succession of montane ridges culminating between 2000 to 3000 m and longitudinal basins filled with 2000 to 3000 m thick alluvial production occur throughout this vast region (Fig. 52). Reliefs of variable length and about 30 km in width, sometimes correspond to the classical horst and graben structures but most often represent half-horsts and half-grabens with elevation differences between them reaching 5000 to 6000 m, at least in the case of tilted blocks. These blocks are bounded by typical normal listric faults (concave upwards). The extension they produce has been estimated to vary between 10 and 35% of the original width, and locally at 100%.

Moreover, these tilted blocks lie flat on a Precambrian basement certainly belonging to the middle or lower crust (amphibolite facies). Outcrops of this basement exposed between these blocks form the 'metamorphic core complex' (MCC) which, locally, may present slight domes. The overlying blocks, pertaining to the upper crust, may also consist of metamorphic Precambrians, Palaeozoics and Mesozoics as well as considerable volcanic material of Tertiary age. These different members are truncated by a low-dip fault which separates them from the basement (Figs. 53, 54) and which is associated with a certain thickness of mylonites. This fault is manifestly a subhorizontal decollement surface which nevertheless intersects the listric faults separating the blocks. It represents the most remarkable feature of the

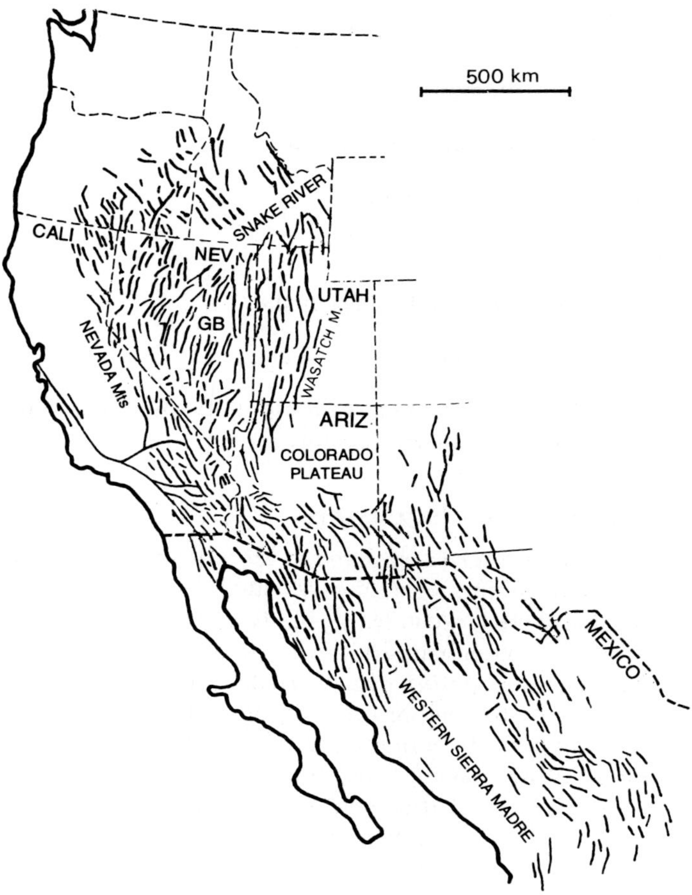

Fig. 51. Geographic distribution of Basin and Range type of extension structures (GB: Great Basin).

region, especially due to its extent (up to 10,000 km^2) and the magnitude of displacement (40 km in Western Arizona).

Geophysical data: The crust is slightly thinned out and yet isostatic equilibrium is established, representing a corresponding rise of the upper mantle. The seismic velocity in the latter is 7.9 km/s, indicating probable partial melting, which is also compatible with double-the-normal heat flux (2 microcal/cm^2/s). Thus the characteristics of extension zones are observed here.

Age of extension: Even though the extensional state has prevailed since the beginning of the Tertiary, in which volcanism appeared, the present structure developed in the Miocene, when sedimentary filling of the basins

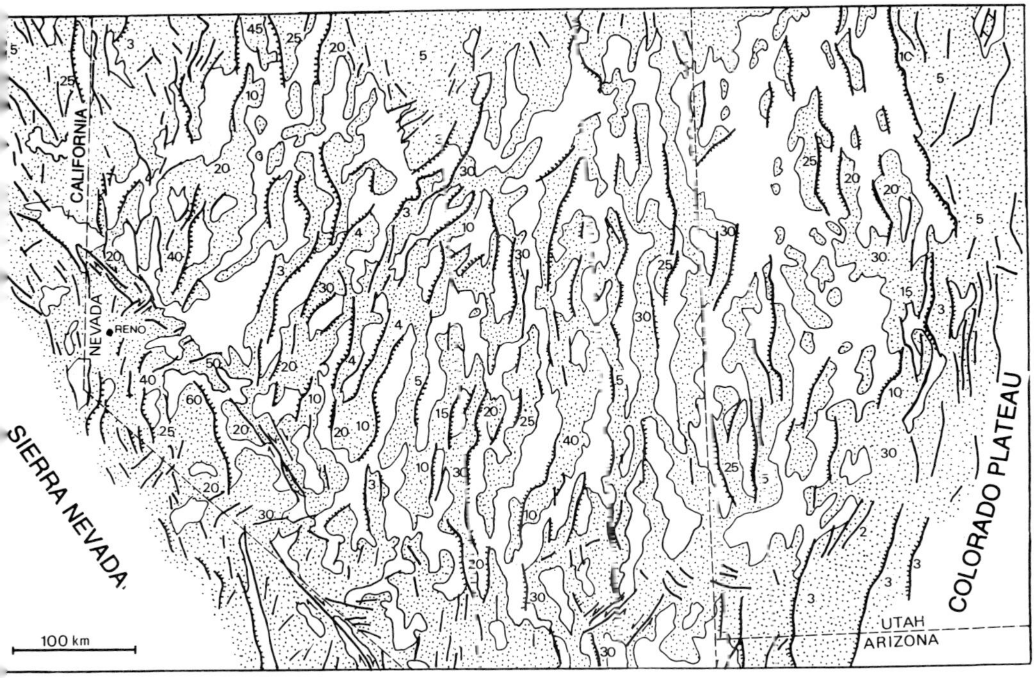

Fig. 52. Detailed structure of the Great Basin (Basin and Range Province) between Sierra Nevada and the Colorado plateau (after Stewart, 1978).
Montane massifs indicated in grey. Normal faults shown by hatched black lines. Numbers indicate the mean slopes of tilted blocks.

commenced. In this epoch the nature of volcanism also changed. Hitherto calco-alkaline (influence of Pacific subduction), it became alkaline.

Extensional mechanism: This has been highly debated to such an extent that these enigmatic horizontal faults were even considered to be the ancient Laramian thrust planes, reactivated into normal faults. Only extensional movement is now suggested, the latter leading to an extreme stretching of the crust during which the upper part may have been subjected to brittle deformation whereas the lower part may have undergone a more plastic (ductile) movement. The result may hence have been a total rupture of the upper crust into tilted blocks which have thus come to rest on their edge over the subhorizontal fault mentioned above, marked by mylonites that indicate a deep intracrustal cleavage.

However, radiometric ages of granites affected by the phenomenon show that the superposed parts of the crust were not deformed at the same time, deformation of the upper crust being later to that of the lower crust. The mechanism is thus more complex and the spectacular fault is certainly not the primitive decollement level. The model suggested in Fig. 53 shows extension shear surfaces more or less obliquely intersecting the crust, surfaces known as 'detachment faults', the possibility of which in another extension structure was mentioned earlier (Fig. 46, p. 72). Views differ on whether these faults continue at depth across the entire lithosphere (Wernicke's model which we shall recall while discussing the continental margins, p. 146) or attenuate horizontally in the ductile formations of the middle crust (model of Lister and Davis, Fig. 53). In any case, for the problem of the fault surface under discussion, the mechanism practically remains the same and explains the presence of mylonites, indicating a deep ductile milieu below the tilted blocks, an evidence of surficial and ductile tectonics.

In the first stage (Fig. 53.1), a ductile shear plane, φ, marked by mylonites, is formed more or less horizontally in the crust, probably between the upper and lower crusts. This is joined by listric normal faults (f) separating the first blocks tilted by extension.

Subsequently, under the effect of persistent extension, listric faults that are flatter and nearer to the surface ('detachment' faults F1) appear and complicate the breaking of the tilted blocks (stage 2).

In stage 3, the extension ends by causing a thinning of the crust sufficient to trigger uplift of the upper mantle and hence of the thinned-out lower crust. The detachment faults F1 are warped and brought nearer to the surface along with the mylonites of stage 1 that are generated in a ductile milieu and henceforth are fossils. This phenomenon leads to sliding of blocks of the overlying surficial crust. In other words, starting from a certain stage (stage 4), a new detachment fault (F2) is formed separating these surface blocks from the underlying basement always topped by their ancient mylonites that are 'inactive' (it thus represents too shallow a regime to form new mylonites). The horizontal fault, characteristic of the present Basin and Range type structures (F2), thus gets its final form whereas the old metamorphic basement comes to outcrop (MCC).

The behaviour at depth of the extending shear plane (F2) (inherited from φ) is beyond the limits of the Figure and is much more conjectural.

The extension phenomenon gives rise to a significant rise in temperatures, leading to crustal melting and hence to emplacement of syntectonic granites whose low density also contributes to uplift of the basement. Their syntectonic nature is manifest by extensional shear deformation which affects them in the vicinity of the detachment fault.

This mechanism ends up in thinning of a thick crust.

Structural framework of Basin and Range extension: The structural set-up of rifting is not very clear and several explanations have been advanced.

1. The first hypothesis implies an extension in an oblique direction relative to the San Andreas fault (Fig. 55). This explanation accords well with the fact that formation of the Basin and Range province—or the Great Basin—is contemporary to the San Andreas faulting and its geographic extent is exactly the same as that of the fault. However, this view cannot explain the circular form of the Great Basin which moulds the volcanic arc of the Sierra Nevada and Cascade Mountains.

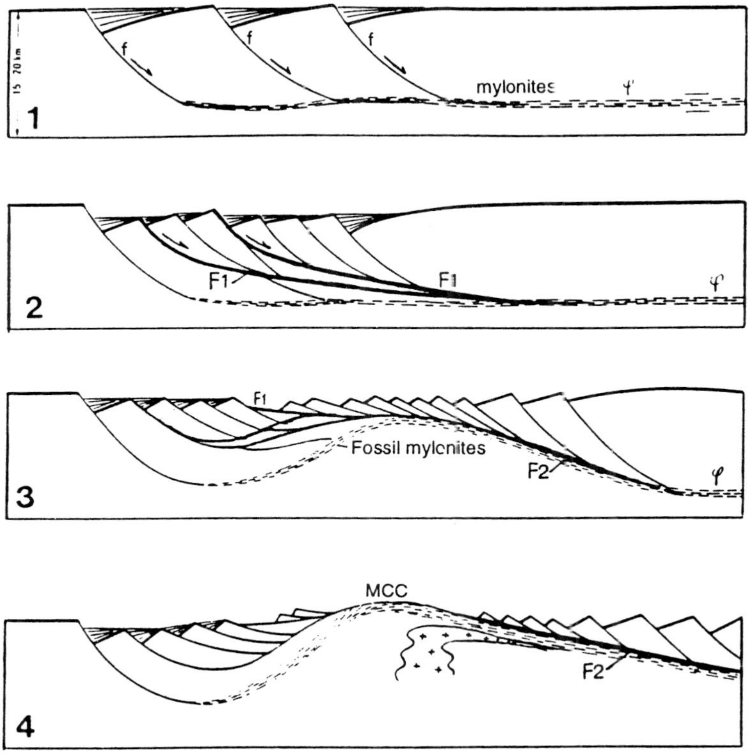

Fig. 53. Model of formation of Basin and Range type extensional structures (after Lister and Davis, 1989, simplified).

2. From the latter emerges the idea that the Great Basin could as well be a back-arc basin (see p. 161), a typical structure on the active margins, immediately behind the volcanic arc and probably associated with asthenospheric convection above the subduction plane underlying the active margin in question (see Fig. 20). This explanation conforms with the calco-alkaline nature of the volcanism, at least at its beginning, high heat flow, crustal thinning and abnormal character of the uplifted upper mantle. The difficulty lies in the chronology: the Great Basin extension in fact commenced when functioning of the said subduction zone ceased.

It may thus be considered that the back-arc basin stage initiated the extension associated with movement of the San Andreas fault. This association is indicated by some quantitative data. As a matter of fact, based on a precise stratigraphic and tectonic analysis of the blocks in the Las Vegas region, Wernicke et al. (1988) were able to calculate that an extension of 247 ± 56 km occurred between the Sierra Nevada and the Colorado plateau in the last 15 Ma, which amounts to an approximate rate of 2 cm/yr. During the same period, dextral extension of the San Andreas fault was 214 ± 48 km, a value of the same order of magnitude as that of the preceding one.

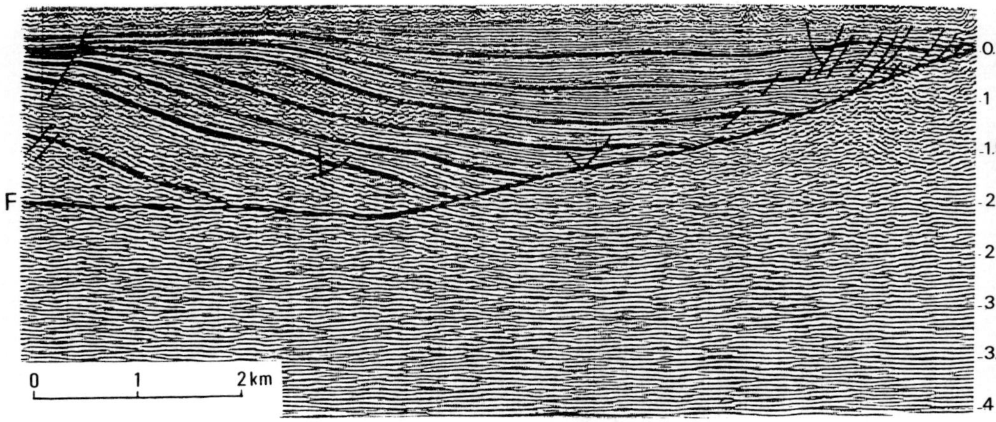

Fig. 54. Detachment fault in a seismic reflection profile below the 'Basin and Range Province' basin (Marys River valley, north of Nevada, SW of Snake River plateau) (after A.W. Bally (1983), Seismic expression of structural style, *Amer. Assoc. Petr. Geol.*, sec. 2-2-2, p. 16).
Vertical scale: two-way time in seconds.

3. A third possible explanation was recently proposed (Molnar and Ménard (1988), *Nature,* 334: 236), namely an extensional spread, by gravity, of an earlier orogenic structure (here Laramian), after which contraction ceased. We shall see later (pp. 278) that most folded mountain belts are accompanied by a thickened crust. The latter can maintain itself only if its margins are blocked, as in the case of the Tibetan plateau. Otherwise, the material tends to creep under the effect of gravity and regains a thickness such that the viscosity of the material maintains cohesion of the whole. Such a state is produced for a thickness between 30 and 35 km. Hence it may be assumed that the Basin and Range type of structure represents the ultimate stage of such spreading of an orogenic plateau, in which 'detachment faults' play a major role.

BASINS PROPER (CRATONIC BASINS)

Simple cases: These basins are oval or circular depressions in which the sediments extend as a continuation of those of neighbouring platforms, which are much thinner with stratigraphic gaps. The maximum thickness remains around the same axis throughout the evolution of the basin. The depth/diameter ratio is of the order of 1:100 to 1:50.

Paris Basin (Figs. 56 to 58)

This basin contains nearly 2500 m thick Mesozoic and Tertiary sediments (as against a few hundred metres of the same formations in its periphery). This

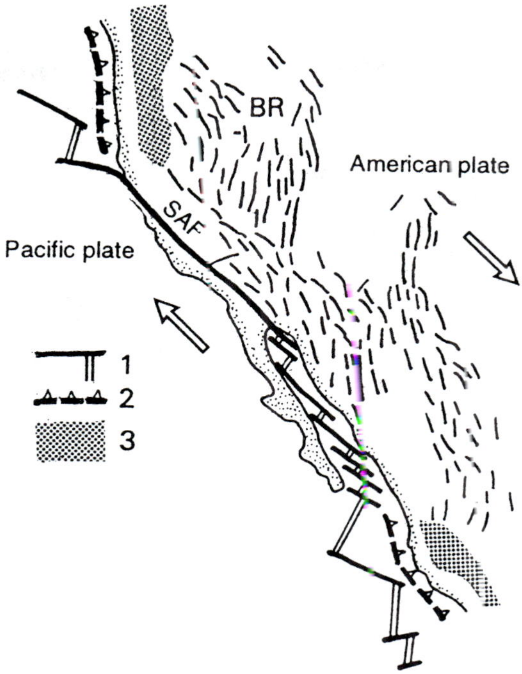

Fig. 55. Tectonic framework of Basin and Range structures (after Stewart, 1978).
One can observe here:
— either the consequence of San Andreas faulting (SAF) (with which these structures are contemporary and which they mark);
— or the effect of subduction of the North American plate (extension of a back-arc basin).
1. Transform fault and ridge; 2. Subduction zone; 3. Neogene calco-alkaline volcanism; BR: Basin and Range province.

thickness corresponds to an average sedimentation rate of 1/10 mm/yr. In fact, this figure is not significant because the evolution of the basin had to proceed in a quite irregular rhythm, even with some positive oscillations (one at the end of the Jurassic and another at the end of the Cretaceous, just to cite two principal ones) that resulted from regressions and a general emergence. These uplifts (sometimes denoted as positive epeirogeny, subsidence being a negative epeirogeny) do not represent a tectonic characteristic of the basin or of the subsidence. They are, in fact, contemporary to the major global tectonic episodes during which the plates were subjected, at least locally, to compression resulting in the formation of salient structures.

The pole of subsidence was displaced slightly during the Mesozoic (Fig. 57) and more strongly in the Tertiary under the evident influence of alpine orogeny.

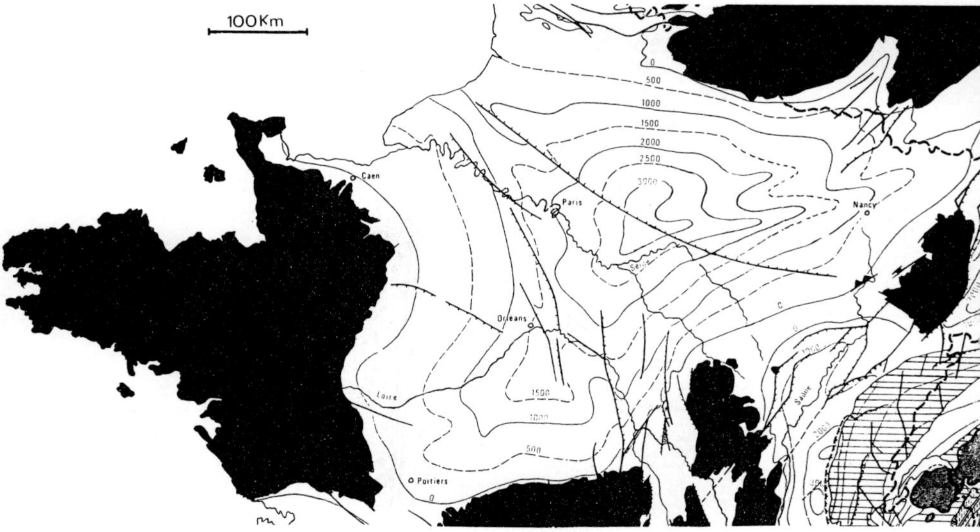

Fig. 56. Paris basin.
Isobaths of the top of the basement and major faults.

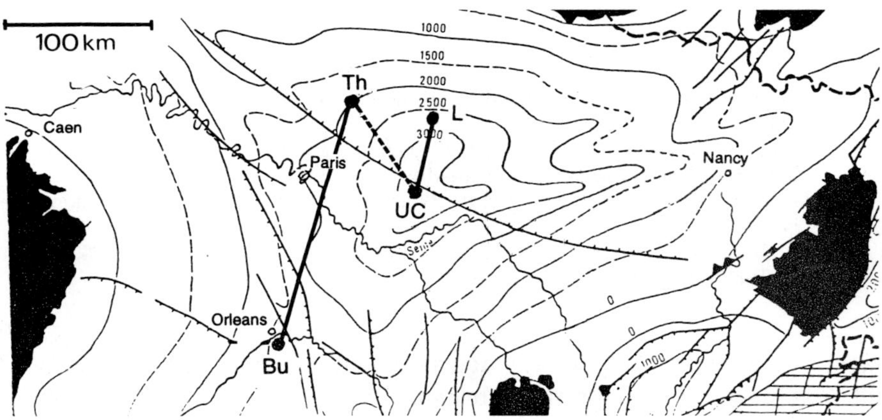

Fig. 57. Migration of the pole of subsidence of the Paris basin between the Liassic (L) and the Upper Cretaceous (UC) by 50 km, and between the Thanetian (Th) and the Burdigalian (Bu) by 150 km. Note that the effect of the early alpine folding has strongly pushed the maximum subsidence of 70 km NW (after Pomerol, 1980).

It may also be noted that all the transgressions have not been due to the subsidence of the basement. *Eustatic variations* of the sea level should also be taken into account. For example, the Upper Cretaceous transgression was caused due to a rise in sea level by 400 m throughout the world.

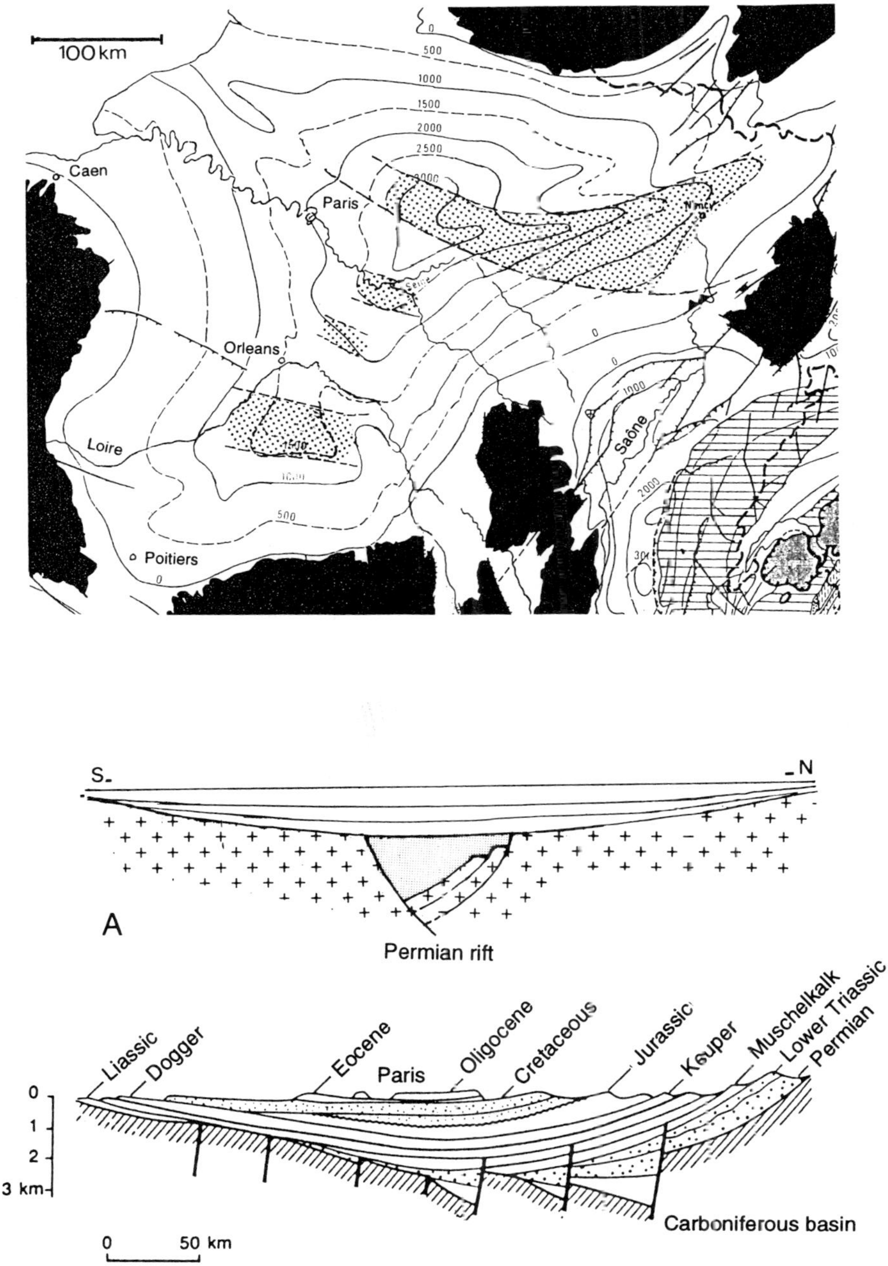

Fig 58. Permian rifts below the Paris basin (reconstructed from results of deep drilling). Note their coincidence with the maximum subsidence zones of the final basin.
A: Theoretical cross-section showing how the existence of a Permian rift leads, by thermal debulging, to the origin of subsidence in its vertical section.

Subsidence Model: The basin obviously involves an extension phenomenon but its *beginning* and its *maintenance* need to be explained.

The origin of the process becomes distinct when the map of Permian subsidence zones is superposed on the map of the present-day basin (Fig. 58). During the Permian, while fragmentation of the Pangaea got underway, the post-Variscan peneplain was subjected to an intensive tensile regime, giving rise to two elongated grabens along the Variscan directions, in which are deposited 500 to 1000 m of Permian sediments. Hence the maximum subsidence of the Paris basin follows the geometry of these grabens, beyond which it greatly extends. It thus follows that after the Permian rifting, *thermal debulging* occurred and the latter led to the formation of a vaster zone of depression in which the Mesozoic subsidence took place (Fig. 58A). Commencement of the subsidence phenomenon is thus an extension applied to crust weakened by older structures which themselves were extensional. This can be readily verified on other examples such as the *North Sea basin* which follows a Permo-Triassic rifting (Fig. 59).

However, it is now necessary to explain why the phenomenon proceeded with a certain regularity and the reasons for this are not very clear. It may be possible to find them in the horizontal displacement of the lithosphere associated with global tectonic movements. This displacement in fact affects a heterogeneous lithosphere consisting of 'hard cores' (which in this case are ancient massifs, such as the Massif Central, Armorican Massif and Vosgian mountains) with a crust thickened by Variscan folding and undergoing a continuous isostatic uplift, and of more supple intermediate zones with a thin and probably fractured crust. Coming together of the hard cores results in the formation of intracontinental mountain chains; their moving away from one another produces a subsidence basin provided the rhythm is sufficiently slow so that there is no brittle subsidence or even fracture (crustal fractures, p. 112).

Finally, this type of basin, situated at a low altitude, has always been very sensitive to eustatic variations. Their stratigraphy faithfully reflects major transgressions and regressions.

OTHER EXAMPLES

Palaeozoic basins of Michigan, Illinois and Oklahoma in North America, of Tindouf and Karroo in Africa and Mesozoic to Tertiary basins of Chad (Fig. 63), Fezzan and South Libya in Africa and of Arabia, China, Orinoco, etc.

Complex Basins

Complications of structures are produced, especially:

1. *When the surface of the basin is very large and its history is long,* because it is then divided into secondary basins in which the maximum subsidence zone is displaced, resulting in tilted basins, one fitting into the other. The Sahara basin is an example (Fig. 60).

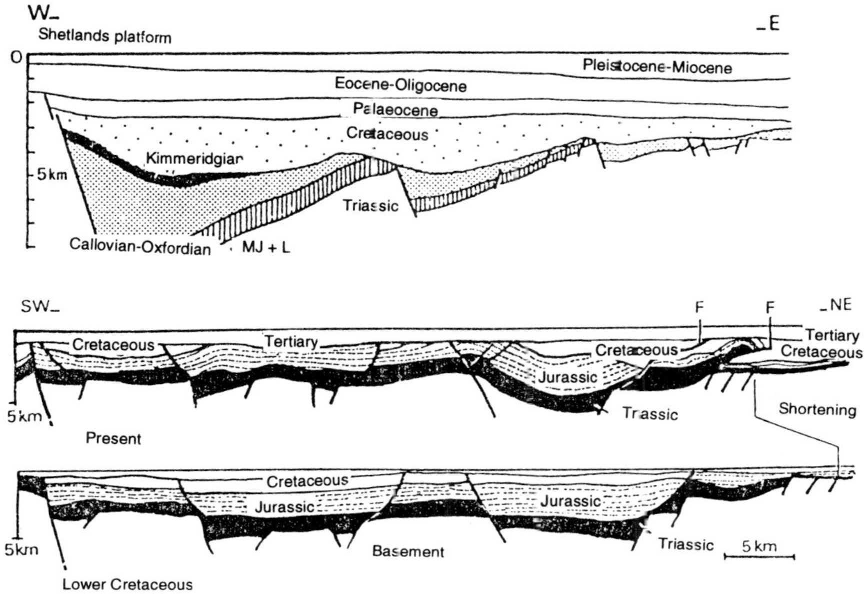

Fig. 59. Cross-sections of the North Sea basin.
Top: Between Shetland Islands and Norway (after Ferrodon, 1983).
Bottom: Between England and Holland (after Huyghe, 1992).
At the end of the Cretaceous, a contraction phase affected the basin which rejuvenated the normal faults into reverse faults (F) and produced some folds. Extensional activity recurred in the Tertiary.

2. *When the subsidence is accompanied by extensional faulting* and is broken by contraction phases that can induce tectonic inversion with folding and even local overthrusting (North Sea basin, Fig. 59).

3. *When basins contain beds of evaporites,* because, in this case, salt tectonics may develop there resulting in salt domes or diapiric folds (Aquitaine Basin, N. Germany-Poland basin, Gulf of Mexico, etc.) or synsedimentary slidings that appear as local thrusts.

4. *When the thickness of the sediments becomes very large,* reaching tens of kilometres and even more, which implies a thinning out or local ruptures of the crust. This is the case of the Aquitaine and Tindouf basins. An extreme case, difficult to interpret, is that of the Precaspian depression west of the Caspian Sea. In its central part it shows an 18 to 25 km thick Ordovician-Neogene sequence with little or no folding. Such a disposition implies considerable thinning and even disappearance of the underlying continental crust, but the mechanism remains unknown. Is it lateral creep or subcrustal erosion by asthenospheric currents? digestion by mantle magmas? or simply an extreme stretching accompanying a crust-mantle decoupling?

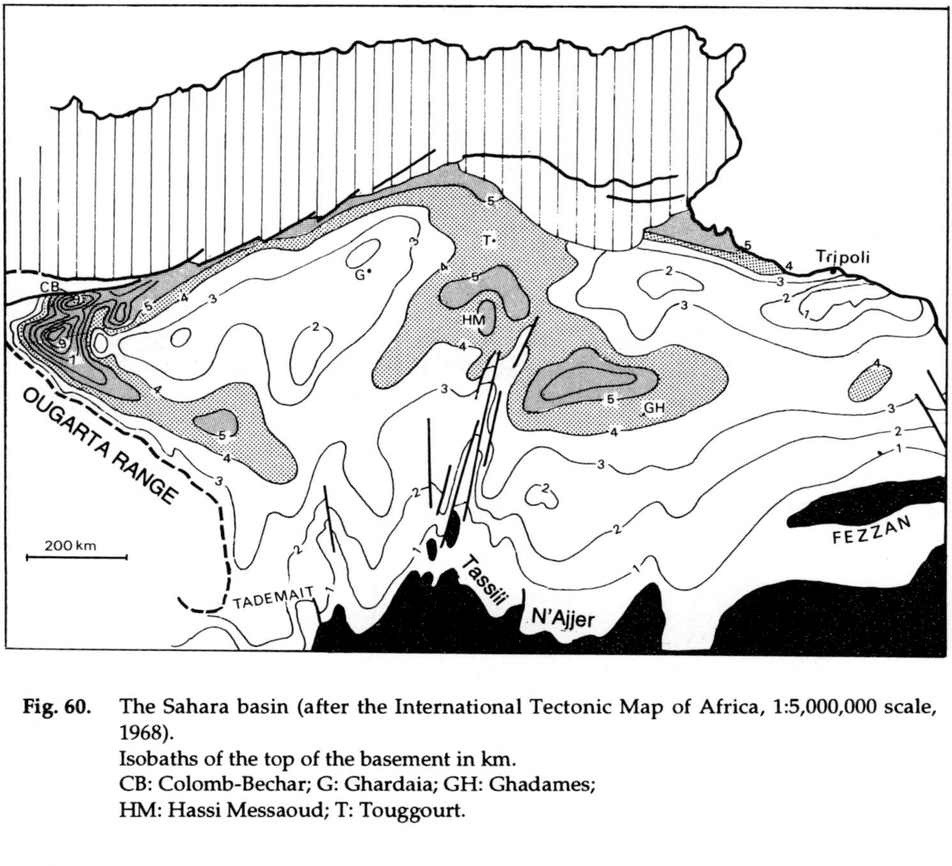

Fig. 60. The Sahara basin (after the International Tectonic Map of Africa, 1:5,000,000 scale, 1968).
Isobaths of the top of the basement in km.
CB: Colomb-Bechar; G: Ghardaia; GH: Ghadames;
HM: Hassi Messaoud; T: Touggourt.

MIXED BASINS

Classical subsidence in this case is actually associated with collapse by faulting, at least locally and temporarily. Moreover, alternation or juxtaposition of extension and compression can also be observed.

The structural sites in which the basins are developed may be highly varied, as the following examples show, viz., aulacogens, strike-slip fault corridor basins and molassic basins of orogenic zones.

Aulacogens[2]

These are sedimentary basins developed on a continental margin perpendicular to its opening into the ocean. The other end of the basin gradually

[2]Term introduced in 1940 by the Russian geologist Shatzki, based on the Greek word *aulax* (trough).

closes in the continental mass. Moreover, aulacogens develop at places where a distinct re-entrant angle exists in a coastline. In fact, there exists in this place a particular zone, a sort of junction between two lithospheric blocks of different surface areas. This zone thus functions as a 'dilatation joint' between these blocks and follows the tectonic regime which the lithospheric blocks in question are subjected to. It may hence operate in compression or extension wherein the latter may even be terminated by the appearance of a crustal fracture (see p. 112).

EXAMPLES

Aquitaine basin in the Cretaceous (Figs. 61, 62)

This basin appears in the Triassic in the form of a large rift valley bounded in the north by a flexure zone (Arcachon-Toulouse flexure) and in the south by a series of major faults of Variscan origin, parallel to the future Pyrenees belt. The North Pyrenean fault is the most important of these faults (see p. 177 and Fig. 125). In the Jurassic, this regime continued with shallow-water sedimentation (evaporites, dolomitic limestones) and ceased with emergence at the end of the Jurassic-beginning of the Cretaceous.

Starting from the Aptian-Albian, following the opening of the Atlantic, the Aquitaine basin becomes an aulacogen open towards the west, i.e., towards the newborn Atlantic, at the junction of the European plate and the Iberian microplate (Fig. 62). Sedimentation then is very active, maximum not only at the emplacement of the ancient rift zone, but also beyond its margins. It is accompanied by light volcanic activity (submarine alkaline basalts) which indicates the existence of fractures at the base of the aulacogen (probably horst-graben structure?).

Like all aulacogens, the Aquitaine aulacogen underwent a true crustal fracture in the North Pyrenean zone during the Aptian and, conversely, several compressive episodes in the Albian, the Upper Cretaceous and the Eocene.

This evolution will be mentioned again with regard to the opening of the Gulf of Gascogne (p. 127) and the origin of the Pyrenees (p. 180).

Benue basin (Nigeria) (Figs. 63, 64)

This 1000 km long and 50 to 100 km wide basin extends from the Gulf of Guinea to the southern part of Lake Chad where it terminates in a Y-shape on either side of the Biu volcanic plateau. This basin, with Cretaceous sedimentation, is directly encrusted in the crystalline basement and offers a sedimentary sequence 1000 to 5000 m thick at some points. This sequence is essentially continental with lacustrine sediments, except in the Turonian when the great and classic marine transgression invaded it, joining the Atlantic with the Tethys across the Sahara.

Subsidence commenced in the Aptian, manifestly associated with the opening of the Atlantic. It continued up to the end of the Senonian (locally

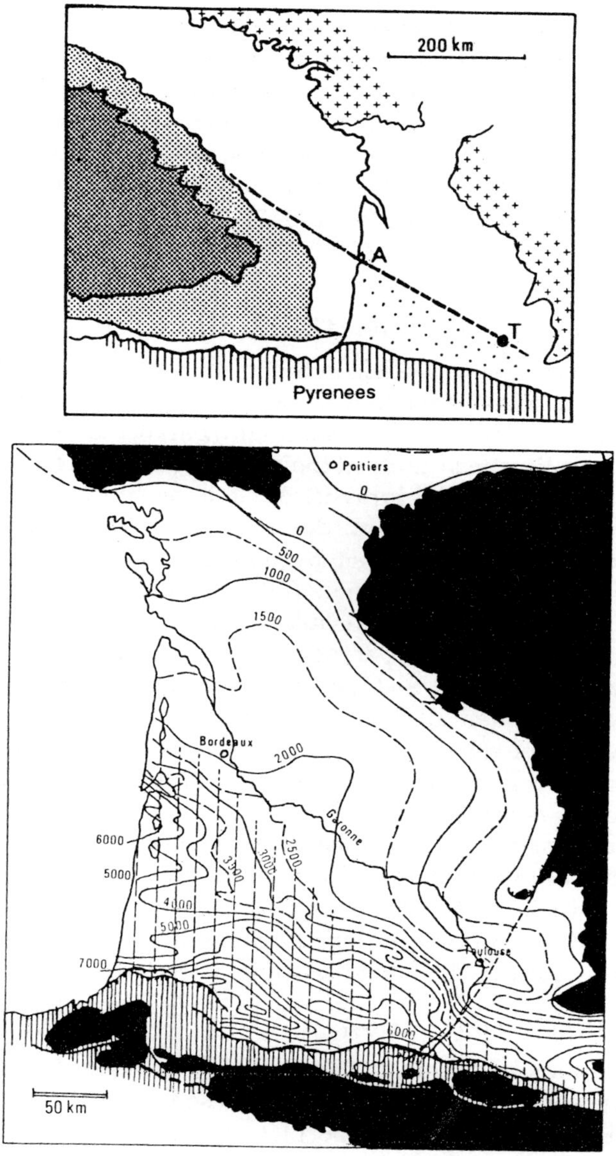

Fig. 61. Aquitaine basin and its relation to the Gulf of Gascogne (dark grey: oceanic crust; light grey: continental slope).
AT: Arcachon-Toulouse flexure, forming the northern boundary of the basin of Triassic evaporites and extended by the South Armorican margin. This flexure separates the Aquitaine basin into two parts, of which the southern one has the nature of an aulacogen.
Note the coincidence between this subsiding southern part of the basin and the zone of Triassic evaporites. The part of the basin affected by Pyrenean folding is indicated by vertical wide-spaced broken lines.

Eocene ?) interrupted by two phases of folding, one in the Santonian and the other at the end of the Cretaceous. These phases gave rise to elongated folds, parallel to the axis of the basin, with a north-westerly vergence, and also rejuvenated the existing longitudinal fractures into sinistral strike-slip faults. The second phase was followed by eruptions of alkaline basalts which continued in the Tertiary and Quaternary in the northern part of the basin.

The Bouguer gravity anomaly map shows that the region on the whole is in isostatic equilibrium. However, a small positive anomaly is observed on the axis of the basin and a negative anomaly away from it. This is interpreted as due to the thinning of the continental crust below the axis of the basin, accompanied by magmatic intrusions, whereas the sedimentation is thickened in the marginal troughs where it may locally reach 4000 to 6000 m.

Other geophysical data suggests that the basement of the Benue basin is faulted below its Cretaceous cover. Based on this, it is deduced that the basin started as a rift but the extension zone was enlarged later with the slowing down of subsidence, before being interrupted by folding and rejuvenation of fractures.

This example clearly illustrates the 'dilatation joint' behaviour of aulacogens in which extension and compression may succeed or alternate with one another.

Aulacogen of Athapuscow or Great Slave Lake, Canada (Fig. 65)

Of Precambrian age, this aulacogen intersects the western boundary of the North American Archaean craton. Its outlet into the ocean of that period is no longer clearly visible, as the ocean in question has disappeared. The Western Canadian Cordillera intersects it at its western extremity.

The time duration of deformations recorded in it extends between 2200 and 1800 Ma. It consists of a classical rifting followed by a geographically enlarged subsidence phase accompanied by volcanism, then a compression phase (Hudsonian orogeny at 1700 Ma). A new stage of extensional subsidence occurs around 1400 Ma. The Variscan and Laramian episodes are mainly accompanied by strike-slip faulting because of the E-W orientation of the primitive structure.

The final structure obtained is exactly identical to those described in the preceding section.

Transverse-Fault Basins

Examples

Small Neogene basins of the Betica region (Spain)

These basins are situated on a swarm of SW-NE faults traversing the eastern part of the Betica mountain chains and developed from earlier structures. During the Miocene this swarm was subjected to a submeridional compres-

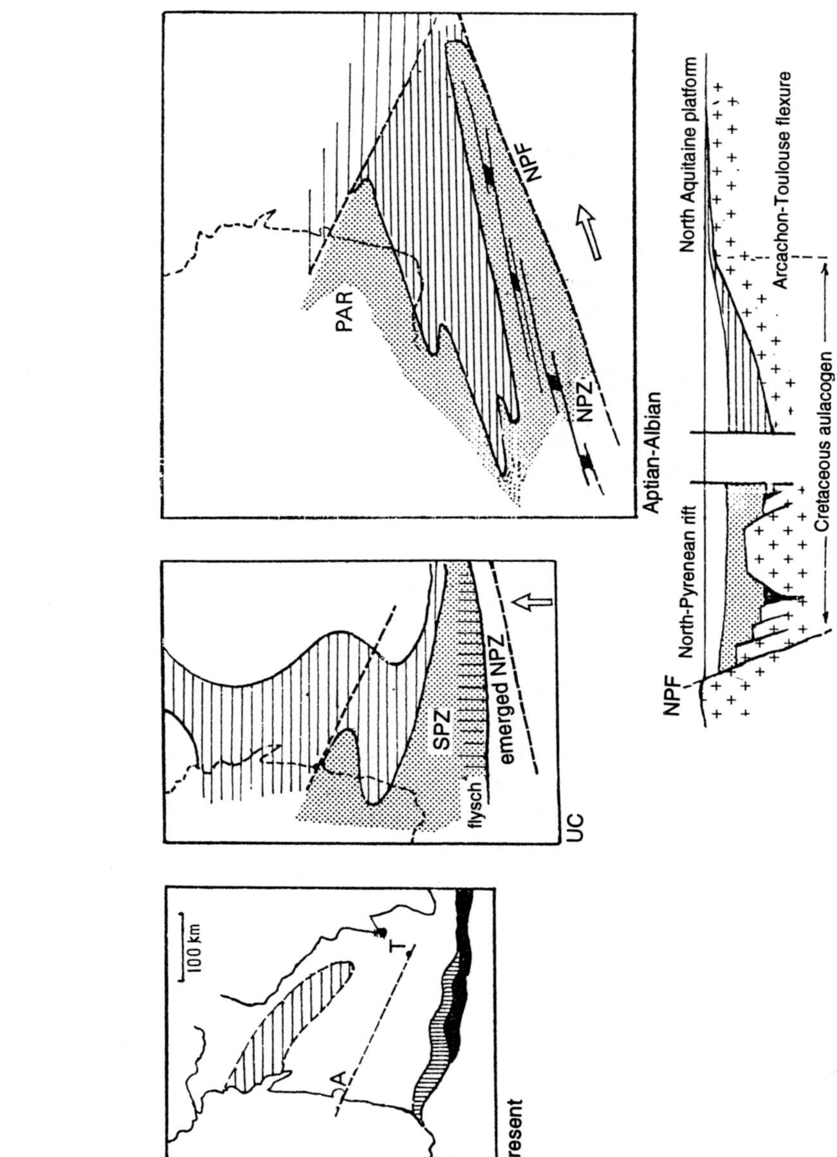

Fig. 62. Palinspatic reconstruction of the Aquitaine basin in the Cretaceous showing its functioning as an aulacogen. AT: Arcachon-Toulouse flexure; NPF: North Pyrenean fault; NPZ: North Pyrenean zone; PAR: Parentis basin; SPZ: Sub-Pyrenean zone.

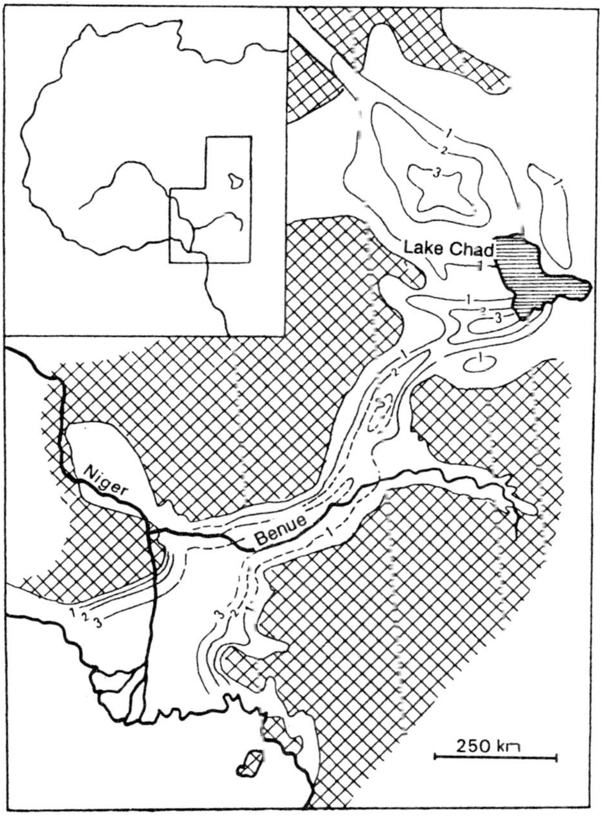

Fig. 63. The Benue Basin (Nigeria) and its relation with the Chad basin.
Isobaths of the top of the basement in km (after the International Tectonic Map of Africa in 1:5,000,000 scale, 1968).

sion regime as a result of convergence of the Africa-Iberian plates. Two types of sedimentary basins have thus appeared (Figs. 66, 67).

— *Extension basins:* Sometimes as half-grabens (Cartagena, Murcia) and sometimes as true pull-apart structures (Lorca), relatively less subsident, poor in detrital material but better endowed with organogenic limestones, indicating a relative stability of their margins. They hence represent surficial structures.

— *Compression basins:* These involve true synclinal folds facilitating sedimentation by the hollow effect. They are strongly subsident with a detrital sedimentation in which turbidites dominate, indicating uplifted and intensely eroded margins that are often overthrust. Synsedimentary deformations are invariably observed, especially in the vicinity of the strike-slip faults: vertical-axis folds, reverse faults, conjugate transverse faults, syn-

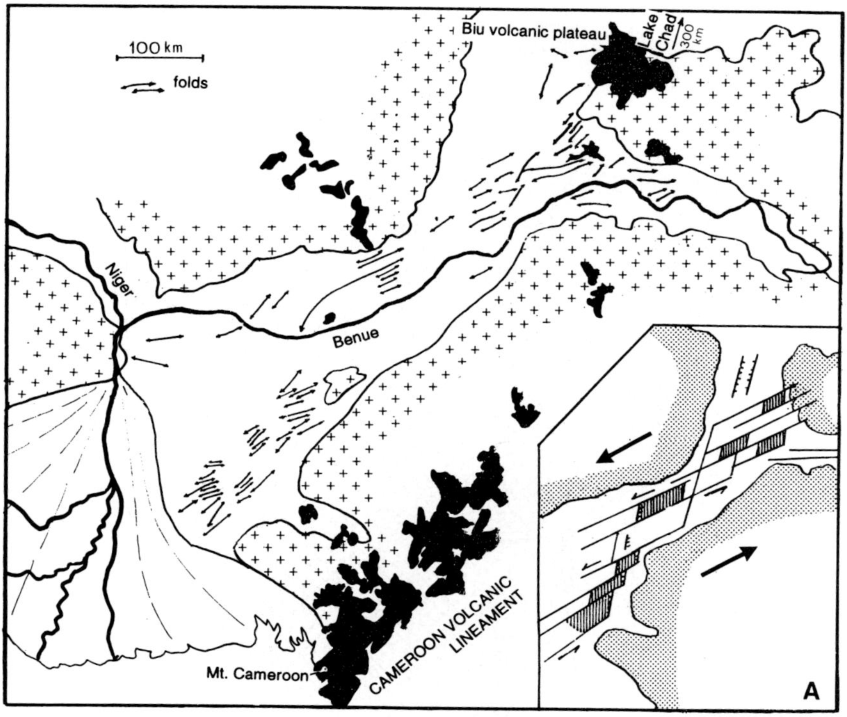

Fig. 64. Simplified structural map of the Benue Basin (adapted from Benkhelil, 1986). A: mechanism of formation of Albo-Aptian pull-apart sedimentary basins.

sedimentary en echelon folds (sometimes topped by algal structures emerging from the marl-turbiditic sedimentation). Subsidence here is obviously associated with compression since 2000 m of sediments were deposited during the Upper Tortonian period in less than 3 Ma, at the moment of the compressive episode.

In spite of the compression regime, the major faults are marked by *high volcanic activity* (notably Cabo de Gata), which attained maximum in the Tortonian period itself. This volcanism is calco-alkaline (dacites and rhyolites) with explosive type products (breccia, cinerites, dominant ignimbrites) often shoshonitic. However, the presence of a number of xenoliths with garnet, cordierite, andalusite, sillimanite, etc. and some veins of leucogranites suggest that this volcanism is produced from the anatexis of the basement gneisses and hence has nothing to do with any subduction plane (Fig. 68).

Volcanic activity was terminated in the Pliocene towards Cartagena by the common alkaline basalts of mantle origin.

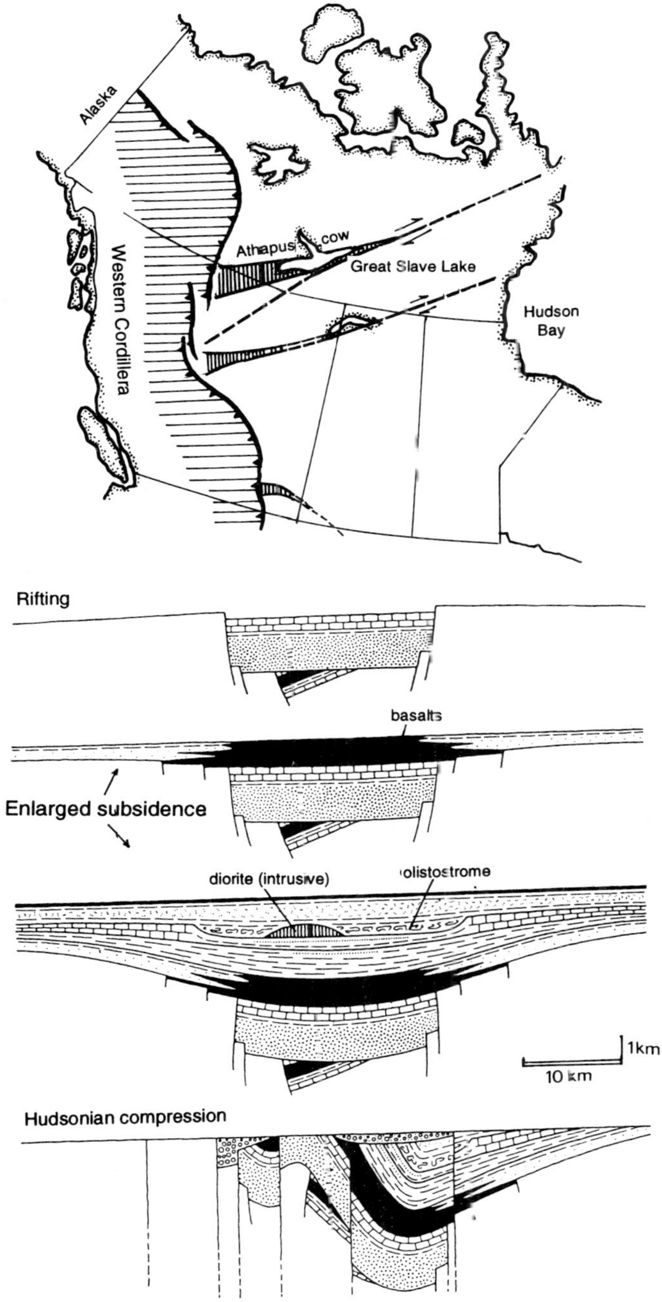

Fig. 65. The Precambrian aulacogen of Athapuscow (Western Canada) (after Hoffmann et al., 1977, simplified).
Location plan and evolutionary sections. Facies are represented by their conventional figures in the latter.

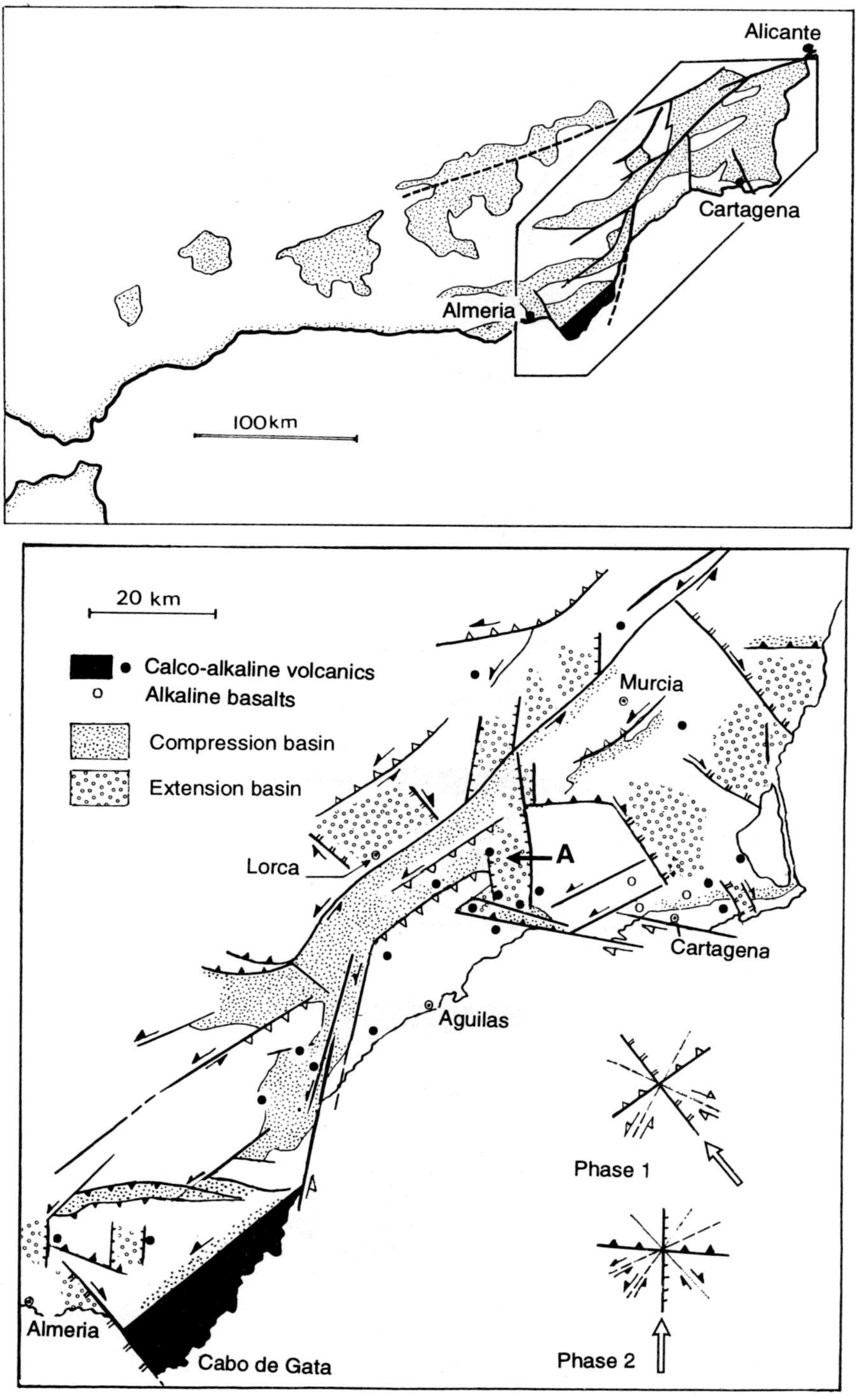

Fig. 66. Eastern Betica molassic basins (after Montenat et al., 1987, simplified).

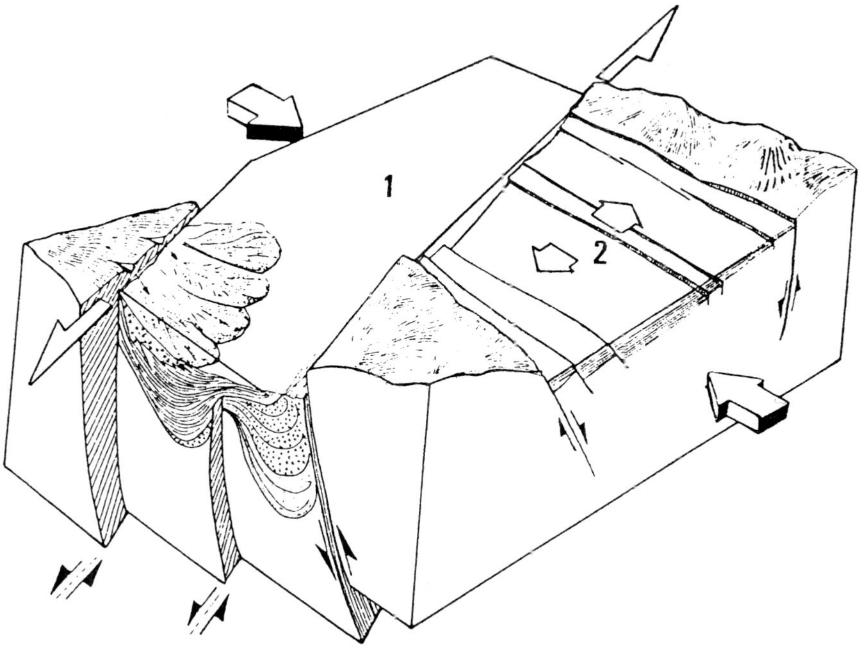

Fig. 67. Different types of Eastern Betica molassic basins (after Montenat et al., 1987).
1. compression basin 2. extension basin

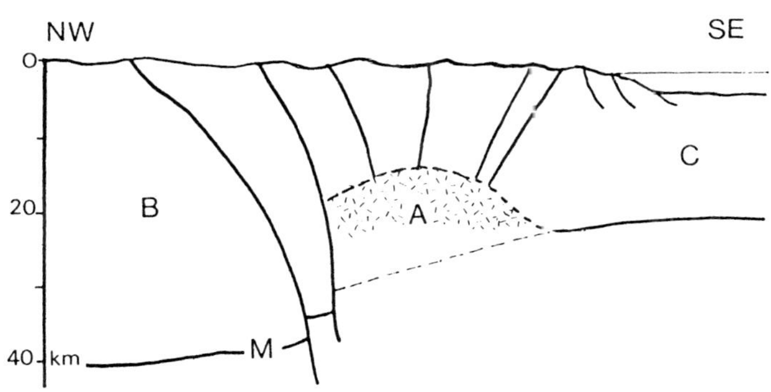

Fig. 68. Structural model of the continental crust under the Eastern Betica molassic basins (after Montenat et al., 1987, highly simplified).
An anatectic zone (A) appears between the thick crust of the Betica mountains (B) and the thinner crust of the Alboran seacoast (C). M: Moho.

Geophysical data show that the strike-slip fault zone corresponds to a strong uplift of the Moho (–22 km) and hence of the upper mantle, the latter with the usual abnormal seismic velocities. The geothermal gradient here is very high (41 to 56°C/km instead of 30–33). Anatexis, normally occurring at great depths, is here produced at shallow depth.

San Joaquin basin

This basin is situated on the border of the San Andreas fault, NWof Los Angeles, California. It is a good example of a transpression basin in which en echelon folds are developed at the contact of the fault (Fig. 69).

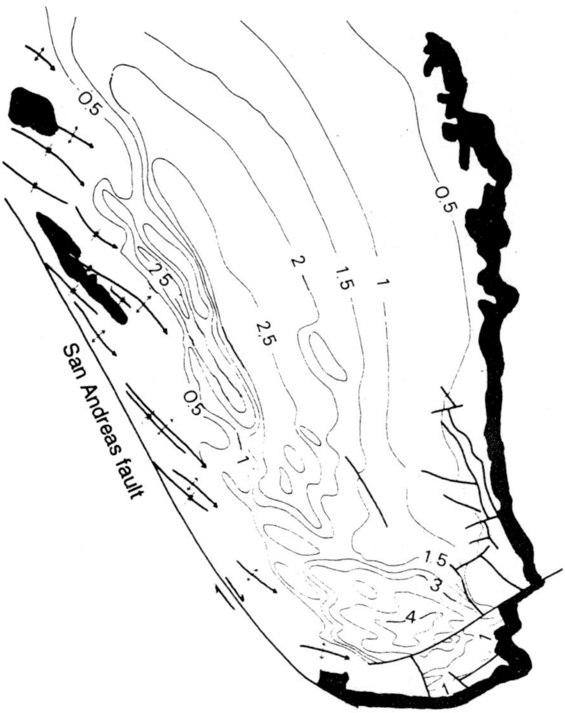

Fig. 69. San Joaquin basin, north of Los Angeles: Example of a transpression basin (after Wilcox et al., 1973).
In black: outcrop of the basement. Isobaths in km. The en echelon folds are indicated by arrows (synclines and anticlines distinguished by usual symbols).
The transpression hypothesis for the origin of folds is debatable, however. Some authors (see Van S. Mount and J. Suppe (1987), *Geology*, pp. 1143–1146) consider that the activity of the fault has no effect on its limbs and that these folds are comparable to those of the Transverse Ranges which are developed to the south of the San Andreas fault (see p. 172 and Fig. 48).

MOLASSIC BASINS

Basins which are situated on the boundary of a mountain chain and receive its disintegration products, or molasse, are known as molassic basins.

That is why molasse representing post-tectonic sediments have long been distinguished from flyschs, which are syntectonic deposits. In fact, syntectonic formations exist which some people call molasse and others flysch. In any case, most of the classic molasse are deposited there, whereas the neighbouring mountain belt is completed to such an extent that part of these formations will be folded by the end of evolution. The distinction between molasse and flysch is thus very vague and local names have emerged more from usage than from precise lithologic characteristics.

Foreland Molassic Basins

A foreland basin is developed in front of a mountain belt undergoing folding parallel to its direction and its tectonic structures on a lithosphere undergoing flexure due to the load represented by the chain, and hence the name 'flexural' basins. Examples are abundant: the Swiss molassic basin in front of the Alps, the Carpathian basin in front of the Carpathians, the Padan basin in front of the Apennines. The Adriatic Sea can be considered a present-day basin of this type in front of the Dinaric Alps.

EXAMPLES

Swiss molassic basin (Figs. 70 to 72)

From Lausanne to Zurich, its extension exactly corresponds to the front of the great ranges of the Swiss Alps (Prealpine and Helvetic nappes). Beyond Lake Constance, it changes to the Bavarian basin and later to the Vienna basin in front of the Eastern Alps consisting of the Austroalpine nappes.

The basin is dissymmetric in the sense that the thickest sediments occur in the vicinity of the mountain chain. Moreover, subsidence episodes are contemporary to the periods of emplacement of nappes.

The isobathic map of the basin (Fig. 71) shows two very distinct periods of subsidence:

—*Oligocene period* which accompanies and succeeds the first emplacement of Helvetic and Pennine nappes, i.e., a 10 to 12 km thick mass according to studies of metamorphism (Frey (1978), *J. Petrol.*, 19: 95–135) and taking 30°C/km as the geothermal gradient (see also Steck, 1984).

—*Miocene period* corresponding to the beginning of the major Neoalpine episode which mainly affected the Helvetic region and resulted in its overthrusting on the molassic basin at the end of the Miocene (and also in intramolassic thrusts) (Fig. 72).

The subsidence corresponding to this period is very different from the preceding one, as indicated by the isobath map (Fig. 71), and further be-

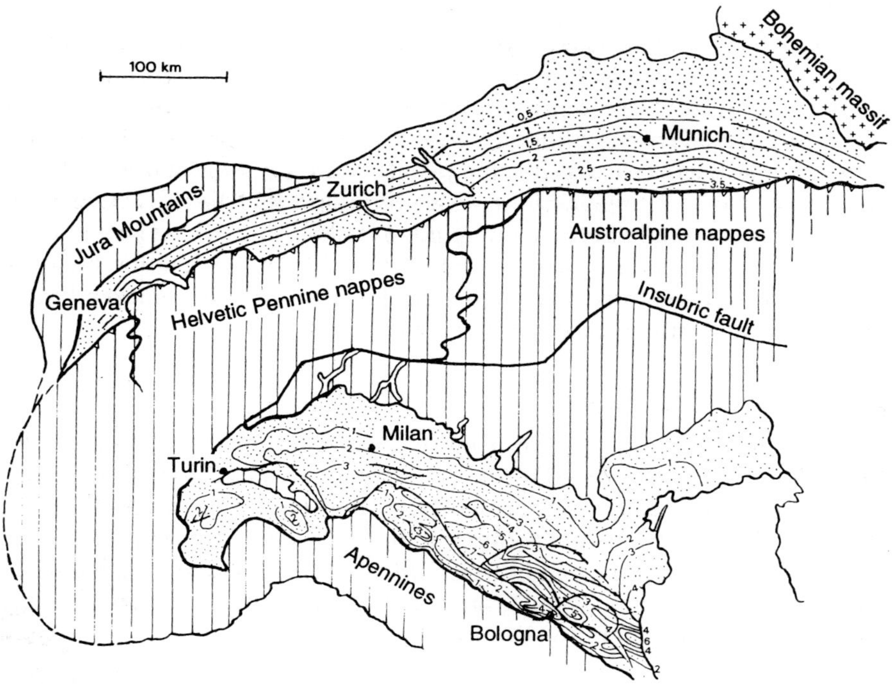

Fig. 70. Swiss and Padan molassic basins in front of the Alps and the Apennines. These two basins are not synchronous, the former being of Miocene age and the latter Plio-Quaternary. Note that in both cases subsidence is maximum where the thickness of the stack of nappes is greatest (Austroalpine nappes, Toscan and Ligurian Apennine nappes). Isobaths (in km) of the Miocene sequence on the top and of the Pliocene horizon at the bottom.

comes parallel to the alignment of external crystalline massifs whose origin is effectively of this period.

These facts enable us to interpret the subsidence of the basin as the result of a flexure of the European lithosphere, thick and hence rigid, under the loading it was subjected to because of the piling up of nappes (Karner et al., 1983)[3]. In fact, the process migrated northwards over time: the axis of maximum subsidence was displaced by 70 km northwards between the beginning of the Oligocene and the end of the Miocene, i.e., at a rate of 2 to 3 mm/yr. The bending of the lithosphere was accompanied by synsedimen-

[3]G. Karner and A. Watts, 1983. Gravity anomalies and flexure of the lithosphere. *J. Geophy. Res.*, BB, B12: 10449–10477.

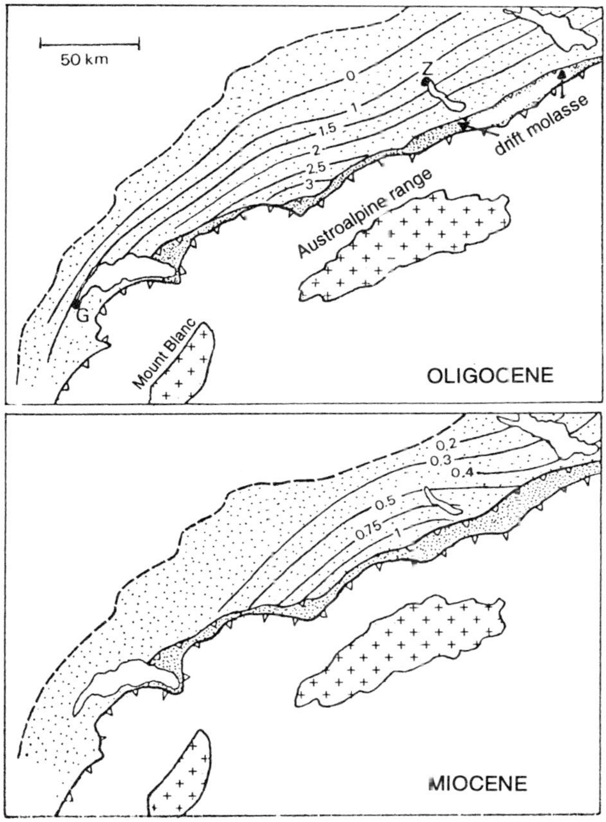

Fig. 71. Thickness (in km) and distribution of deposits in the Swiss molassic basin during the Oligocene and the Miocene (after Menard, (1988), simplified).

tary normal faults which produced sudden changes in the facies and thickness or even abrupt disappearance of some stratigraphic units (Fig. 72).

Considering the thickness and rigidity of the European lithosphere, its elastic flexure should have induced, at its NW or N boundary, an upwelling forming the corresponding bank of the molassic basin. However, it does not appear distinctly in the Tertiary palaeogeography. Some authors have suggested that it was fragmented by extension structures and that it might have been replaced by rift valleys like that of Bresse. However, this interpretation is debatable as this rift was produced by another process.

In fact, this graben, like the Bas-Dauphine basin, though being in a situation of foreland molassic basins, follows a different logic because the latter mark the Germany-Provence extensional axis. This axis consists of a series of rifts created in the Upper Eocene-Lower Oligocene. Subsequently, i.e., in the Neogene, the hollow structures which were nearest to the alpine axis have

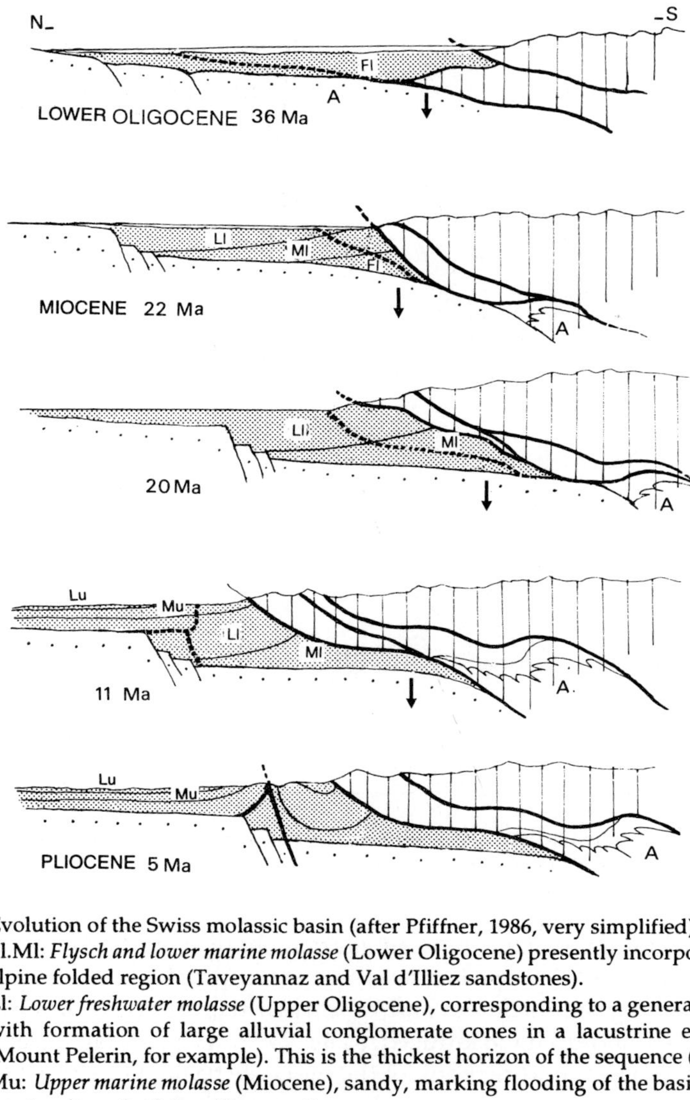

Fig. 72. Evolution of the Swiss molassic basin (after Pfiffner, 1986, very simplified).
Fl.Ml: *Flysch and lower marine molasse* (Lower Oligocene) presently incorporated in the alpine folded region (Taveyannaz and Val d'Illiez sandstones).
Ll: *Lower freshwater molasse* (Upper Oligocene), corresponding to a general regression with formation of large alluvial conglomerate cones in a lacustrine environment (Mount Pelerin, for example). This is the thickest horizon of the sequence (4 km).
Mu: *Upper marine molasse* (Miocene), sandy, marking flooding of the basin by the sea coming from the future Rhone valley.
Lu: *Upper freshwater molasse* (Upper Miocene) representing the end of a marine regime and the arrival of new conglomerates. It is interesting to note that east of Munich (Fig. 70), this member remains marine. This Bavarian and Austrian molassic basin, intensely subsident, is associated with the emplacement of Austroalpine nappes.
A: The Aar massif, serving as a marker. It shows the progression of alpine nappes northwards. The corresponding subsidence of the bottom of the basin (arrow) by flexure in the south and normal faults in the north are also shown. The light dots indicate, for each stage, the trace of cleavage which would give the drift structure of the following stage.

been affected by a subsiding flexural process, certainly less distinct than in the Swiss basin, but sufficient enough so that the area of subsidence in each case exceeds the boundaries of the primitive Palaeogene graben. This process somewhat resembles the one we have mentioned for the origin of the Paris basin. It has contributed to the formation, from the Bas-Dauphine to the Mediterranean, of a kind of subsidence channel consisting of successive subbasins (Valence, Valreas, Digne, Lower Rhone), a channel which facilitated Miocene transgression in reaching the Swiss basin.

Po basin (Padan basin) (Figs. 73 to 75)

Its history and organisation are far more complex than those of the Swiss molassic basin because it was subjected to the successive influences of the Alps and the Apennine orogenies, exerted in two different subsidence zones.

The alpine influence was exerted in the Oligocene and beginning of the Miocene (Aquitanian). A sedimentary sequence of this age is found on the W and NW boundary of the basin, which thickens in the direction of, and parallel to the Alps (Fig. 73). These sediments occur as outcrops in the Como region.

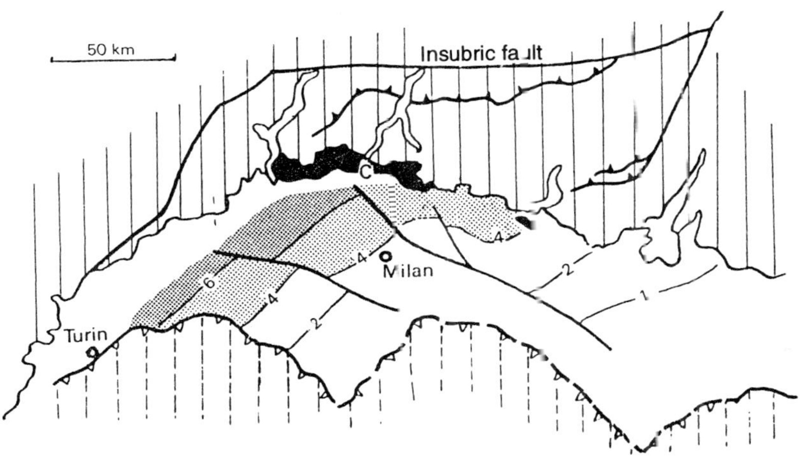

Fig. 73. Oligo-Aquitanian subsidence in the Po basin (after Ménard, 1988, simplified). Directly related to alpine evolution and not to that of the Apennines. Note the rough parallelism between isobaths (4 km) and the Insubric fault.
In black: Present-day outcrops of the Oligo-Aquitanian formations in the Como region (C).

This Oligo-Aquitanian sequence immediately succeeds the folding phase of the inner alpine zones, some of which have a Padan vergence (see p. 266). The molasse of Como thus represents the filling of a common flexural basin.

In the Upper Miocene a new phase affects mainly the outer alpine zone and completes emplacement of the great Penninic and Austroalpine nappes on the western and northern slopes of the Alps, thus initiating subsidence of the Swiss molassic basin mentioned earlier, which took over the subsidence of the Po basin on another slope of the mountain chain.

The influence of the Apennine orogeny is exerted mainly from the Middle-Upper Miocene[4]. The new subsidence, which became manifest in the whole eastern part of the Padan basin, is associated, chronologically and geometrically, with formation of Apennine nappes.

The *chronological* association between the subsidence and the emplacement of the Apennine nappes is evidenced by the fact that subsidence became active only after the Tortonian, starting from which the Padan sediments contain olistoliths of the highest nappes of the Apennine range (Ligurian nappes) (Fig. 74). This regime continued and became more active during the Pliocene in which the sediments lie, moreover, unconformably on the earlier structures (end of the Miocene phase).

The geometrical association is indicated by the maximum subsidence (7 to 8 km only during the Plio-Quaternary period) at those places where the Apennine nappes are the thickest and the best represented, viz., in Emilia and Toscana (Fig. 75).

Thus, since the Pliocene the Padan basin consists of two parts (Fig. 75):

— *a northern part*, the 'Southern alpine monocline', which subsides gradually southwards up to basement depths of 12 km.

— *a southern part*, in which the sediments mask the Apennine nappes, folded as successive arcs (the arcs of Monferrat, Emilia, and Ferrare-Romagna) with a northern vergence and corresponding to equally large piled-up thrust nappes. It is precisely at such places where piling up is maximum that subsidence of the floor of the basin is the most intensive.

In the part outside the Apennine front, subsidence is accompanied by normal faults which are found on the alpine boundary surrounding the basin and which are perhaps still active ('Piemont arc' of alpine earthquakes).

The Padan basin is also significant in giving examples of 'transported' basins, since a part of the sedimentary complex has developed on active overthrusts (FR, Fig. 75, or the basin situated between faults 1 and 2 of Fig. 74 (so called 'piggyback' basins).

Backland Molassic Basins

These are found only in the interior of arcs of certain mountain chains and their mode of formation is associated with the specific origin of the arcs.

---

[4]The older sediments are already incorporated in the Apennine frontal folds.

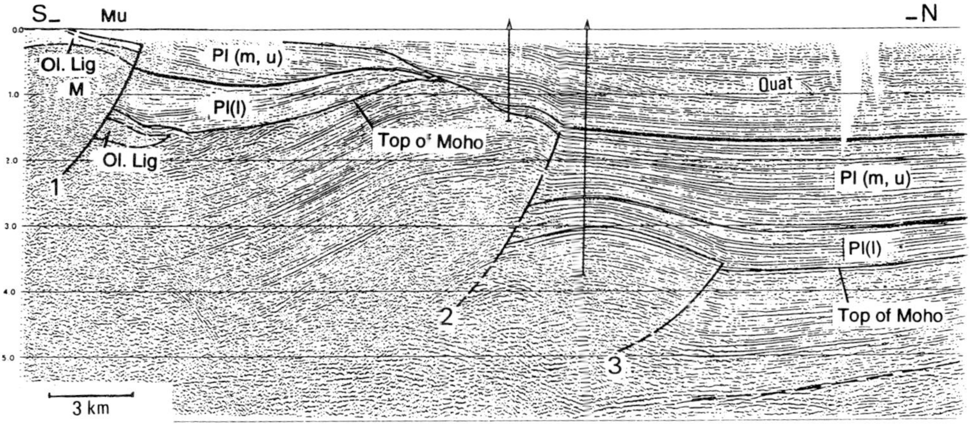

Fig. 74. Seismic reflection profile in the southern Po plains, on the boundary of the Apennine range (after A.W. Bally (1983), Seismic expression of structural style. *Amer. Assoc. Petr. Geol.*, sec. 3-4-1: 15).
The effect of post-Tortonian subsidence is reflected in the thickness of deposits visible on the right side of the Figure. It is accompanied by Apennine thrust movements which deform the Neogene formations.
Note the olistoliths of the highest Apennine nappes (or Ligurian nappes) in the Upper Miocene and the migration of depocentres (Lower Pliocene, in front of fault 1; Middle-Upper Pliocene, in front of thrust zones 2 and 3; Quaternary, north of the preceding ones).
M: Miocene; Ol. Lig: olistoliths of the Ligurian deposits; Pl (l, m, u): Lower, Middle, Upper Pliocene.
Vertical scale: in seconds of two-way time.

EXAMPLES

Pannonic basin, in the centre of the Carpathian arc (Fig. 76)

This basin actually consists of a series of small, nearly coalescent depressions. They developed starting from the Aquitanian and mainly during the Miocene, i.e., during the period when the Carpathian arc acquired its structure and its specific form. Thus an obvious relation exists between the compression prevalent on the periphery of the arc and the extension regime in the interior of the basin.

All these intra-Carpathian basins show about the same history with two periods of successive subsidence:

—In the *Lower Miocene* (Aquitanian and Burdigalian), subsidence is rapid, such that the initial deposits show relatively deepwater facies and are produced in small subsidence basins created by normal faults that are often synsedimentary.

—Later, in the *Middle Miocene*, a much slower process takes place, which is terminated more or less slowly depending on the region, but the subsidence

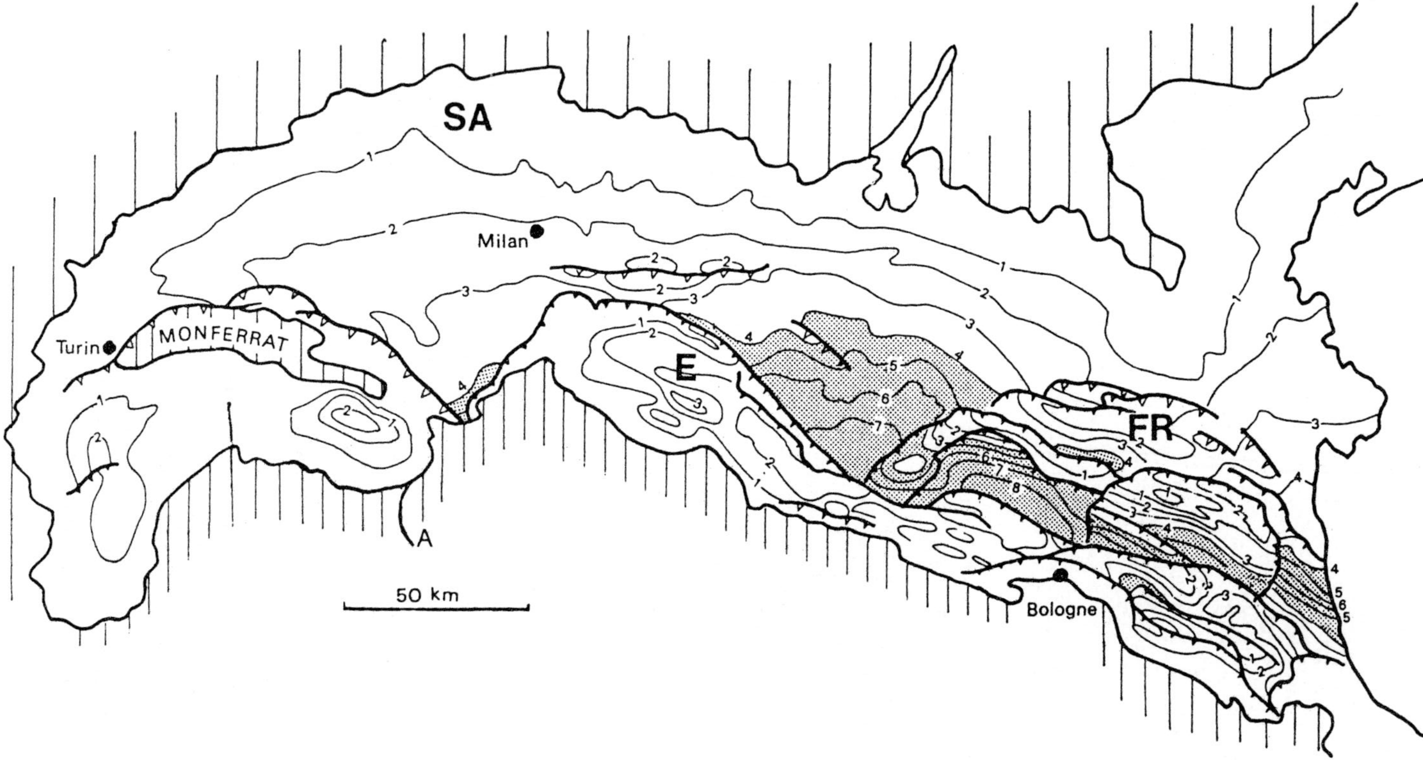

Fig. 75. Plio-Quaternary Padan basin and its structures (adapted from Pieri and Groppi, (1981)* and Castellarin and Vai, 1986).
The isobaths (in km) correspond to the base of the Pliocene sequence.
The basin consists of two parts: in the north, the 'Southern Alpine monocline' (SA), which gradually subsides southwards up to a depth of 12 km for the basement, 7 to 8 km for the base of the Plio-Quaternary sequence.
In the south, the front of the Apennine nappes, masked by the sedimentary formations of the basin. This front is folded: folds of Monferrat, Emilia (E) and Ferrare-Romagna arc (FR) and the corresponding basins are called 'transported'.

*Subsurface geological structures of the Po plain, P.F. Geodinamica Publ. 414, C.N.R., 23 pp.

then affects much larger areas: the Upper Miocene and Pliocene deposits may even directly overlie the pre-Neogene basement. They mask, in any case, the largest part of the older structures. The total thickness of sedimentation, however, remains moderate at 1700 to 4000 m.

The first stage (initial subsidence) reflects the immediate response of the crust to extension, with surficial fracturing and deep ductile stretching, resulting in uplift of the upper mantle and finally of the asthenosphere. The second stage is that of thermal debulging (thermal subsidence, to which is added the effect of sedimentary overloading).

Actually, this evolution varies depending on the basin. In the peripheral depressions, stage 1 is distinct and stage 2 is short (it ceases in the Upper

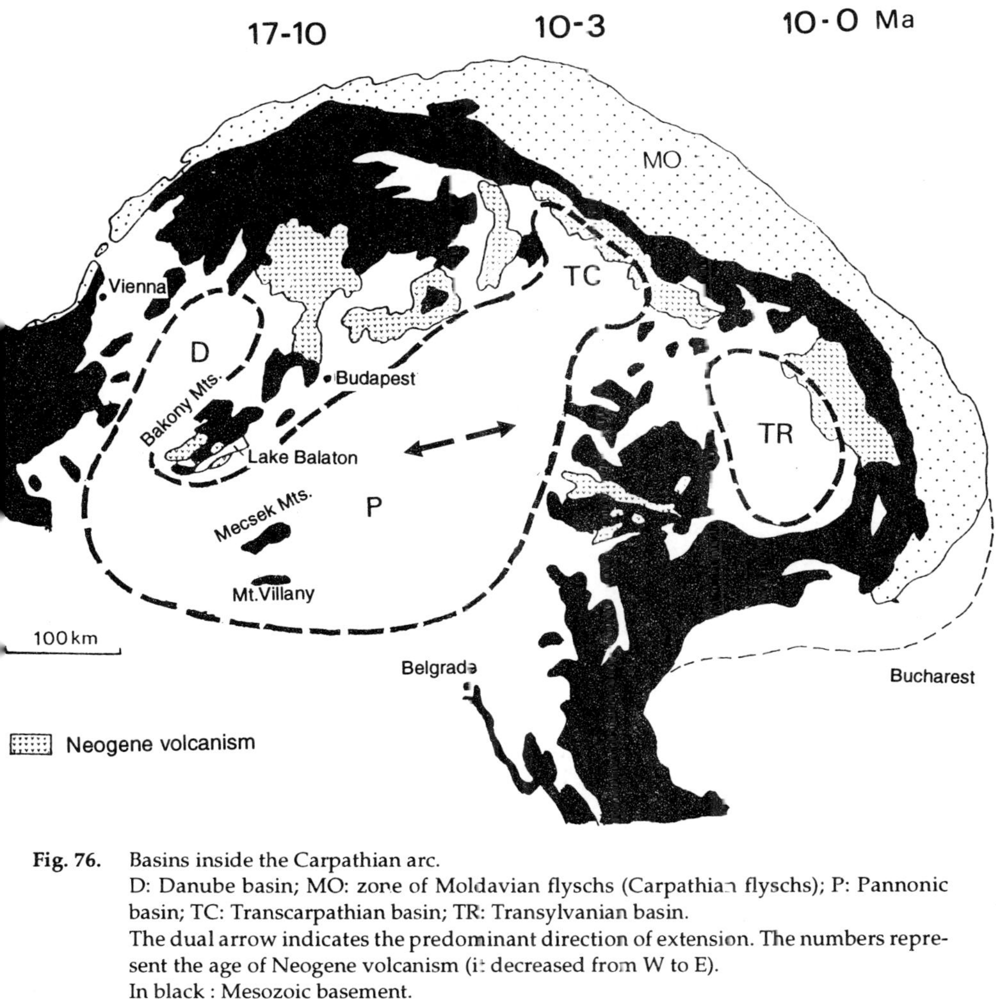

Fig. 76. Basins inside the Carpathian arc.
D: Danube basin; MO: zone of Moldavian flyschs (Carpathian flyschs); P: Pannonic basin; TC: Transcarpathian basin; TR: Transylvanian basin.
The dual arrow indicates the predominant direction of extension. The numbers represent the age of Neogene volcanism (it decreased from W to E).
In black : Mesozoic basement.

Miocene). On the contrary, in the case of the central basin (the Pannonic basin proper), stage 1 is hardly represented in the Miocene and is strongly indicated in the Plio-Quaternary. Undoubtedly, it is not terminated because here the heat flow is double that of the normal and the crustal thickness is lower than the average (20 to 25 km).

However, what is important is that the two extension episodes (Lower Miocene and Middle Miocene) are contemporary to the compression episodes which then affected the outer Carpathians (the zone of Moldavian flysch or Moldavides). This zone is in fact formed by a piling up of nappes following the extreme shortening of their basement or even its disappearance, which most authors attribute to a subduction phenomenon.

The basin of Carpathian flysch might constitute an oceanic crustal region situated at the foot of the inner Carpathian mountains (Dacides), exactly like the present-day Black Sea relative to the north Turkish ranges (Pontides).

The Neogene subduction of the Carpathian basin south or southwestwards might have thus transformed the inner Carpathians into a real active margin. Hence there exists a major volcanism, calco-alkaline and acidic in composition (andesites and mainly dacites; the latter were named after this region, the province of Dacia). This volcanism is also of Neogene age and its association with the folding of the outer Carpathians is confirmed by the fact that both were simultaneously displaced from W to E during the Neogene (Fig. 76).

The Carpathian flysch might thus represent a true accretionary prism as effectively indicated by its numerous piled-up slices, the bottommost being the most recent. Such a disposition is indicative of an active continental margin. As a matter of fact, in a later chapter we shall observe that these active margins are affected by chronic extension phenomena generally manifested by the appearance of one or several basins, parallel to the volcanic arc, known as *back-arc basins*. This might also explain the origin of the Pannonic basin. There extension phenomena are generated in a stress field controlled by the origin of the volcanic arc. But the fracture network resulting from it has been disturbed by the rejuvenation of older structures of any orientation.

Tyrrhenian basin (Fig. 77)

This oceanic basin, triangular in shape, is surrounded by Italy, Sicily , Sardinia and Corsica. It is thus situated in the midst (or at the back) of the *Sicilian-Calabrian arc,* a Tertiary folded belt above a subduction zone of NW plunge. This zone, though almost inactive, is still indicated by some deep earthquakes, as also by the volcanism it generates (Eolie Islands).

The disposition of this basin is thus very similar to that of the Pannonic basin inside the Carpathian arc but the significance of the Tyrrhenian basin is twofold. Firstly, it represents a present-day example of undersea evolution. Moreover, the extension regime prevailing there has caused splitting of the continental crust. A young oceanic crust appears there in two basins of

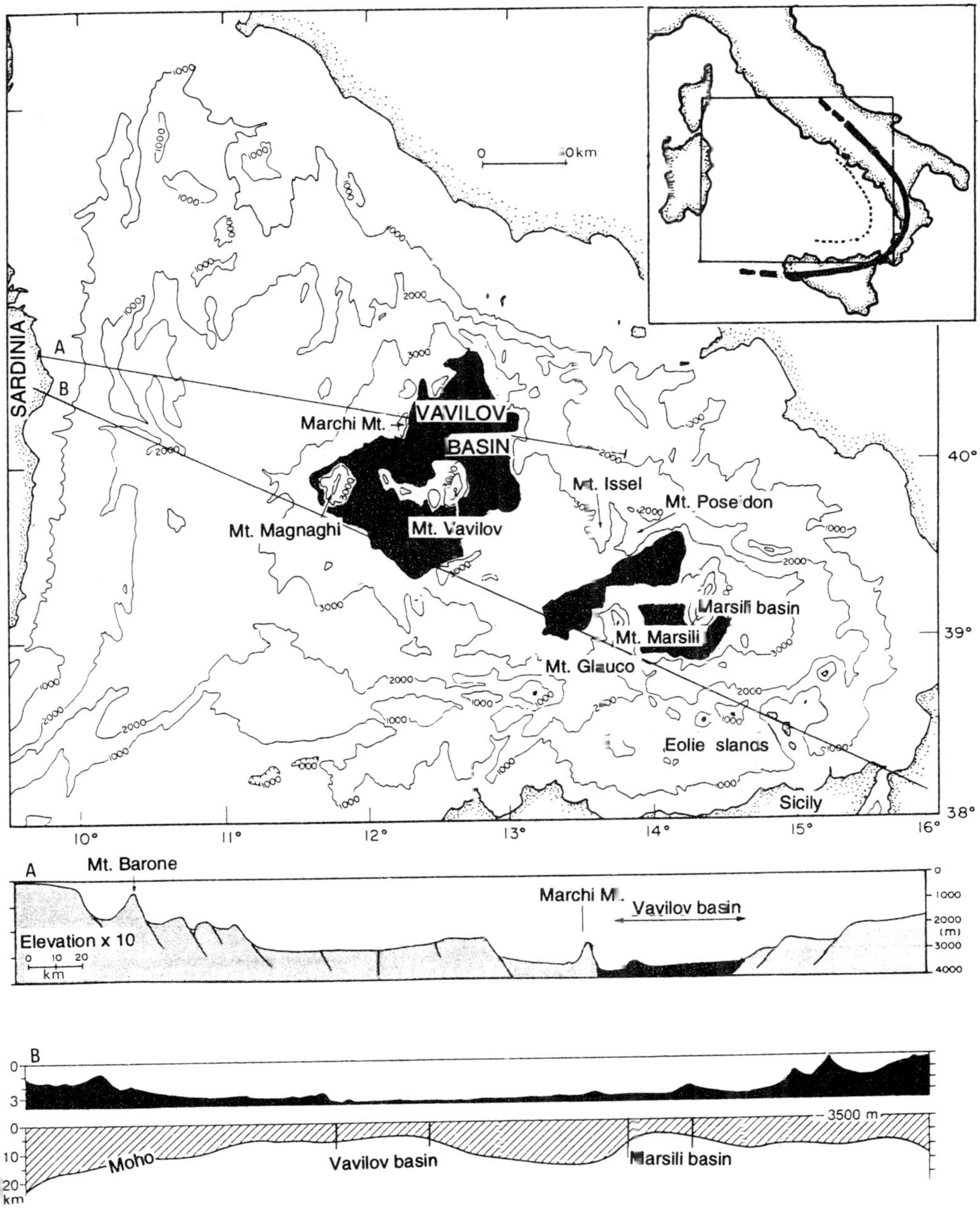

Fig. 77. Bathymetric map of the Tyrrhenian Sea.
In black: depths greater than 3600 m, also corresponding to outcrops of oceanic crust.
Inset: position of the Tyrrhenian Sea in the midst of the Sicilian-Calabrian arc (thick curved line). Volcanic arc: dotted curve.
A—Cross-section showing the relation of the Vavilov basin (oceanic crust, in black) with the neighbouring continental margins (continental crust, in grey).
B—Cross-section of the two oceanic basins showing crustal thickness.

large depths (about 3700 m), viz., the Vavilov and Marsili basins. It exhibits very distinct magnetic anomalies except in regions where the basement is entirely peridotitic or covered by turbidites.

The structure of the Tyrrhenian Sea is clearly seen on a bathymetric map (Fig. 77). Here one can observe a succession of reliefs with orientations of N-S to N 30°, more or less intersected by E-W to N 100° lineaments. In particular, the Sardinian margin shows a series of step topographies representing the transition between emergent region and the deep Vavilov basin. Geophysical and geological data (seismic profiles, underwater observations, drilling) show that they constitute a sequence of tilted blocks of continental crust in the margin of the young oceanic crust of the Vavilov basin.

Evolution of the Tyrrhenian Sea has taken place in several periods. The first episode of stretching was produced in the Upper Miocene in the presently elevated part of the margin (the synrift sequences of the Upper Tortonian-Messinian). It was followed by a second episode in the present lower part of the margin during the Messinian-Lower Pliocene. The oceanic lithosphere appears in the Pliocene in the Vavilov basin and at the end of the Pliocene in the Marsili basin. Note that the maximum extension is displaced in the direction of the arc but not in the opposite direction. Some geologists have suggested that the displacement towards the arc was associated with an increase in slope of the subduction zone (Fig. 78).

Submarine volcanic activity of tholeiitic type exists, giving rise to thick volcanic edifices such as those of Vavilov and Marsili. In the former, being a little older and farther from the arc, the lavas show a more alkaline tendency however. The heat flow is very high.

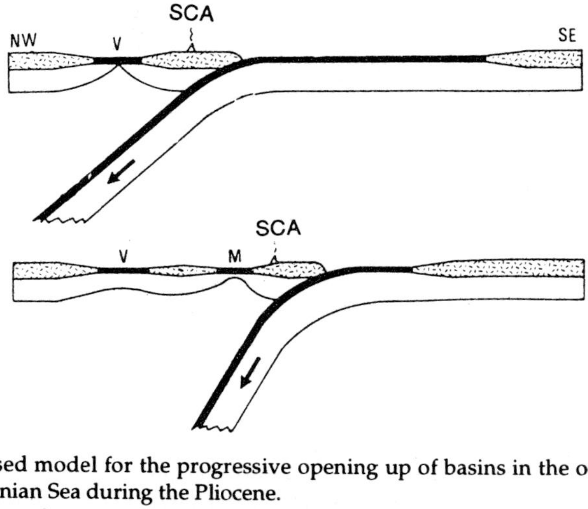

Fig. 78. Proposed model for the progressive opening up of basins in the oceanic crust of the Tyrrhenian Sea during the Pliocene.
Continental crust in grey. Oceanic crust in black.
M: Marsili basin; SCA: Sicilian-Calabrian Arc; V: Vavilov basin.

Thus the origin of the Tyrrhenian basin, like that of the Pannonic basin, implies the presence of an active subduction zone. These basins are definitely back-arc basins, therefore, similar to those of active margins, which we shall describe later (p. 161). They reflect the existence of extension phenomena probably associated with divergent convective currents inside the upper mantle, vertically below the subduction zone (Fig. 20).

Another example would be the *Aegean basin* for which crustal thinning is less severe and evolution is still in progress.

CHAPTER 2

CRUSTAL FRACTURES

These are rifts whose rims are separated from each other such that the continental crust, being stretched, is broken and allows a new oceanic type crust to appear.

EXAMPLES

Red Sea-Afars-Gulf of Aden Group (Fig. 79)

The group of these three regions forms an L-shaped fracture which separates the Arabian platform from the rest of Africa. The bottom of this fracture is an oceanic type crust appearing in the Miocene or the Pliocene because its different segments are not in the same stage of development.

1. Red Sea

This sea protracts the Gulf of Suez as also the Gulf of Aqaba southwards, i.e., the Jordanian transform fault zone whose structural influence will be considered later. Southwards, the Red Sea structure is abruptly interrupted and offset westwards (en echelon pattern). It is found again in the Afars region.

Structural evolution: The first synrift sediments of the margins are of Oligocene age locally and of Lower to Middle Miocene elsewhere. This initial opening was accompanied by emplacement of numerous basaltic and gabbroic dykes parallel to the general alignment of the rift (Fig. 80) and, in the southern part of the latter, by large masses of rhyolites and even small granitic bodies, which indicate that the crust was always continental.

Localisation of this volcanism only on the Arabian side of the rift suggests that the extension which induced the rise of the upper mantle was due either to a dissymmetric process or the effect of a hot spot.

The uplift of the boundaries was still small or negligible because all the detrital sediments are fine grained (it will occur 14 Ma later).

Then occurred a quiescent period of 10 Ma, i.e., up to the end of the Miocene, which witnessed the same evaporitic episode as that of the Gulf of Suez. Until that period, the history of the Red Sea is intertwined with that of the latter.

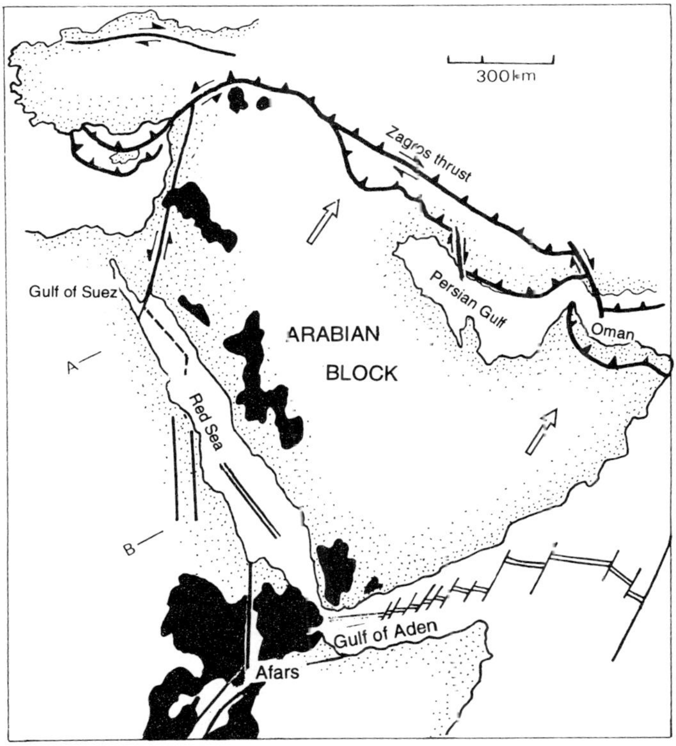

Fig. 79. Red Sea, Afars and Gulf of Aden in the structural framework of the Arabian block (after d'Estévou, 1987).
In black: volcanism; Arrows: direction of displacement of the Arabian plate; A, B: lines of cross-sections of Fig. 81.

In the beginning of the Pliocene an opening was effected in the region of the Gulf of Aqaba (associated with sinistral movement of the Jordanian fault) and the Red Sea, whereas the Gulf of Suez escaped this event (see p. 62). This stage of Plio-Pleistocene extension is clearly marked by the numerous discordant relations in the corresponding sedimentary sequences, as also by the uplift of the rift margins to elevations of 2000 to 3000 m and by a strong volcanism (alkaline basalts with olivine and inclusions of crustal rocks, dated as 0.3 to 0.9 Ma) which is always limited to the eastern margin of the rift (Fig. 80). Microseismicity indicates that the extension is still active.

Present Structure: Only in the southern part of the Red Sea does an oceanic type neocrust appear, 5 to 6 km in width, which is remarkable for its thinness (10 km) and its bands of magnetic anomalies (Fig. 81). The appearance of this crust is contemporary to the Plio-Pleistocene crisis. Between this neo-oceanic axis and the plane continental boundaries exists a 50 to 100 km wide region of detrital sediments or reefal structures that are mostly covered by the sea. The thickness of the crust increases from 10 to 20 km on

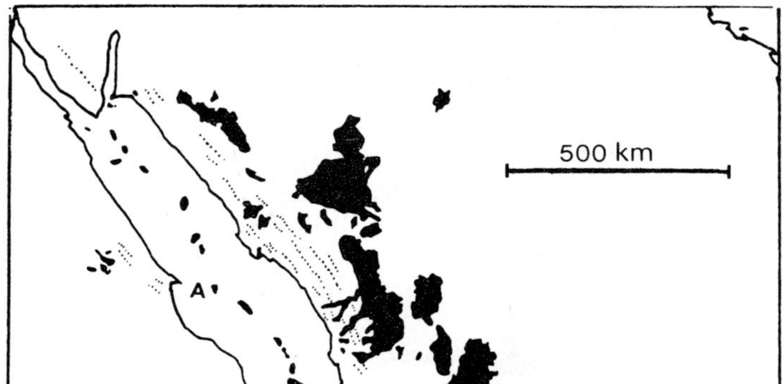

Fig. 80. Volcanism of the Red Sea-Afars-Gulf of Aden group (after Bayer et al., 1989, completed and simplified).
Vertical hatching: alkaline volcanism contemporary to the initial opening of the rift (30-10 Ma). In the Red Sea area it mainly consists of swarms of basaltic and gabbroic dykes dated as 22 Ma (dotted lines).
In black: volcanism contemporary to the opening of the crustal fracture (alkaline to tholeiitic) (5 Ma-Present).
A: Zabargad Island.

moving away from the Red Sea and its nature is disputed. It probably represents an 'intermediate' crust (see p. 6). The rare outcrops (such as Zabargad Island) show mantle peridotites with diabase and gabbroic veins, injected into granulite and amphibolite facies gneisses which represent the base of continental crust thinned by extension. This 'oceanisation' of a continental crust might have started around 13 Ma, whereas the extension phenomenon commenced, as already mentioned, in the Upper Oligocene.

On the margins the structure is that of blocks bounded by normal faults. Their disposition is symmetrical in the northern part of the Red Sea but

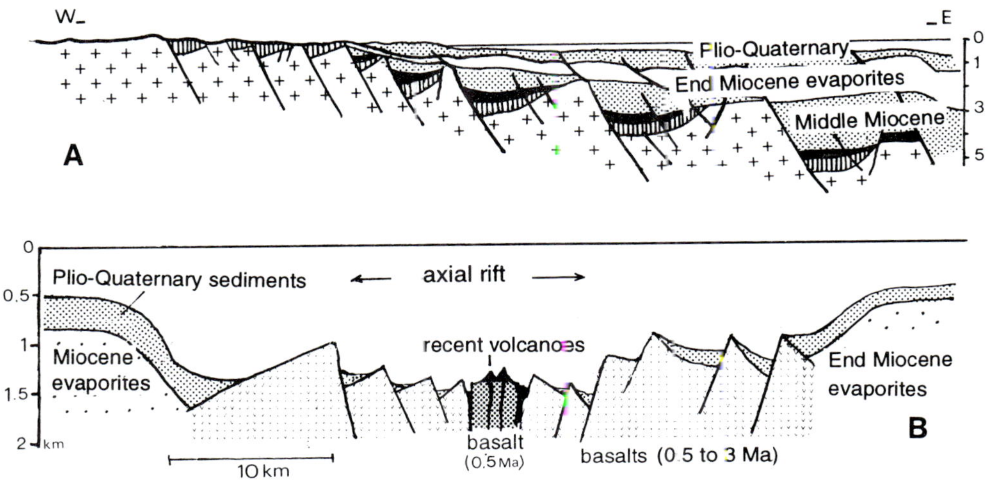

Fig. 81. Structure of the Red Sea.
A—Cross-section of the western margin of the rift, in the Safaja region (A of Fig. 79) (after Ott d'Estévou et al., 1987, simplified).
The tectonic style and sedimentary formations of the Gulf of Suez can be found here (compare with Fig. 39). In black Oligocene; hatched: pre-Oligocene sediments.
B—Axial cross-section of the rift around 19° parallel (B of Fig. 79) (after Juteau, 1983, simplified).
This rift is bounded by two scarps developed on a thick mass of Miocene evaporites, covered by 200 to 300 m of Plio-Quaternary sediments. Nature of the basement not known.
This rift itself consists of nearly symmetric tilted blocks, bounded by major fault planes, reaching elevations of up to 600 m, with dips of 60° towards the axis. The tholeiitic basalts have been dragged onto this plane.
In the rift axis, one can observe a belt of recent basalts (less than 500 years old), 4 to 5 km in width, separated from the nearby scarps by deep linear rifts. Actually, the entire basin is broken by almost-open extension faults, parallel to its alignment, and by transverse faults.
In the centre, the extension zone itself, 0.5 to 1 km in width, is marked by recent small volcanoes showing pillow-lava structures.
The entire axial zone indicates roughly symmetric magnetic anomalies, extending up to 1.5 to 2 Ma. After this, they disappear below the sediments and evaporites. They enabled us to calculate that the average expansion rate has been 1.6 cm/yr for the last two million years, as in the case of the Atlantic ridge (slow ridge). This analogy facilitates comparison of the axial rift of the Red Sea with the rift of the Atlantic ridge.

becomes very asymmetrical southwards, with narrow blocks that are highly tilted in the same direction (Fig. 82).

This structural asymmetry is further accentuated by the redistribution of volcanism (only in the east) and the uplift of the boundaries (somewhat

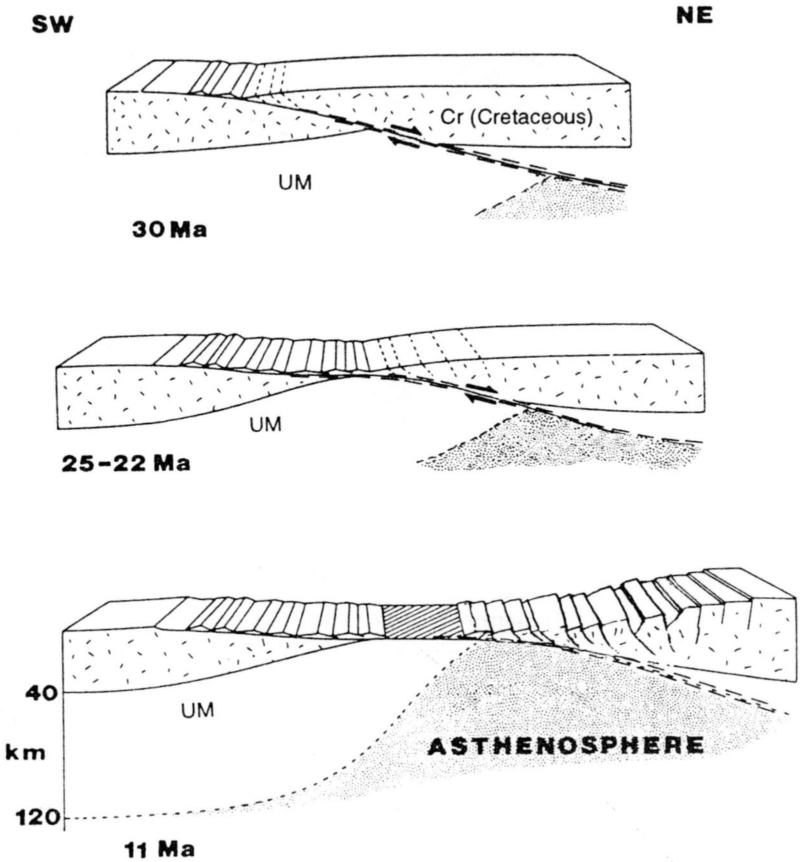

Fig. 82. Opening of the Red Sea, between the Oligocene and the Middle Miocene, by a dissymmetric crustal thinning.
The rise of the upper mantle (UM) below the NE margin explains the concentration of volcanic activity observed there.

more distinct in the east). In these conditions the present interpretations propose an extension by a low-dip shear surface.

2. *Afars region*

This is a complex region in which the oceanic opening occurs over a hot spot. The crustal fracture of the Red Sea disappears around 14° parallel but is further continued westwards by an en echelon fracture of the same orientation, known as the Northern Afar fracture, which is developed between the Ethiopian plateau in the west and a less elevated plateau in the east, known as the 'Danakil horst' (Fig. 83).

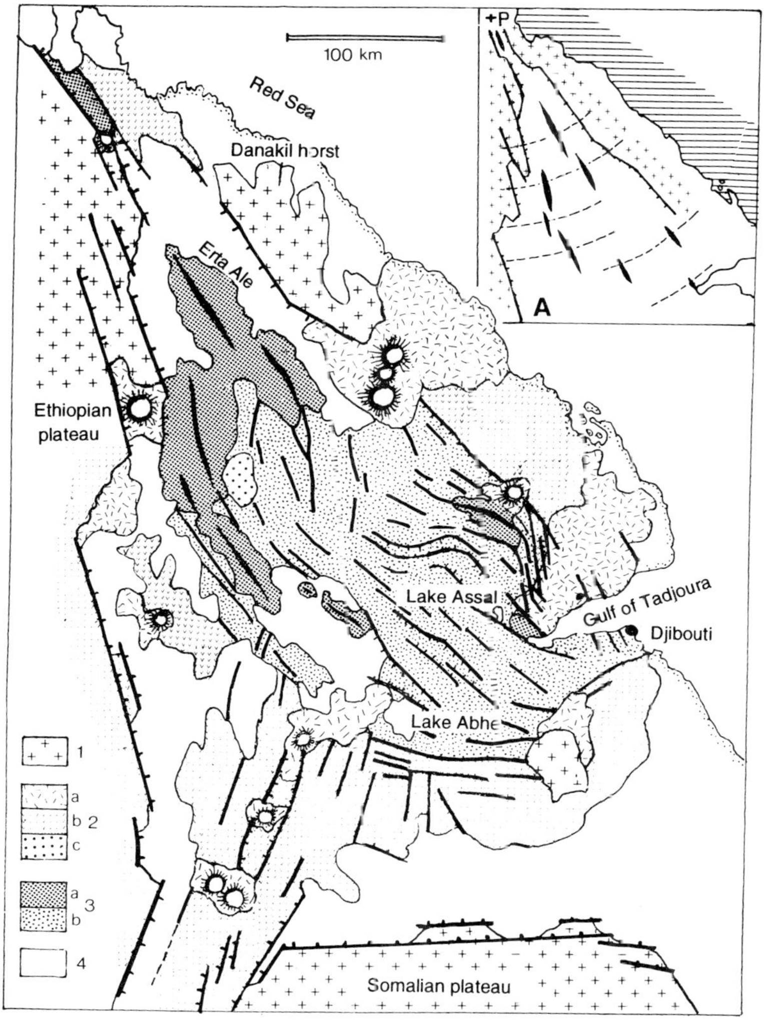

Fig. 83. Structural map of the Afars region (after Varet, 1975, simplified and modified).
1. Precambrian basement.
2. Alkaline volcanism in the plateaus and margins of the rift (a. acidic; b. basic; c. Miocene granite).
3. Tholeiitic volcanism (a. major massifs with their axes on the fractures; b. tabular volcanism, in general without distinct bodies).
4. Mesozoic, Tertiary and Quaternary sediments.
A—Figure showing that the orientation of the principal volcanic massifs of the Afars rift is controlled by an opening controlled by the same rotation pole (P) as that of the Danakil horst.

Rifting and subsidence of the Danakil basin commenced, as in the case of the Red Sea, in the Lower Miocene, around 25 Ma, the age of the lavas emplaced there. These are highly fluidic alkaline basalts (traps), often subaquatic, outcropping on the rift margins. They are associated with continental sediments, which may be lacustrine, marine or evaporitic, and also with an abundant alkaline rhyolitic volcanism (trachytes, rhyolites, ignimbrites, and even granite), evidencing the existence of a continental crust and functioning of a hot spot.

After a certain quiescent phase, a new extension occurred in a later period, which has not been precisely established. The continental crust is broken, giving rise to an axial fracture with an oceanic crust, as revealed by geophysical data (high magnetic anomalies and seismic velocities). This crust is not visible because topped by the large volcanic massif of Erta Ale, which is still active and appears to have been generated around 1.2 Ma, but might be concealing older volcanic material also. The products of this volcano are intermediate between alkaline basalts and tholeiitic basalts, with acidic differentiation products (trachytes and rhyolites): these could suggest the conversion of a continental magma to a ridge magma.

Between this 'oceanic' axial fracture and the border of the Ethiopian or Danakil plateau, of continental crust, exists a 75 to 100 km wide region in which geophysical data indicates, as in the case of the Red Sea, a thin continental crust. However, this zone is buried under recent volcanic formations. This axial volcanism does not hinder the activity of more lateral centres, where basalts and rhyolites always coexist.

Uplift of the margins is quite intense (4000 m in Ethiopia and 3000 m in Yemen).

The *southern part* of the Afar fracture is a more complex region, in which the extension is more diffuse. The axial oceanic zone of the northern Afar is continued by new en echelon troughs, both to the west as well as the east, generally surrounded by Present to Recent volcanic formations with an oceanic composition, functioning parallel to the lateral alkaline centres (basalts and rhyolites).

The tectonic structure, controlled by numerous extension faults (Fig. 84), is also complex, exhibiting a sequence of rifts fitted into each other ('rifts in rifts') whose orientations are curved, becoming E-W in the Gulf of Tadjoura, thus transforming to the structures of the Gulf of Aden. A structural continuity thus exists between Afar and this gulf, but none whatsoever with the East African rift which is intersected by the E-W structures in the region of Lake Abhe (Fig. 83). Nevertheless, its presence is certainly responsible for the staggering and crumbling of the northern Afar crustal fracture into the southern Afar.

3. *Gulf of Aden*

This is no longer a continental structure. It is already an oceanic gulf in which the oceanic crust outcrops with gradually increasing thickness

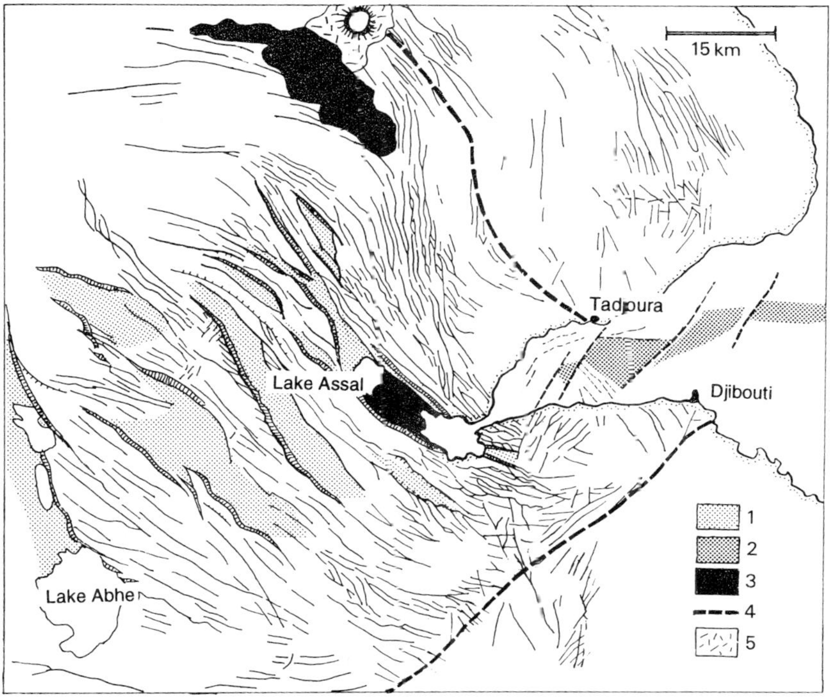

Fig. 84. Structural map of the region around Djibouti (transition from the Afar crustal fracture to the oceanic Gulf of Aden) (after Boucarut et al., 1980, simplified).
1. grabens; 2. oceanic crust of the Gulf of Aden; 3. volcanic massifs with tholeiitic tendency; 4. boundary of the Afar rift; 5. alkaline and acidic volcanism of the margins.

eastwards until the Carlsberg ridge (Indian Ocean Ridge) finally appears. Opening of the Gulf of Aden took place earlier than that of the structures mentioned above, as evidenced by the fact that oceanic crust appears there in the Miocene itself (10 Ma).

In the Gulf of Tadjoura, bathymetric, magnetic and gravity surveys indicate several narrow troughs of oceanic nature, offset in sinistral fashion by NE transform faults (whose westernmost members experienced earthquakes in 1973) (Fig. 84). The first outlines of a rift in a ridge can be observed here.

On the land, and offset in the same direction, the 'axial' rift of Lake Assal is observed, marked by very recent volcanism, which appears to be one of the last manifestations of the Afar crustal fracture south-eastwards. Thus transition to the oceanic Gulf of Aden occurs here under the effect of a series of transform faults.

Gulf of California (Fig. 85)

Unlike the Red Sea, the Gulf of California does not correspond to a rift in a ridge. It is steeply inclined because produced by the sliding of a fragment of the North American plate, of continental material, along the transform faults of the East Pacific ridge, which just touches the American continent here.

So there is a dextral strike-slip faulting of the peninsula relative to the continent along the well-known San Francisco (or San Andreas) fault, which can be considered as the north-easternmost element of this transform fault cluster.

A study of the striae and tension fractures associated with the faults in the gulf margins revealed two stages in their evolution, Miocene and Plio-Quaternary, confirmed by a discordance between the corresponding formations.

1. Initially, a strong SW-NE extension in the Lower Miocene gave rise to the classical NW-SE (N 125°) rift, whose margins are intersected by more or less dislocated blocks and pinchouts. The stretching is large, as indicated by graphic constructions, giving a value of β (for definition see p. 59) as 1.4 (an extension of 40%).

2. This is followed by a more complex Plio-Quaternary phase, dominated by a NW-SE (N 125°) transform fault movement in the interior of the gulf, which is thus obviously controlled by the normal faults of the preceding stage. The oceanic gap is opened up along this direction. On land the same faults induce dextral strike-slip movements. However, there also appears a new generation of normal faults, roughly N-S, parallel to the faults of the Basin and Range Province, which indicates that extension has become nearly E-W.

Some marker horizons enable us to verify and quantify the displacement. For example, an outlier of Pliocene conglomerates with calcareous pebbles containing fusilinides is found here, which is broken into two parts and the two halves are located on either side of the rift, in the two extremities of a line parallel to the transform faults, separated by 240 km from each other. The ancient Colorado delta is offset by 300 km (Fig. 48).

Seismic velocities and the thinness of the crust below the gulf indicate that a strip of oceanic crust occurs there. It shows magnetic anomalies which reveal that divergence commenced in the Pliocene and is of the order of 6 cm/yr since 4 Ma ago; this also gives a displacement of 240 km, as mentioned above. However, geodesy shows only a displacement of the order of 3 cm/yr at present. The reason for this is due to the fact that the displacement is in fact distributed over all the faults of the swarm as indicated by the seismicity map (Wallace, 1990).

The total displacement since the Lower Miocene is 850 km parallel to the fault (or 4 cm/yr on average) and 180 km perpendicular to the latter, i.e., moving away from North America. However, since 5 Ma, there has been a new tendency of coming nearer. The regime thus has been that of transtension between 20 and 5 Ma, and that of transpression since 5 Ma.

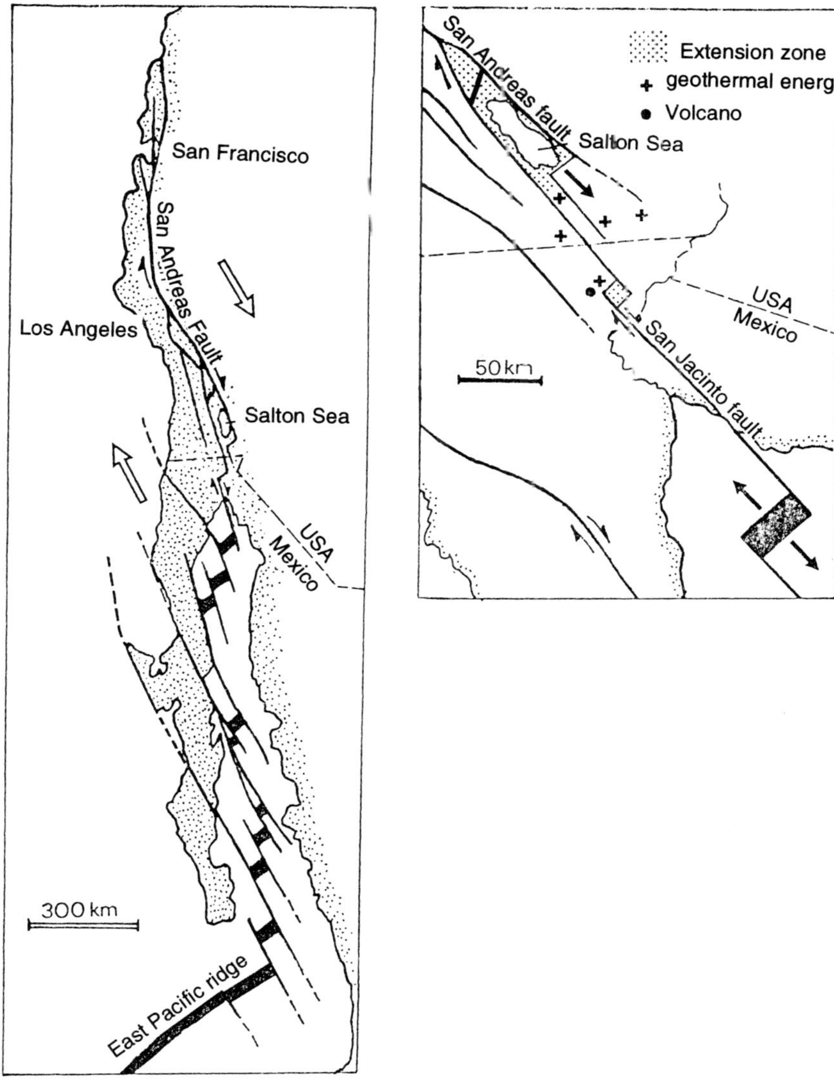

Fig. 85. Gulf of California (after Elders et al., 1972, simplified).

Heat flow is high, as in all extension axes, explaining the occurrence of former geyser fields, now utilised as sources of geothermal energy in the northern extension of the gulf.

Gulf of Genoa (Fig. 86)

With this gulf we enter into the description of crustal fractures which open up by the rotation of their two rims around a point, like two branches of a

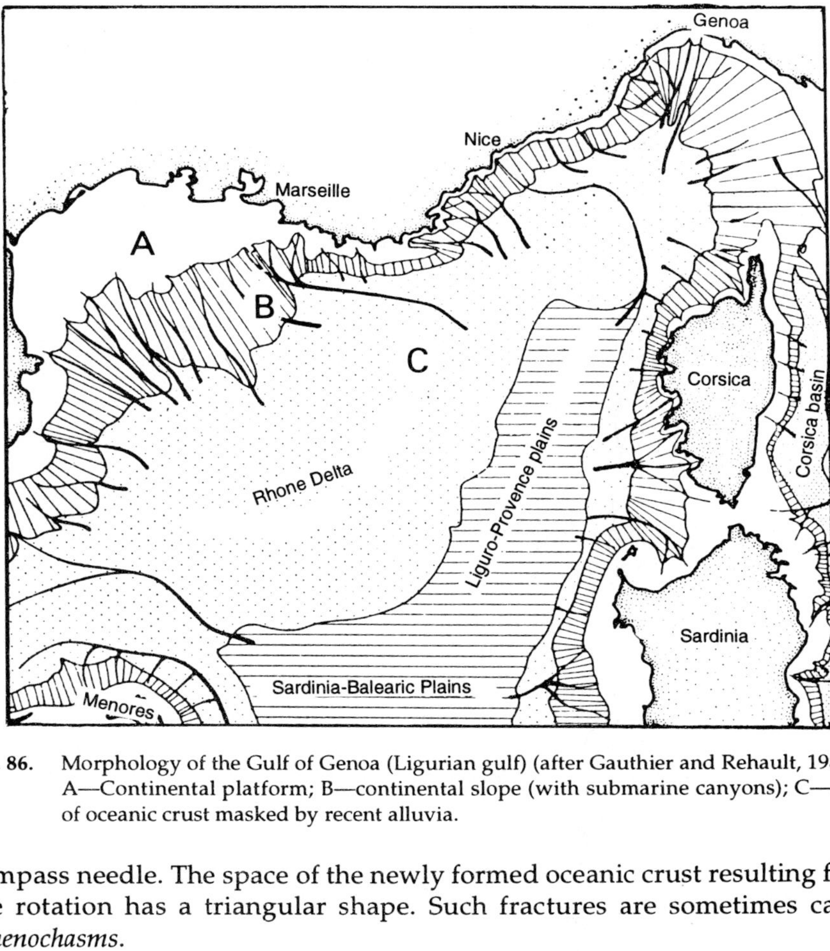

Fig. 86. Morphology of the Gulf of Genoa (Ligurian gulf) (after Gauthier and Rehault, 1986). A—Continental platform; B—continental slope (with submarine canyons); C—Zone of oceanic crust masked by recent alluvia.

compass needle. The space of the newly formed oceanic crust resulting from the rotation has a triangular shape. Such fractures are sometimes called *sphenochasms*.

The Gulf of Genoa is centred around this region. It is developed between the Corsica-Sardinian axis and the Provence-Languedoc coast. The base of this gulf presents all the characteristics of an oceanic crust (high seismic velocities, smaller thickness, bands of magnetic anomalies). Continental crust appears only near the coasts, affected by numerous extension fractures which blur its exact margin, especially on the coast of Provence.

Details of sedimentary filling are known only from an offshore well drilled north of Baleares, since none of the wells in the remaining part of the Gulf of Genoa has reached the basaltic crust. Data from the Baleares well, supplemented by seismic reflection data off the Provence coast, show that the first sediments to be deposited are Aquitanian to Burdigalian. It has thus been thought that the Gulf of Genoa was opened up at the very beginning of the Miocene by a sinistral (anticlockwise) rotation of the Corsica-Sardinian axis, centred in the Genoa region.

This hypothesis is based on several arguments:

1. *Volcanic evidence*: Sardinia shows a high volcanic activity in the Oligo-Miocene period, obviously related to the opening of the gulf. The beginning of this activity is dated as 29 Ma (Middle Oligocene), the maximum occurring between 20 and 13 Ma (Middle Miocene).

However, this volcanism is *calco-alkaline*, whereas one would expect it to be alkaline or tholeiitic. A more widely held view is that the opening of the gulf is combined with a phenomenon of active margin: displacement of the Corsica-Sardinian axis eastwards would be accompanied by subduction below it of a 'pre-Tyrrhenian' oceanic crust.

2. *Stratigraphic evidence*: The Palaeozoic deposits of Sardinia resemble those of the French Montagne Noire (south of the Variscan Massif Central), and the Permian volcanic formation of Monte Cinto (Corsica) is similar to that of the Esterel massif (near Nice) and finds itself opposite the latter if the Corsica-Sardinian axis is put back in its primitive position (Fig. 87). The Mesozoic formations in the NW of Sardinia are a continuation of the deposits in the Pyrenean-Provencal region, and the Eocene of the Balagne region (NW Corsica) is the same as that of the maritime Alps (the famous Annot sandstones, for instance).

It is thus evident that the Corsica-Sardinian block, terminated on the European coast, fits in perfectly here. Along with the latter, it formed the Corsica-Sardinian domain whose erosion supplied the material for the Annot sandstones during the Upper Eocene and Lower Oligocene.

3. *Palaeomagnetic evidence*: The Permian rhyolites of Monte Cinto and the NW part of Sardinia on the one hand, and those of Esterel on the other, show a difference in the remanent magnetic declinations of the order of 30 to 60°, which is precisely the angle of rotation derived from palaeogeographic studies. Measurements on Sardinian Miocene lavas also show that the major part of rotation was produced during the Lower Miocene, between 20 and 18 Ma. Previously the extension had been only of intracontinental type; this agrees well with the stratigraphic data on the submarine sedimentation of the gulf.

A difficulty arises in the fact that the total component measurements give a rotation of 25 to 30° for Corsica and 60° for Sardinia. The Sardinian block hence might have had a behaviour different from that of Corsica. However, field data show that the Corsican structures pass onto those of Sardinia with no deviation on either side of the Strait of Bonifacio. To date, this contradiction has not been resolved. Hence other models assume that the Corsica-Sardinian axis is not separated and that the angle of rotation is the same (30°) for both islands (Fig. 88).

4. *Structural evidence* (Fig. 89): The lineations of Variscan stretching and late Variscan faults of Maures region may readily be made to coincide with the continuations of their Corsica-Sardinian homologs if the latter block is rotated by an angle of 30°.

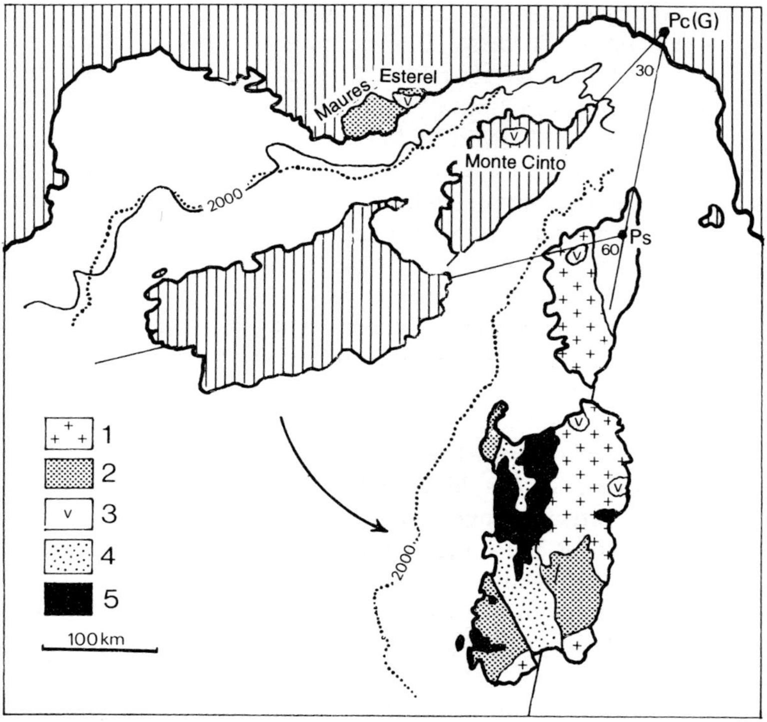

Fig. 87. Sphenochasm of the Gulf of Genoa (after Westphal et al., 1976).
The Corsican and Sardinian blocks are represented in their Present and palinspatic (pre-Miocene) positions. The second position has been obtained by coinciding the 2000 m isobaths of the Languedoc-Provencal coast and the Corsica-Sardinian submarine plateau. These isobaths are considered to represent the boundary of the respective continental platforms. This reconstruction shows the deviation between the rotation of the Corsican block (30°) and that of the Sardinian block (60°), as deduced from palaeomagnetic measurements of Permian lavas. But this deviation does not conform with the exact continuation of the Corsican granitic massif into its Sardinian equivalent.
1. Variscan granites; 2. partly metamorphosed Palaeozoics; 3. Permian volcanism; 4. Central Sardinian graben, of Tertiary age, and associated volcanism (5). Pc, Ps: poles of rotation of Corsican and Sardinian blocks. G: Genoa

Consequences of Ligurian sphenochasms on land: The opening of the Gulf of Genoa must have induced complications on the land, such as an angular gap on one side of the pole of rotation and overthrusting on the opposite side. This cannot be conclusively established however; there are clearly overthrusts dependent on Apennine folding and they have a precisely NW vergence, as can be expected there (Fig. 90). However, this folding continued after cessation of the opening of the gulf and, by superposition of new structures, masks the effect desired.

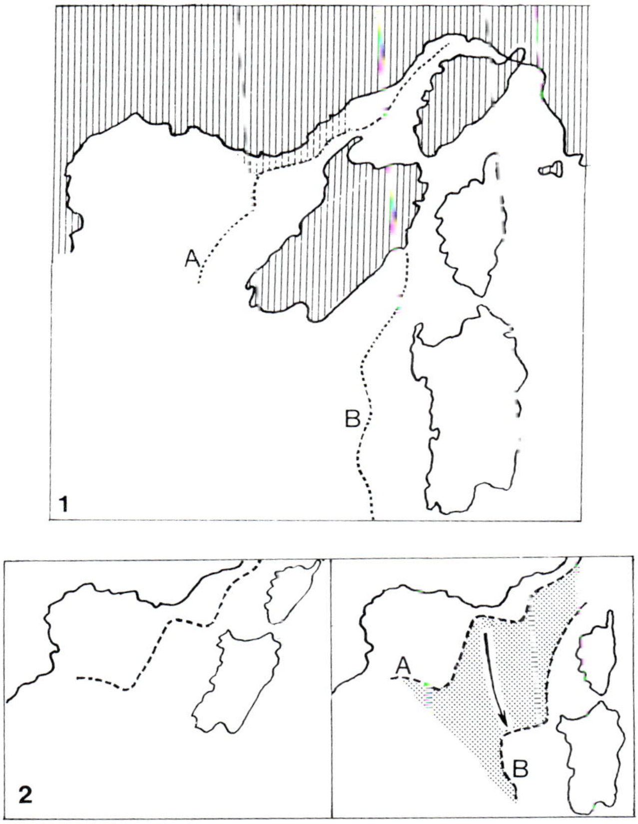

Fig. 88. Other possible models for reconstructing the Corsica-Sardinian axis (1. after Chabrier and Mascle, 1975; 2. after Mattauer, 1973 , *Coll. Cagliari*, pp. 297–300).
Unlike the model of Fig. 87, these models do not separate the Corsican block from the Sardinian and hence assume the same angle of rotation (30°) for both. The differences between the two models lie in the determination of the continental-crust boundaries of the Languedoc-Provencal (A) and Corsica-Sardinian (B) blocks. These boundaries still serve as reference but are masked by the Rhone alluvial fan (Fig. 86).

Consequences of Ligurian sphenochasm in the sea: Development of submarine canyons of the coast of Provence (Fig. 91). The opening of the Gulf of Genoa aids in understanding the origin of these canyons which affect the boundary of the Provençal platform in the continuation of some present-day water courses. The process occurs in four stages:

1. Immediately before rifting, the Corsica-Sardinian block, united with the European plate, was uplifted by Provençal folding and was an elevated zone subjected to erosion that supplied the future Annot sandstones.

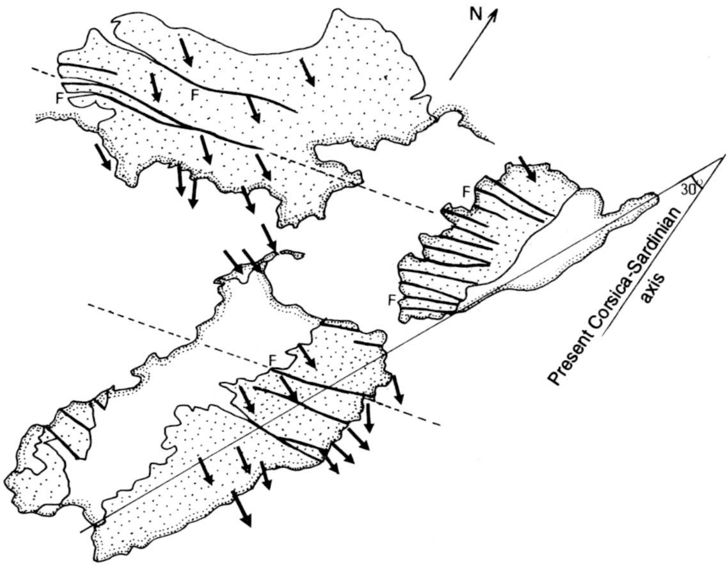

Fig. 89. Correspondence between lineations of Variscan extension (arrows) and late Variscan faults (F) in the Maures and the Corsica-Sardinian blocks (after Chabrier and Mascle, 1975). In order to relocate them as a continuation of each other, the Corsica-Sardinian axis needs to be rotated by 30°. This model is not a palinspatic reconstruction (which is given in Fig. 88).

2. This block was cut in two by an extension axis which became a rift. These two limbs were uplifted like any other rift margins after their formation (Oligocene-Lower Miocene).

3. By the end of Miocene the crustal fracture had widened. The boundaries moved away from the zone of high heat flow which marked the rift axis. They cooled down, became denser and subsided, resulting in *thermal debulging*. The Provençal margin hence became inclined towards the oceanic gulf and the continental waters started flowing southwards, into their ancient valley whose slope is found to be reversed.

Drainage southwards was also facilitated by the general regression of the Upper Miocene which transformed the Mediterranean into a gigantic lagoon, thus lowering the base level of the water courses that reach there.

4. Pliocene transgression restored the sea level and submerged the canyons produced in the preceding stage. Their slope is further accentuated by two other factors:

— under the weight of water, the floor of the Mediterranean subsided due to isostasy;

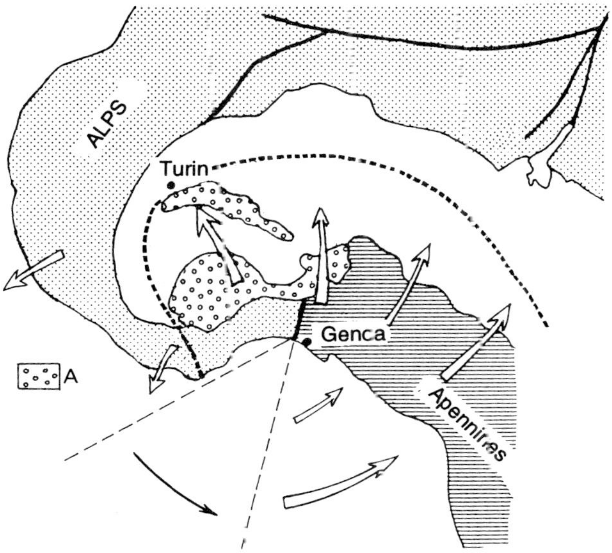

Fig. 90. Gulf of Genoa and the Ligurian Apennines.
Their disposition might suggest that the opening of the Gulf of Genoa is associated with folding of the Ligurian Apennines. However, these two are offset in time, the latter being later.
A. Ligurian Miocene thrust.

— on land, the Pliocene compressive movements induced, at least locally, the uplift of some sectors of the Provençal margin (Maures region, for example), resulting in the present-day morphology.

Gulf of Gascogne (Bay of Biscay) (Fig. 92)

A deep triangular basin, known as the abyssal plain of Biscay, extends between the southern Armorican continental margin and the northern Spanish coast (Cantabrian margin), which, like the Gulf of Genoa, has all the characteristics (seismic velocities, thinness of crust and magnetic anomaly bands) of an oceanic crust. The boundaries of this small oceanic basin are very dissymmetric. The Armorican margin is wide with numerous step faults (typical passive margin), whereas the Cantabrian margin is narrow and complicated by thrust faults.

The data of offshore wells, combined with that of seismic reflection profiles, indicates that sedimentation commenced at the end of the Lower Cretaceous (which is observed only in the NW of the gulf). The gulf was completed in the Upper Cretaceous as the sediments of this period are found everywhere.

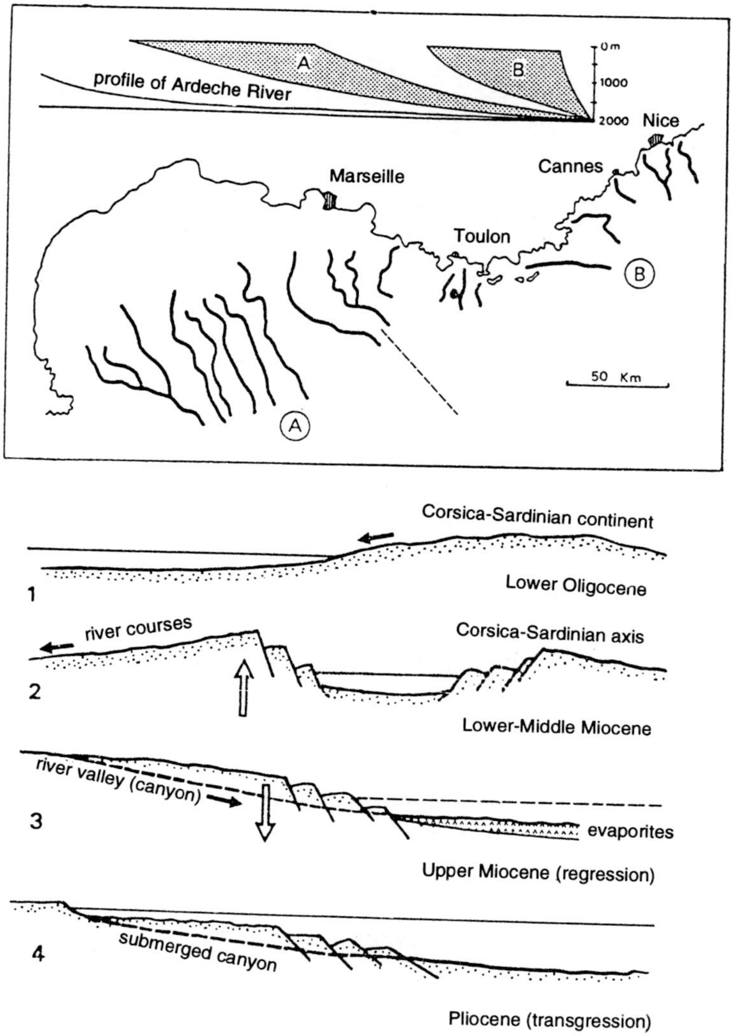

Fig. 91. Submarine canyons of Languedoc (A) and Provence (B). Their average slopes are included in the corresponding grey areas.
Below, their origin as related to the opening of the Gulf of Genoa.

These facts accord with the palaeogeography of the Aquitaine basin. In fact, after the major regression at the end of the Jurassic and the beginning of the Cretaceous, the sea returned during the Aptian from the newborn Atlantic, resulting in the formation of two gulfs (Parentis and Adour) which gradually opened up mostly westwards.

It has hence been thought that a typical sphenochasm exists here in which the Iberian block has rotated by 30 to 35° relative to the rest of

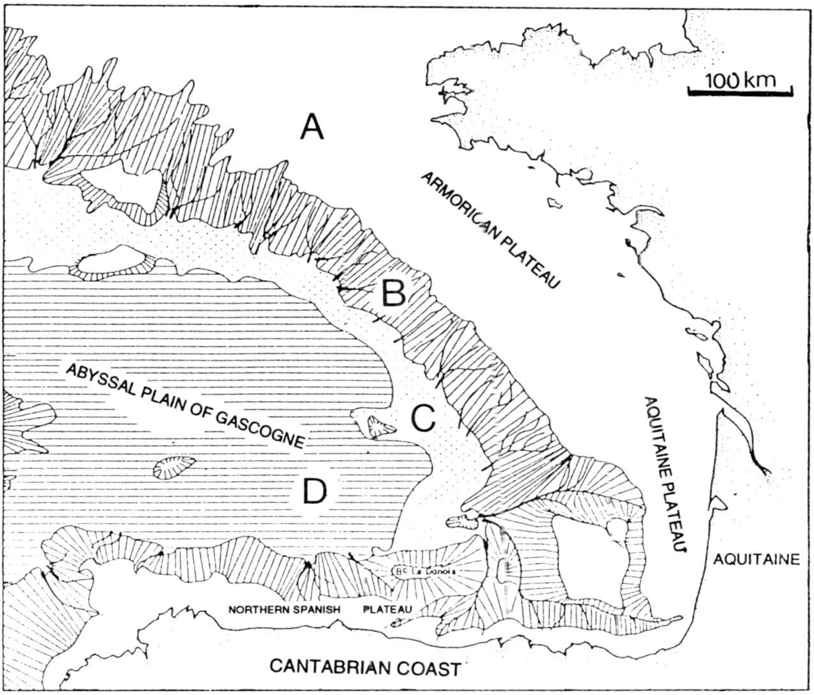

Fig. 92. Morphology of the Gulf of Gascogne (after Rehault, *in:* Boillot, 1984). A—continental plateau; B—continental slope; C—sedimentary glacis masking the contact between continental and oceanic crusts; D—oceanic crust.

Europe. This theory also accords with the fact that the Variscan structural zones of the Armorican massif extend into Spain without a gap, on either side of the Gulf of Gascogne which did not exist then (Iberian-Armorican virgation) (Fig. 93).

This model is also corroborated by palaeomagnetic data. The Carboniferous and Permian lavas of Spain on the one hand, and those of the Armorican massif on the other, show magnetic declinations deviating from each other by 35° following a sinistral (anticlockwise) rotation. On the contrary, the lavas of the Upper Cretaceous and Eocene give identical values in both cases. Thus a rotation has clearly taken place between the Permian and the Cretaceous.

The foregoing explanation is based on palaeogeographic reconstructions of the Atlantic from magnetic anomalies (Fig. 94).

In the gulf itself only one pair of anomalies, viz., no. 34 (Santonian, 80 Ma) is observed. The next one (no. 33, Campanian, 75 Ma) does not pass through the gulf, indicating that accretion there has been completed.

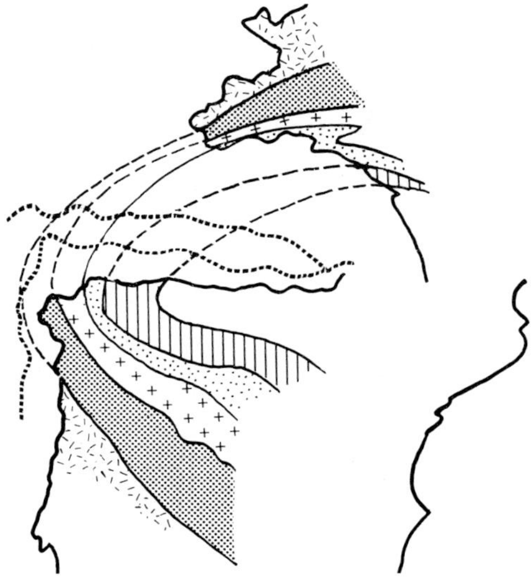

Fig. 93. Variscan virgation of the Iberian-Armorican block reconstructed after closure of the Gulf of Gascogne and rotation of Spain relative to France.
Broken lines: continental margins of Iberia and Armorica. The different shaded regions represent Variscan structural zones. Their names are of no interest here.

To reconstruct the shape of the gulf of this period, it suffices to make anomaly pairs 33 of the two coasts of Atlantic coincide with each other (they are presently separated by 1800 km). In other words, the ridge of this epoch is reconstructed and the continents are relocated in their palinspatic positions of the Campanian period. It is thus found that in this epoch the Spanish continent was at 100 km to the SSE of Europe. It was Pyrenean folding during the Tertiary which closed these 100 km.

To know the position of Spain before the opening of the gulf in the Cretaceous, one can use anomaly J (= Mo) which is observed off the Spanish coast and which is 110 Ma in age (Upper Aptian), hence immediately before the opening. Following the same procedure as earlier, we find that Spain was situated farther away 200 km NW.

In other words, Spain had a dual movement:

— a rotational sliding SE by 200 km (Albian-Campanian);

— a post-Campanian migration northwards by 100 km.

The *first movement* is contemporary to the opening of the gulf: the sliding movement of Spain thus occurred on a transform fault, which must have been curved. The pole of rotation was initially situated in the north of France and then *moved gradually southwards, which is the only possible manner of opening of the gulf into a sphenochasm* (Fig. 95).

The fault, or rather the transform zone in question, is again found, highly deformed by subsequent tectonic movements, in the 'North-Pyrenean

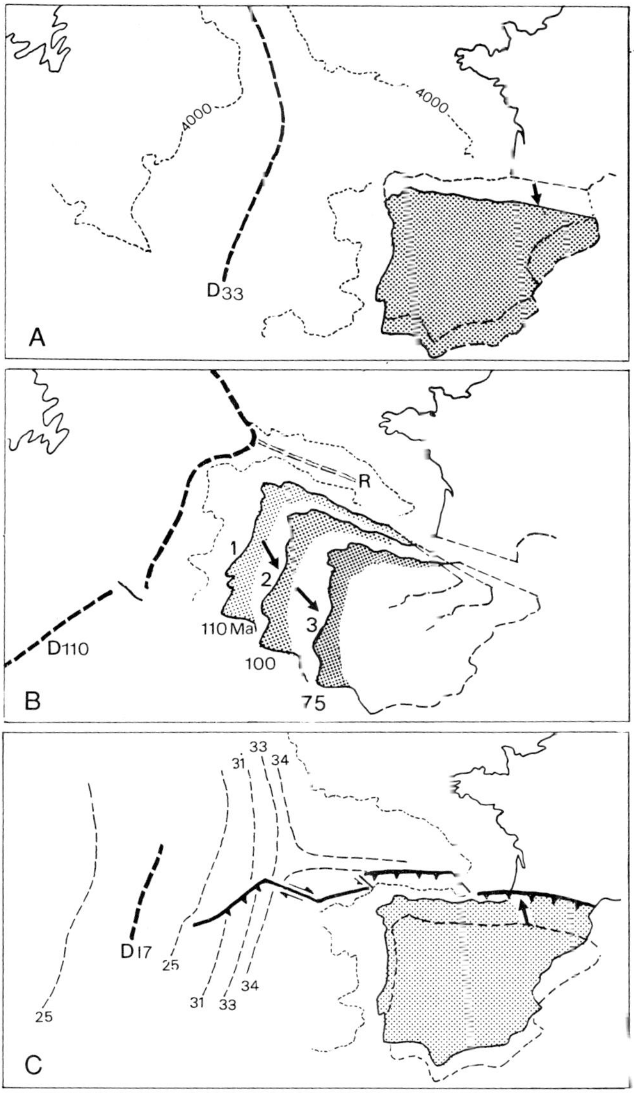

Fig. 94. Relative movements of the Iberian block with respect to Europe (after Boillot, 1984, simplified).
A—Reconstruction of the Atlantic during the period of anomaly 33 (75 Ma, Campanian). For this, anomalies 33 situated on either side of the present-day ridge are made to coincide with each other. Thus the Campanian ridge (D 33) is found. The Spanish block is displaced by 100 km southwards.
B—Reconstruction of the Atlantic during the period of anomaly J (110 Ma, Aptian). In a manner similar to the earlier case, the ridge of this period is reconstructed, which gives the location of Spain in position 1. The trajectory from 1 to 3 (nearly 200 km) is covered between the Aptian and the Campanian. R: Probable intracontinental rift preceding the oceanic opening of the Gulf of Gascogne.
C—Reconstruction of the Atlantic during the time of anomaly 17 (40 Ma, Upper Eocene). Between the Campanian and the Eocene, the Iberian block has moved up over 100 km in the direction NNW, thus causing Pyrenean folding.

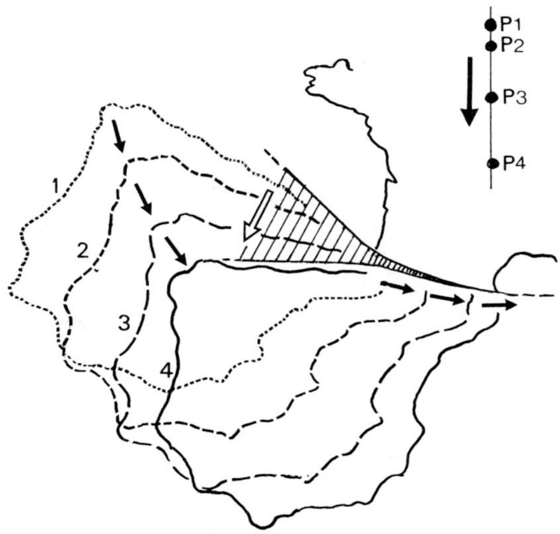

Fig. 95. Opening of the Gulf of Gascogne in the Upper Cretaceous by anticlockwise rotation of the Iberian block (after Choukroune et al., 1973).
The pole of rotation was situated in northern France but gradually moved southwards during the Upper Cretaceous (P1, P2, P3). The Iberian block rotated relative to Europe at the same time as it moved (black arrows), resulting in the opening of the gulf (white arrow). This model accords with the position of the Iberian block in the Upper Cretaceous (Fig. 94). The slip surface is actually a transform fault.

zone', which is effectively a very complex region as we shall see later (p. 176).

The *second movement*, contemporary with folding of the Pyrenees, makes the entire southern part of the Gulf of Gascogne disappear by compression and subduction below the Cantabrian margin. This can readily explain the very specific style of this margin, quite different from that of the southern Armorican margin. The two coasts of the Gulf of Gascogne are thus highly dissymmetric.

The two branches of anomaly 34 facilitate determination of the ancient axis of symmetry of the sphenochasm. Further, they are closer to the Spanish coast: 70 km of oceanic crust is missing in the south. This is the part subducted below the Cantabrian margin. It is too short to have transformed the latter into an active margin with developed volcanism.

CONCLUSIONS

If all the brittle extension type structures described above are juxtaposed, a coherent model of evolution can be obtained (Fig. 96).

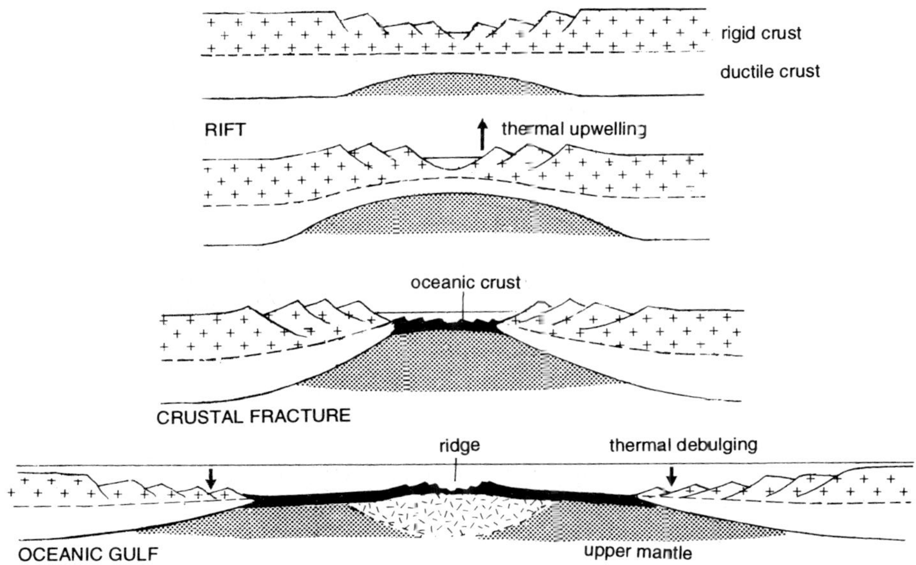

Fig. 96. Evolution of rift-crustal fracture-oceanic gulf.

1. *Rift stage:* Under the effect of a severe extension, the rigid upper crust is split into several fragments by normal faults. These blocks then undergo subsidence. At depth, the crust is subjected to ductile stretching, resulting in the rise of the mantle. This generates a strong thermal flux which heats up the rift margins and produces their uplift (stage of *thermal intumescence*).

An alkaline volcanism is manifest, nourished not by the upper mantle as one would first expect, but by a deeper mantle (at least 100 km).

2. *Stage of crustal fracture*: If stretching continued, the continental crust, gradually thinned out, is injected by basic products of tholeiitic origin, emerging from the upper mantle, closer to the surface this time. Thus an oceanic type crust formed which replaced the older continental crust. At this stage the sea invades the rift (the Red Sea).

3. *Stage of oceanic gulf*: If the extension process continued, the crustal fracture is enlarged into an oceanic gulf, triangular or otherwise; this implies an oceanic accretion with the development of a ridge. At this stage the rims of the fracture are sufficiently far off from the thermal axis, such that they are cooled and subside (stage of *thermal debulging*).

The oceanic gulf may then become an ocean and its rims will become totally submerged *continental margins*, the subject matter of the next chapter.

Figure 96 shows a model of rifting and the symmetric structures of the two opposite rims. However, this disposition is rarely observed. This can be explained by a dissymmetric extension per se, caused by 'detachment faults' (see Figs 44, 46). These symmetric or dissymmetric evolutions in fact cor-

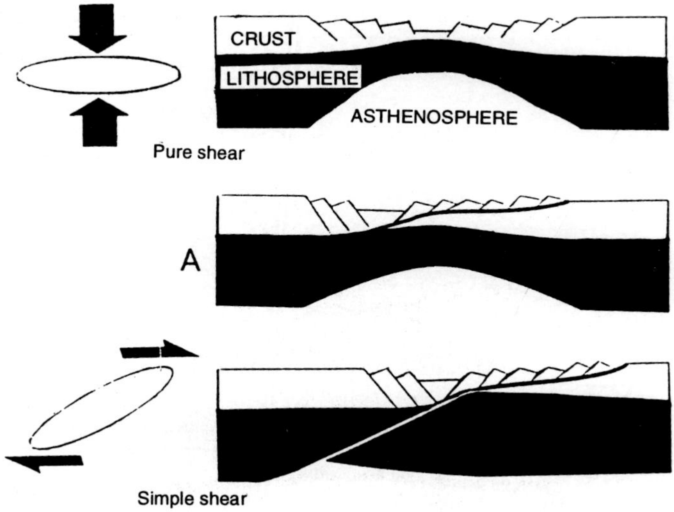

Fig. 97. Symmetric or dissymmetric extension depending on the type of shear of the crust (after Jolivet, 1995). A. Intermediate case.

respond to classic processes of pure shear and simple shear (Fig. 97). The former is caused by homogeneous stretching of layers and the latter by tilting, rotation and stretching of blocks juxtaposed above a detachment fault. This will be discussed in detail while describing the continental margins (p. 145).

CHAPTER 3

CONTINENTAL MARGINS

We saw in the preceding chapter that the continental margins are the rims of highly opened ancient crustal fractures. The two margins thus formed moved away from each other and are invariably broken into blocks and compartments by numerous extension faults. Some fragments of the continental crust, themselves separated, are now found isolated in the midst of young oceanic crust. Such is the case, for example, of the Rockall islet, off the coast of Scotland and the Seychelles in the Indian Ocean.

Such margins are known as *atlantic* or *passive*. However, sooner or later, the passive migration of the continent ceases following modifications associated with sea-floor spreading and intercontinental collisions. This blockage leads to severe folding of part of the continent or, more often, subduction of the oceanic crust below the continental crust at the boundary of the margin in question. The margin then becomes an *active* or *pacific* type.

PASSIVE MARGINS

It is known that the continents extend below the sea along the *continental plateau*, with an average depth of 200 m, where the Moho is 25–30 km deep. The crust there is thus normal. Then it passes in to the zone of *continental slope* (even though the slope there hardly exceeds 3 to 5° in general), which ends at the abyssal plains of the oceanic crust.

Structure

The large amount of geophysical data available on these margins reveals the following facts:

The continental crust *thins out gradually* even though it is often difficult to distinguish the oceanic crust in the slope region (Fig. 98A). This situation has given rise to the concept of *intermediate crust* commonly employed by geophysicists, and already mentioned earlier (p. 6). The crust here is sometimes compounded by the intrusion of basic magmatic products into the extension zones, which are frequent in this region.

If these products are abundant, they are referred to as *passive volcanic margins* (Fig. 98B). These margins may have resulted from the opening up of the sea floor above a hot spot.

Thinning of passive margins, in combination with thermal debulging and the weight of sediments, explains the progressive subsidence of these margins up to the depth of abyssal plains and thus makes it difficult to detect the continent-ocean boundary on a bathymetric map, especially when sedimentation is abundant.

The crust is fragmented into separate blocks by normal faults, a fact which is clearly indicated by geophysical profiles, even below a certain thickness of sediments (Fig. 98A), but only in their upper part or trajectory. Hence these faults become less distinct in deeper parts of the crust. Their trajectory nonetheless appears to become concave towards the surface (listric faults), in which case the movement of the blocks they demarcate forms a rotation, which does not exclude additional fractures along conjugate faults.

Termination of these listric faults at depth still remains an unresolved problem. It is thought that they become tangential to the boundary between the rigid crust and the ductile crust (S-surface) (Fig. 98A).

However, more complex structural dispositions may occur locally. For example, investigations carried out in the Galicia zone, off the western Spanish coast (Fig. 99), have shown that below the partly tilted blocks of the continental crust exists a thick layer of serpentinised peridotites on which the listric faults have been terminated. But the 'seismic' Moho is much deeper. This disposition is interpreted as the consequence of complete disappearance of the ductile lower crust, due to stretching, at the time of origin of the margin. This extension results in the outcropping of the mantle in the axis of the primitive crustal fracture and its direct contact with the upper crust on either side of the fracture.

Moreover, it is thought that the thinning and fracturing of the continental margin were sufficient to produce hydrothermal alteration *per descensum* of the upper mantle. Due to this alteration, the latter acquires seismic velocities of 6.7 to 6.8 km/s, which are the same as those of continental crust. So the 'seismic' Moho is situated deeper, indicating that it moves to the interior of the upper mantle.

On the surface, the system of normal faults is discontinuous and irregular (Fig. 100). These faults are far from perfect step-fault structures, as cross-sections show. This disposition, moreover, is similar to that observed in rift margins (see Fig. 34).

Further, the 'slip surface' of the listric faults does not always face the ocean. Due to the presence of conjugate faults, the blocks of the margin may also occur as nearly symmetric horsts, forming the boundaries of grabens or intramarginal rifts (Fig. 101).

Finally, when evaporite layers occur prior to the development of fracture, they may serve as amortising horizons for listric faults and facilitate the

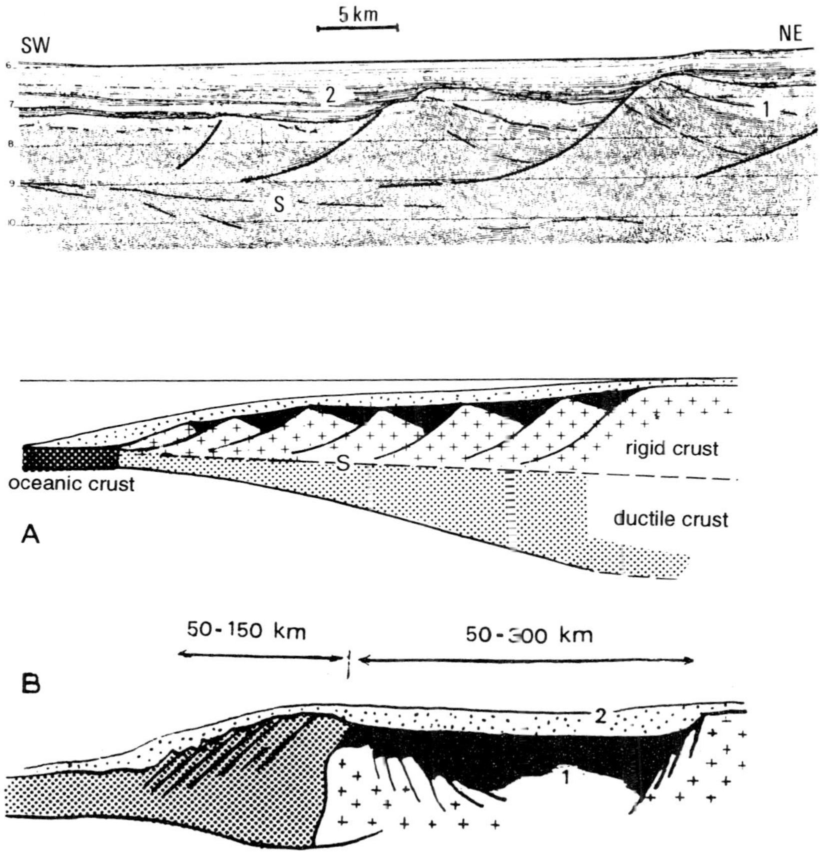

Fig. 98. Passive continental margins.
Top: seismic reflection profile across a passive margin showing the tilted block structure and the S-surface (inlet of the Manche, 300 km west of Finistere) (after Montadert, *in* Boillot et al., 1984, simplified).
1. prerift sediments, 2. post-rift sediments (Upper Aptian-Eocene).
Elevation scale: two-way time, seconds.
A—Mechanism of thinning by stretching of the ductile crust (S: rigid crust-ductile crust boundary).
B—Example of passive volcanic margin, Voringe plateau (Norway) (after White and MacKenzie (1989) *J.G.R.*, 94: 7685–7729). The continent-ocean transition is marked by a thick basaltic sequence (10 to 15 km) showing reflectors dipping towards the ocean and overlying a zone of the same thickness comprising dense materials of high seismic velocities (7.2 to 7.5 km/s), probably gabbros, plated at the base by a very thin continental crust supporting the basalts or directly below the latter. Emplacement of these margins may have resulted from the presence of a deep hot spot vertically below them (here the Icelandic hot spot).
In black: pre- and synrift sedimentation (1). Dotted: post-rift sedimentation (2).

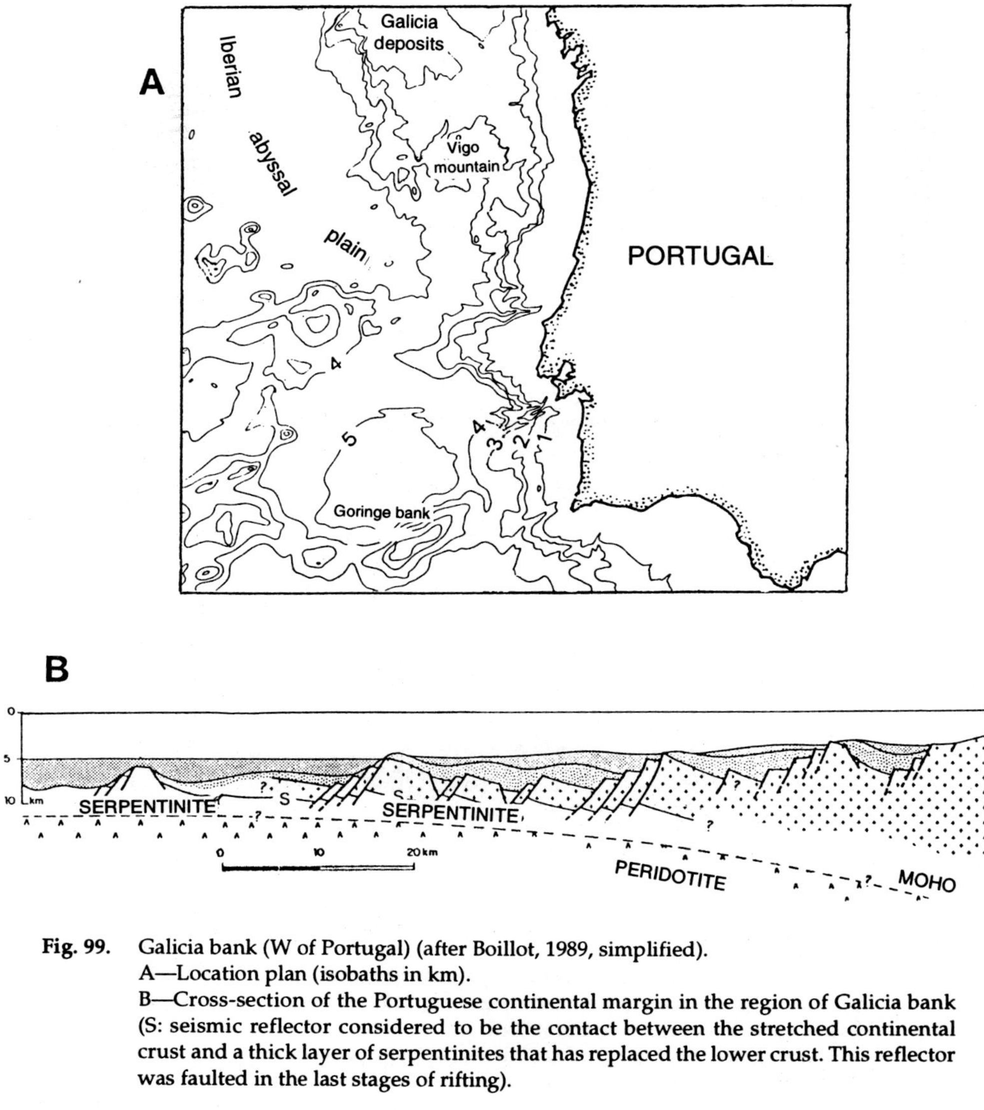

Fig. 99. Galicia bank (W of Portugal) (after Boillot, 1989, simplified).
A—Location plan (isobaths in km).
B—Cross-section of the Portuguese continental margin in the region of Galicia bank (S: seismic reflector considered to be the contact between the stretched continental crust and a thick layer of serpentinites that has replaced the lower crust. This reflector was faulted in the last stages of rifting).

rotation or tilting of the blocks, thus complicating the interpretation of the diagrams.

Passive Margins and Sedimentation

Sedimentation on a margin may occur either prior to the rifting ('prerift' sedimentation), or contemporary to it ('synrift' sediments with fan-shaped geometry, offset by the faults but containing olistoliths released from the

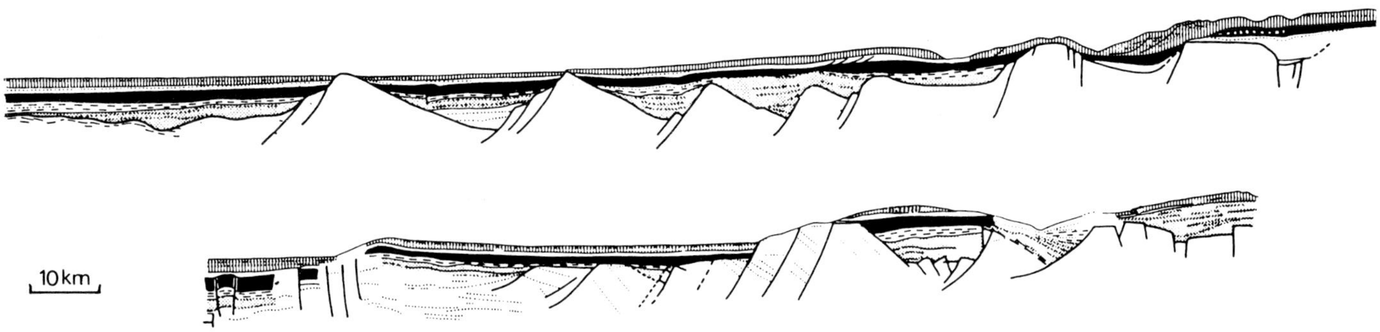

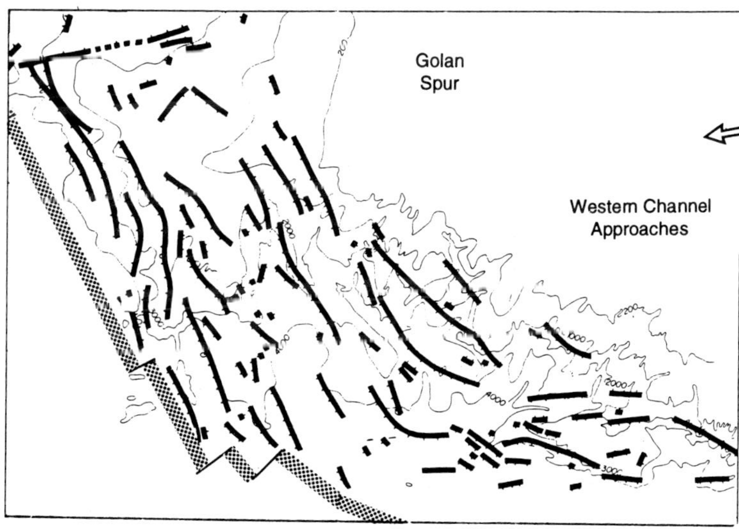

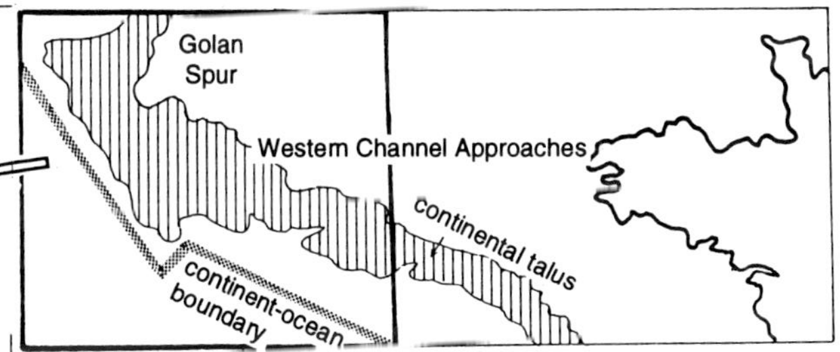

Fig. 100. Example of starved margins: the Golan Spur and the Western Approaches of the English Channel (after Montadert et al., 1979, simplified, *in:* Boillot et al., 1984).
In the cross-section, the synrift sediments (Lower Cretaceous) are shown in grey. The overlying beds range from Upper Aptian to the Present.
On the map, the normal faults (of Lower Cretaceous age) are indicated by thick lines hatched towards the fault plane. The grey band is the probable boundary between the continental and oceanic crusts.

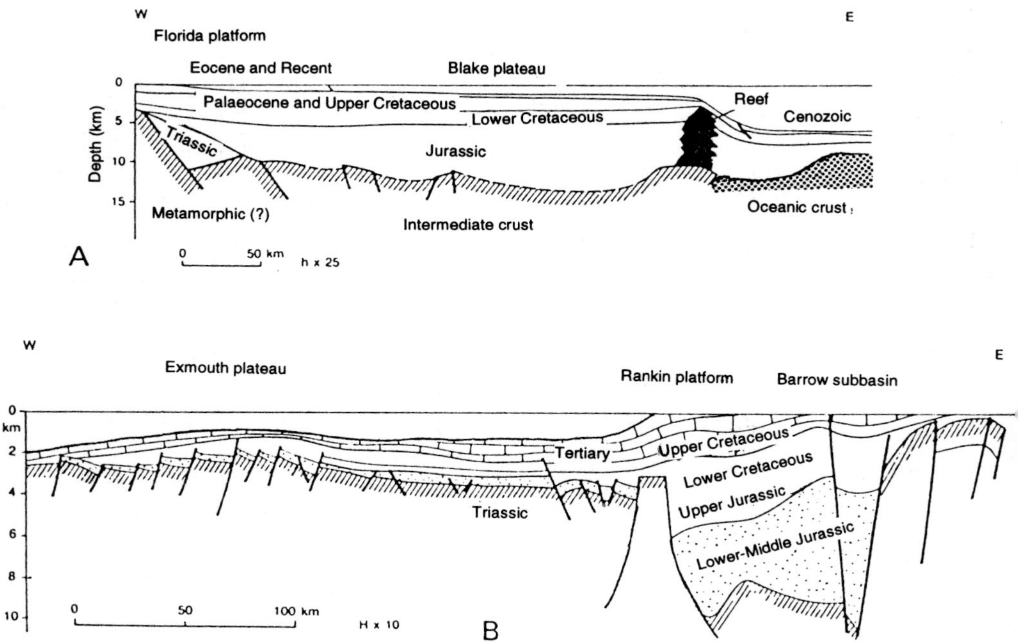

Fig. 101. Passive margin basins.
A—by reefal barrier (Blake plateau, Eastern USA); B—by graben (Exmouth plateau, Western Australia) (after Perrodon, 1988, slightly modified).

fault walls and often showing internal discordances) or, finally, later to rifting ('post-rift' sediments, unaffected by faults).

Since post-rift sedimentation always exists, it can be inferred that fracturing and rotation of the blocks ceases after some time. It is generally considered that this cessation of movement occurs in the beginning of oceanic accretion, i.e., when there is cooling of the margin and thermal debulging.

Further, depending on the abundance of sediments, margins may be distinguished as *starved* (2 to 4 km of sediments) or *sedimented* (greater than 4 km).

Starved continental margins have a varied topography because their tilted-block structure or offset disposition remains visible. A good example is that of the Armorican margin of the Gulf of Gascogne (Fig. 100). These margins show a succession of ridges and plateaus separating small basins in which sedimentation is gradually reduced towards the sea since the ridges act as barriers to the inflows from the continents. In the case of lowering of the sea level, these basins may evolve as lagoons with evaporites.

Sedimented margins have a more monotonous surface topography, similar to that of the sedimentary cover wrapping the continental block. The accumulation of sediments may be attributable to various causes, such as blockage by coral reefs (Blake plateau on the eastern coast of the USA,

Fig. 101A) or a graben (Exmouth plateau in Western Australia, Fig. 101B). Beyond these barriers, the thickness of sediments rapidly decreases except in 'prograding shorelines' where the sediments overflow the continental plateau and advance onto the oceanic crust as a progradation talus which masks its boundary with the continental region (Fig. 102).

The internal structure of such talus zones is sometimes made complex by the effect of tectonic or eustatic oscillations which cause superposition or grooving of successive talus deposits, or by gravitational sliding of mud or

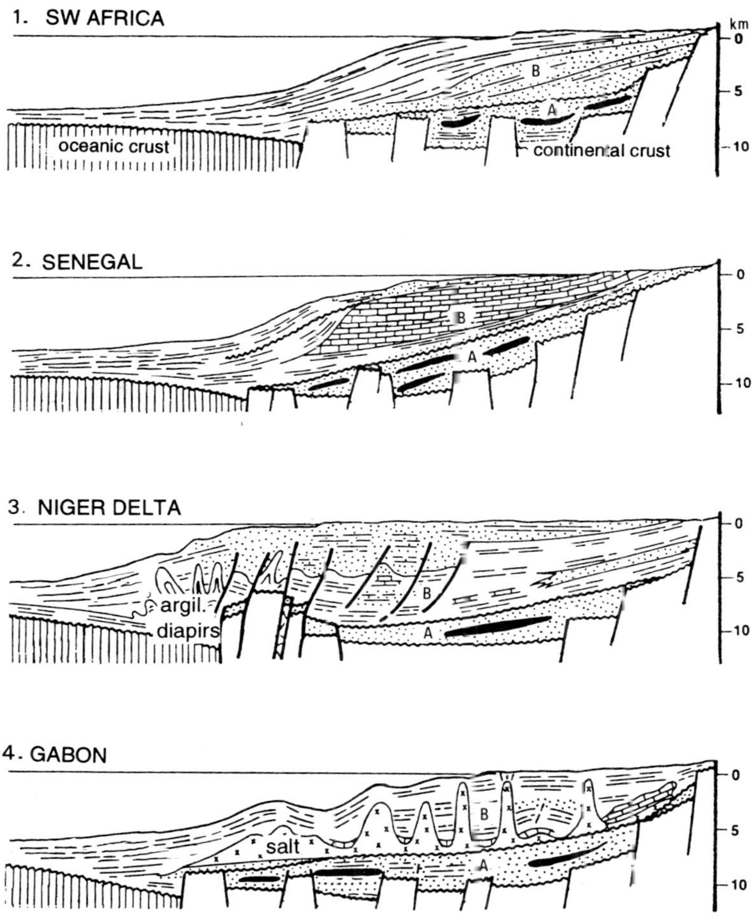

Fig. 102. Some examples of sedimented margins (Western Africa) (after Kingston et al., 1983, simplified).
In 1, dominant detrital sedimentation. In 2, dominant calcareous sedimentation. In 3, deltaic sedimentation. In 4, argillo-evaporitic sedimentation.
A—prerift sediments; B—post-rift sediments.

turbidite flows produced on gentle slopes (2° in the case of the submarine delta of the Niger).

A starved margin may naturally become a sedimented one. Such an evolution is favourable for the origin of petroleum basins. In fact, in the first stage there may be formation of confined basins with high organic charge, subsequently trapped below a sufficiently thick protective sedimentary cap rock. This is the case of basins along the Atlantic coast of Africa: Congo, Gabon, Cameroon and Angola.

Figure 102 gives some examples of these basins in the present-day passive margins, where all the elements of complications described above are superposed.

Fossil Passive Margins

In most great mountain belts one or more ophiolitic suites are found to exist, which are interpreted as the remnants of an ancient oceanic domain pinched between the continental margins. By unfolding these belts, we can reconstruct the margins in question in which all the characteristics of the present margins can be observed.

The western (or European) margin of the alpine ocean in the Upper Jurassic is a good example. Figure 103 gives an interpretation of this margin in which large alpine tectonic units are found to correspond to blocks or basins separated by synthetic or antithetic faults. It can also be seen that the order of magnitude of the Mesozoic blocks thus reconstructed is the same as that of the starved margins of the present Atlantic. Moreover, when alpine folding disturbed the disposition of these blocks by compressing them from east to west, the normal faults often functioned as reverse faults. This fact explains the parallelism, long known in the Alps, between the palaeogeographic units and structural units.

Transform Fault Margins

This is a specific type of passive margin in which the margin is bounded by a transform or transcurrent fault. In this case, neither the characteristic thinning nor the juxtaposition of partly tilted blocks is observed, unless transverse faulting has taken place on older normal faults. Figure 104 shows the origin of such margins. They are recognised from the fact that the magnetic anomalies of the neighbouring ocean are perpendicular to the margin instead of parallel to it.

The northern coast of the Gulf of Guinea is a simple example of such a margin (Fig. 104). A more complex case is the SW coast of Spitsbergen (Fig. 105). The Spitsbergen block is in effect only the emerged part of the European margin on the boundary of the North Atlantic and the strait, also oceanic, which connects it with the Arctic glacial ocean. The SW coast of Spitsbergen is rectilinear, bounded by strike-slip faults, between which exist

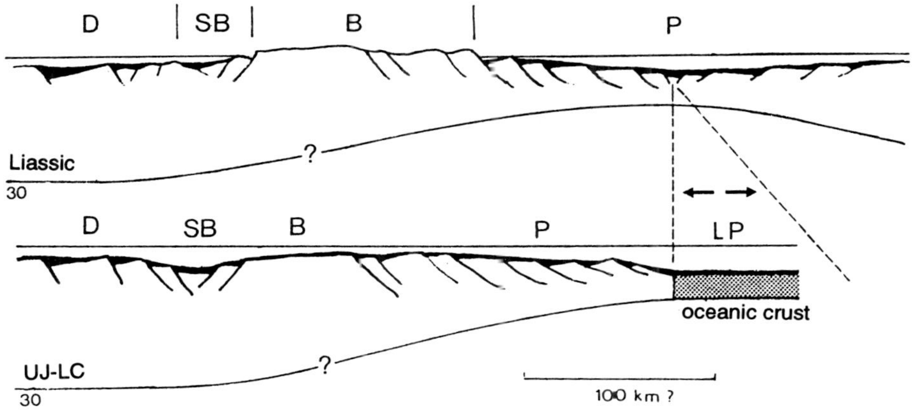

Fig. 103. Hypothetical reconstruction of the European continental margin in the border of the alpine mini-ocean in the Upper Jurassic-Lower Cretaceous, across the Dauphinois (French-Italian) Alps. It represents a typical starved margin.
During the Liassic (top section) this margin is still only an intracontinental rift boundary. Oceanic crust appears only at the end of the Middle Jurassic.
B: Briançonnais domain; D: Dauphinois domain; LP: Liguria-Piemont domain; P: Piemont domain; SB: sub-Briançonnais domain;
UJ-LC: Upper Jurassic-Lower Cretaceous.

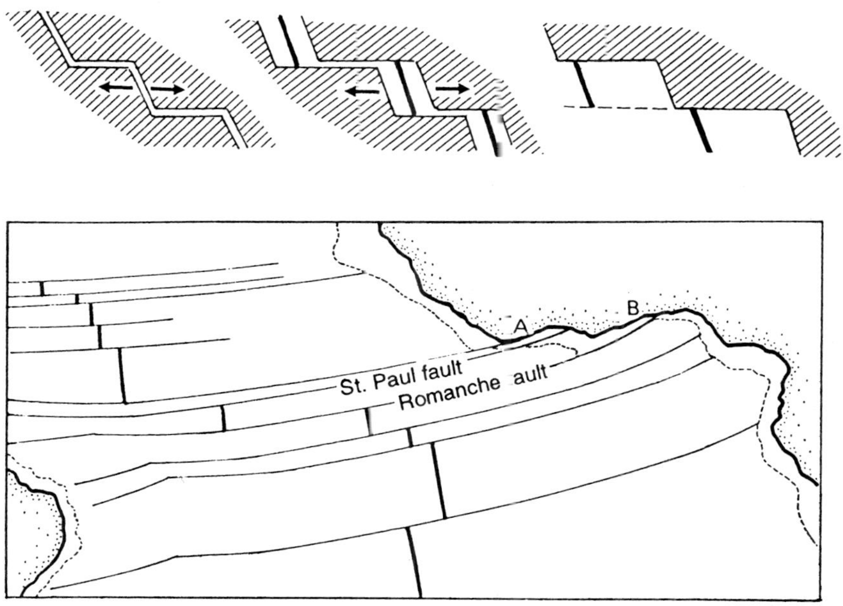

Fig. 104. Transform fault margins in equatorial Africa.
A. Ivory Coast, B. Ghana
Top: Mechanism of their origin.

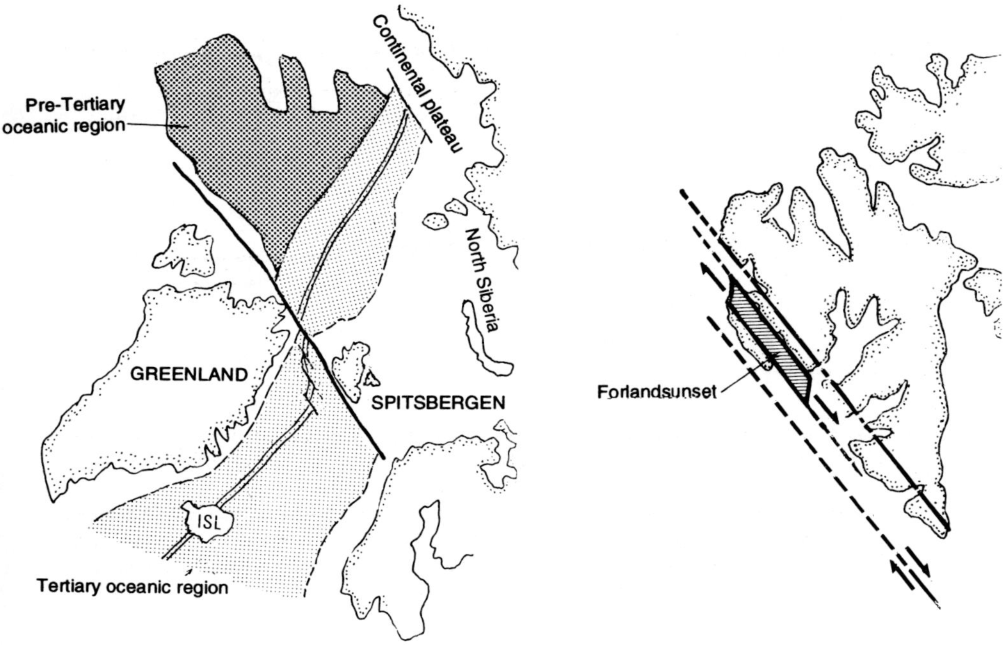

Fig. 105. Transform fault margin in SW Spitsbergen (after Lepvrier and Geyssant, 1985).
Left: position of Spitsbergen in the structural framework of the North Atlantic. Its SW coast is determined by a major transform fault separating it from the Greenland block.
Right: The small tertiary basin of Forlandsunset is a pull-apart structure associated with the transform faulting of the Spitsbergen block relative to the Greenland block.

pull-apart structures that have given rise to the small subsident basins of the Eocene, sometimes containing coal. The Eocene is the epoch of the opening of the North Atlantic into the Arctic glacial ocean. A section across this margin clearly shows a series of juxtaposed blocks, but they indicate neither thinning nor oriented tilting.

Dissymmetric Passive Margins

In the preceding sections it was assumed that the margins result from the opening of a nearly symmetrical rift. However, the opposite margins in the present-day oceans rarely exhibit such symmetry, neither in structure nor in origin. Moreover, geophysical data have shown that in some cases the uplift of the Moho and the zone of maximum crustal thinning do not coincide. Finally, offshore wells have revealed that even inside a passive margin, the mantle may occur directly below the sediments, indicating an irregular and discontinuous spreading.

It can obviously be inferred that such dispositions are the result of chance, as the continental rupture is not produced forcibly in the axis of symmetry of the extension system, and the fragmentation of blocks may also be influenced by older structures. However, other mechanisms may also be possible, such as spreading, which triggers rifting, itself a dissymmetric process.

The model proposed by Wernicke, (*Nature* (1981) p. 291) (Fig. 106), uses for this dissymmetric extension a low-dip (7 to 15°) shear plane or 'detachment fault' traversing the entire lithosphere[1]. Thinning is produced from the sliding of one of the blocks on this fault, resulting in its movement. Figure 106 shows that a significant offset occurs between the zone of maximum thinning of the lithosphere (X) and that of the crust (Y). Further, the disposition of the blocks on the margins A and B is not at all symmetrical. In part A the tilted blocks rest on the mantle but in part B, on the deeper zones of continental crust. This model also explains the occurrence of tectonic denudation of upper mantle in the rift axis before the commencement of oceanic accretion, as well as an offset thermal dome, and hence the uplift of only one of the boundaries (and even the ultimate appearance of a subsidiary rift on this uplifted margin, eccentric relative to the first one). Finally, it also explains the dissymmetric distribution of volcanic activity on the margins or the primitive rift.

The origin of these very flat detachment faults can probably be explained by the layered structure of the crust and the gradual modification of its mineralogical composition from top to bottom. Consequently, all the superposed formations do not react to the extension in the same way. In other words, differences in behaviour are manifest on either side of the horizontal or slightly inclined shear planes.

ACTIVE MARGINS AND DERIVED ISLAND ARCS

Continental active margins are situated vertically above a subduction zone[2] and hence show a chronic seismic and volcanic activity, which is manifest by the existence of a volcanic (or magmatic) arc.

We can, however, see that this statement should be considered with caution because several active margins exist which show no volcanic activity whatsoever; they are associated with low-dip subduction zones. These margins also indicate a compressive tectonic regime. We shall discuss such margins in the chapter on subduction belts.

---

[1]It may be recalled that this model was originally proposed to explain the structure of the Basin and Range in western America, but not for dissymmetric continental margins.

[2]Sometimes called the 'Benioff' or 'Benioff-Wadati plane'. It is a B type of subduction, in contrast to the alpine type or A type subduction, which is 'intracontinental' (see p. 184).

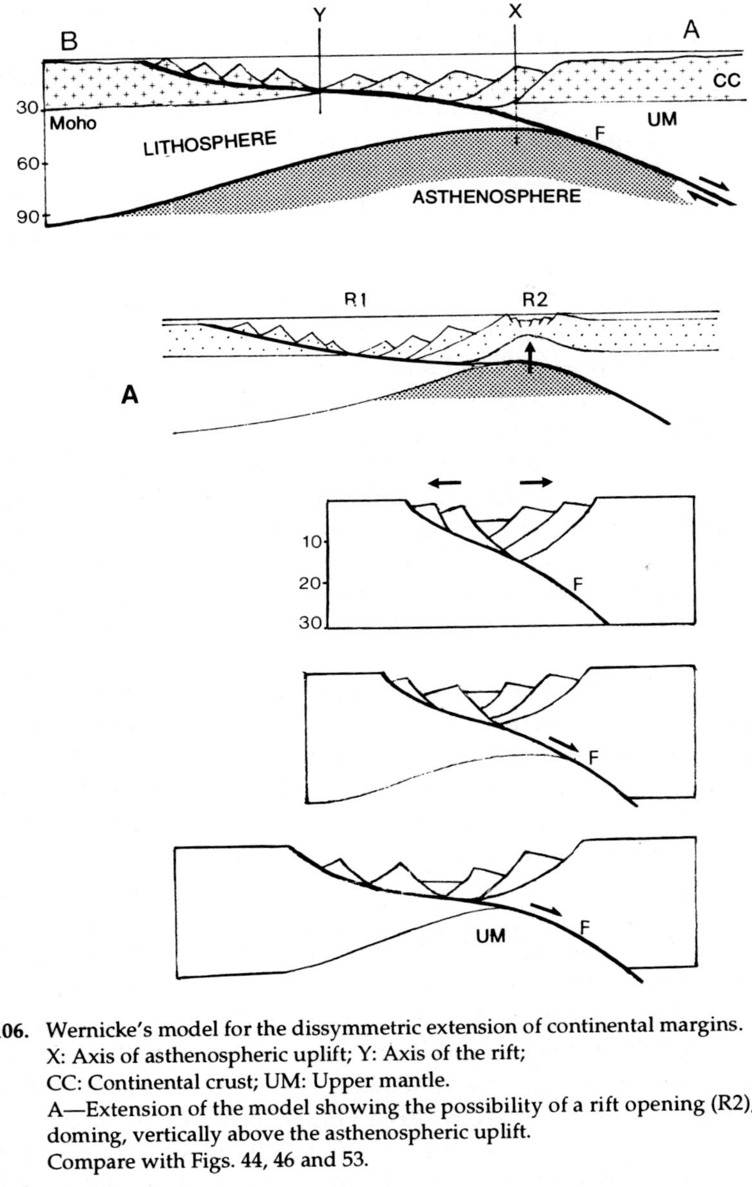

Fig. 106. Wernicke's model for the dissymmetric extension of continental margins.
X: Axis of asthenospheric uplift; Y: Axis of the rift;
CC: Continental crust; UM: Upper mantle.
A—Extension of the model showing the possibility of a rift opening (R2), by thermal doming, vertically above the asthenospheric uplift.
Compare with Figs. 44, 46 and 53.

Active margins are subjected to incessant tectonic forces because they constitute plate boundaries. Some are in the process of approaching a neighbouring plate (dominant compression regime), while others are divergent, i.e., in a dominant extension regime resulting in the disintegration of the margin, and transform to an *island arc* separated from the continent by a *back-arc basin* or a *marginal sea* with an oceanic crust (Fig. 20). Thus there are two types of active margins: those essentially forming the boundaries of a continent on the one hand, and island arcs with a basement of continental crust on the other.

The two coasts, eastern and western, of the Pacific clearly illustrate this distinction. The reason for such a contrasting difference is disputed. The age, i.e., the weight of the concerned oceanic crust, has been mentioned as a cause. The eastern Asiatic coast has an older (Upper Jurassic-Cretaceous) crust, which is denser and readily sinkable, whereas the American coast has a more recent (Tertiary) and lighter one.

Whether the margin be a true continental one or an island arc, two basic elements are invariably found associated with both: a *submarine trench*, the point from which the subduction zone departs, and a *volcanic arc*. These two elements are sometimes separated by an accretionary prism (= *sedimentary arc*) and a *forearc basin*. At the rear of the volcanic arc the observable structures depend on the extensional (or compressive) regime to which the margin in question is subjected (Fig. 107).

Subduction Trench

Being narrow and elongated, these trenches present the largest known oceanic depths, greater than 10,000 m.

Their *outer slope* (oceanic side) is gentle (2 to 5°), but forms a sort of salient which joins with the surrounding abyssal plain (for its definition, see p. 7), a salient affected by extension faults that make the oceanic crust appear below the abyssal sediments. Such faults have been observed in a number of places (Japan, Peru and Chile, Central America, Puerto Rico, Timor, etc.). These faults may also be absent as in the case of the Eastern Aleutians but, in this case, a very large accretionary prism is observed.

Besides the escarpments produced by these normal faults, the outer slope of the trench may exhibit, as in all areas of oceanic crust, varied reliefs such as isolated submarine volcanoes and even a ridge sometimes, active or inactive.

These reliefs will be either totally engulfed by the subduction zone or decapitated, the part thus isolated being incorporated into the structures of the inner slope of the trench (accretionary prism). Either way, the absorption of these reliefs produces modifications in the morphology of this slope. Such a classical example is observed in the trench of Japan, close to its transition to the Kuril trench (Fig. 108).

Base of the trench: It is in this zone that the oceanic crust, more or less covered with sediments, commences to be engulfed by the subduction zone. This base is flat, wide or filled with turbidites over a variable thickness.

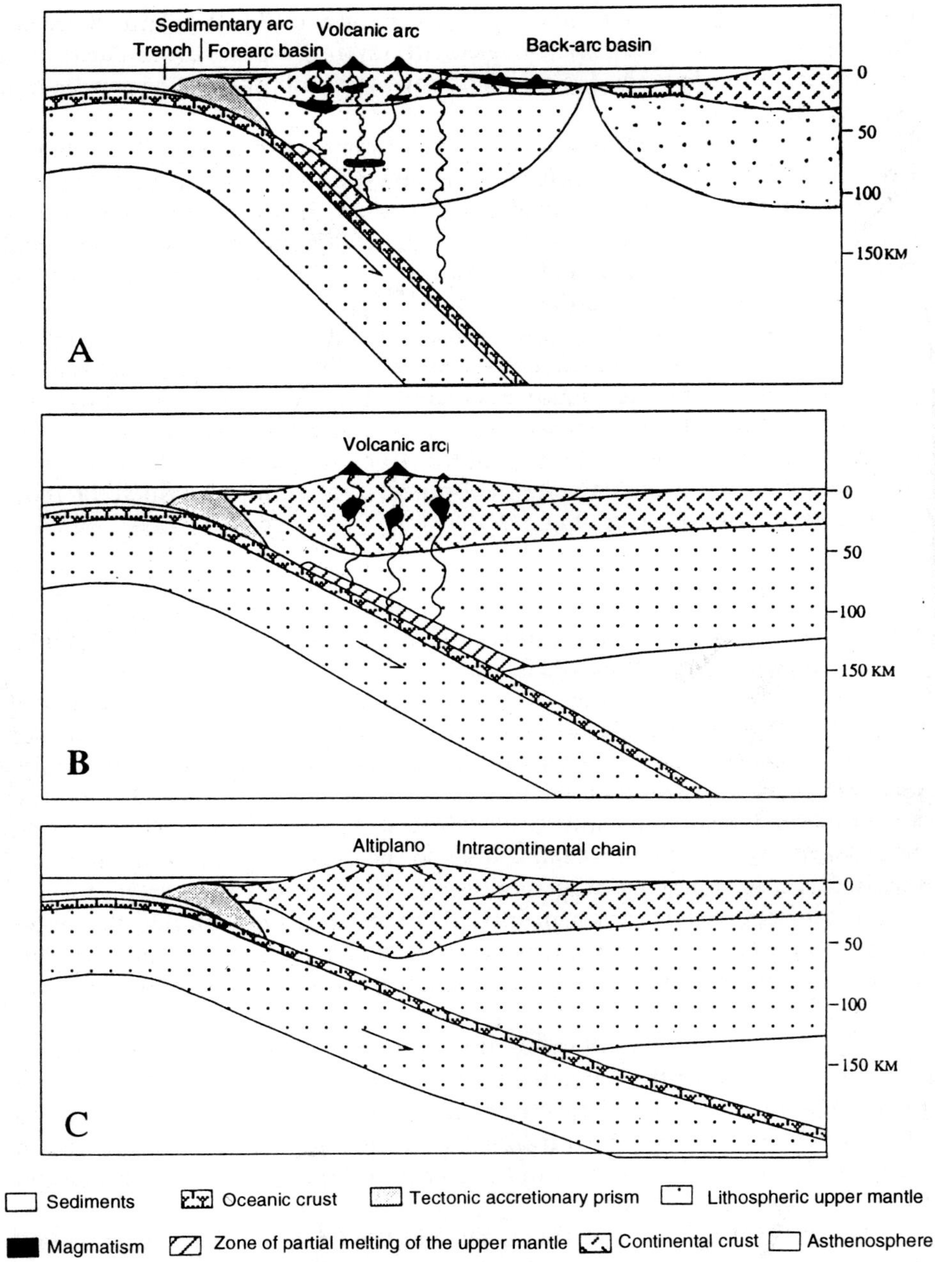

Fig. 107. Theoretical cross-section of an active margin.
Top: extension regime (eastern Asiatic type margin). The cross-section corresponds to the Indonesian arc.
Middle and bottom: Compression regime (western American type margin). The cross-sections are based on those of the Andes of South Peru-Chile-Bolivia (B) and Central Peru-Bolivia (C).

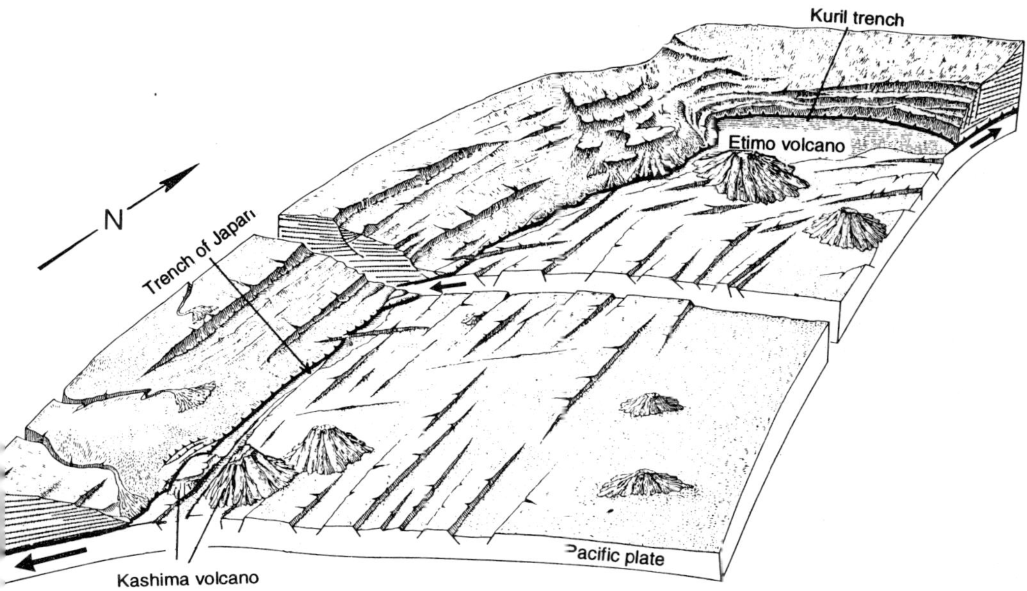

Fig. 108. Oceanic trench of Japan (after Cadet et al., 1987, simplified).
Note especially the extension faults affecting the oceanic crust in the boundaries of the trench. This crust also supports some submarine volcanoes of which the Kashima volcano is split into two parts by these faults; the western half is gradually buried in the subduction zone.

In the Guatemala trench, the pressure of pore fluids in the subduction plane below the trench has been measured. The figures obtained (40 kg/cm^2) are of the same order of magnitude as those of lithostatic pressure.

In the case of Barbados prism, the fluids are hot (80 to 120°). They escape in the form of mud volcanoes associated with an intense biological activity.

Inner slope (continental side): The investigation of this slope has revealed two types of disposition:

— *The inner slope is the boundary of the basement of the volcanic arc*, i.e., the continental crust with its older structures. Such is the case of the Central American trench off the coast of Guatemala. Extension faults are more or less common in this region. Such active margins are called extension margins. The sediments brought to the trench are totally engulfed along with the plunging oceanic lithosphere.

— *The inner slope of the trench exhibits a complex build-up of layers consisting of abyssal sediments of oceanic origin*, and even with small sheets of oceanic crustal material, totally covered by a pellicle of sediments trapped in the irregularities of the slope. Hence the materials brought in by the plunging oceanic lithosphere are not buried in the subduction zone, at least not totally, but piled up as successive layers below the older sediments. Such margins are known as *compression* margins (Java and Aleutian Islands). The structure produced in this case is a *tectonic accretionary prism*, almost similar to the one described earlier for the Barbados Islands in the context of intraoceanic arcs.

Accretionary Prism (Sedimentary Arc)

When such a sedimentary arc exists, which is the case in 56% of active margins, it develops between the trench and the volcanic arc. In the areas where its structure is known, it is found to exhibit thrust sheets, with oceanic vergence (Figs. 109, 110, 138). However, below the sea this disposition is often masked by surficial slidings or by mud flows.

From the sedimentological point of view, these prisms are mainly characterised by the presence of 'melanges' due to the combined action of tectonic brecciation and disorganisation of sediments by submarine sliding. According to many European geologists, these melanges are nothing but partly tectonised olistostromes.

In any case, whatever be their origin, these melanges represent a characteristic of accretionary prisms. One can observe there, side by side,

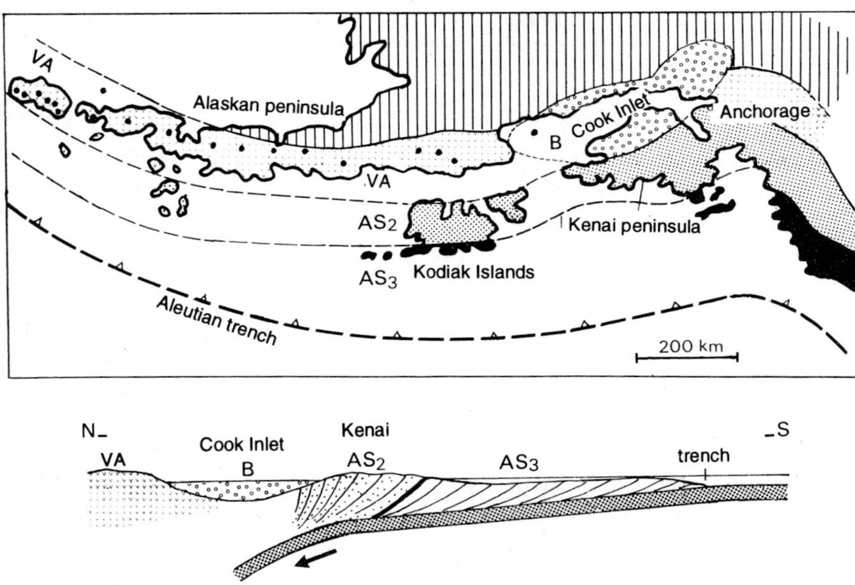

Fig. 109. Partly submerged active margin: the Alaskan peninsula (adapted from Dickinson and Seely, 1979).
VA: volcanic arc (major volcanoes are indicated); B: frontal basin ('forearc basin'); AS2: Mesozoic accretionary prism; AS3: Tertiary accretionary prism.

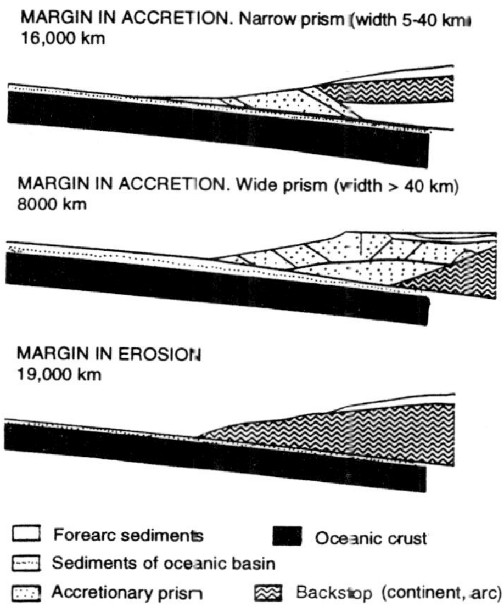

Fig. 110. Different types of margins (after Von Huene and Scholl, 1991, *Rev. Geophysics*, 29: 279–316).

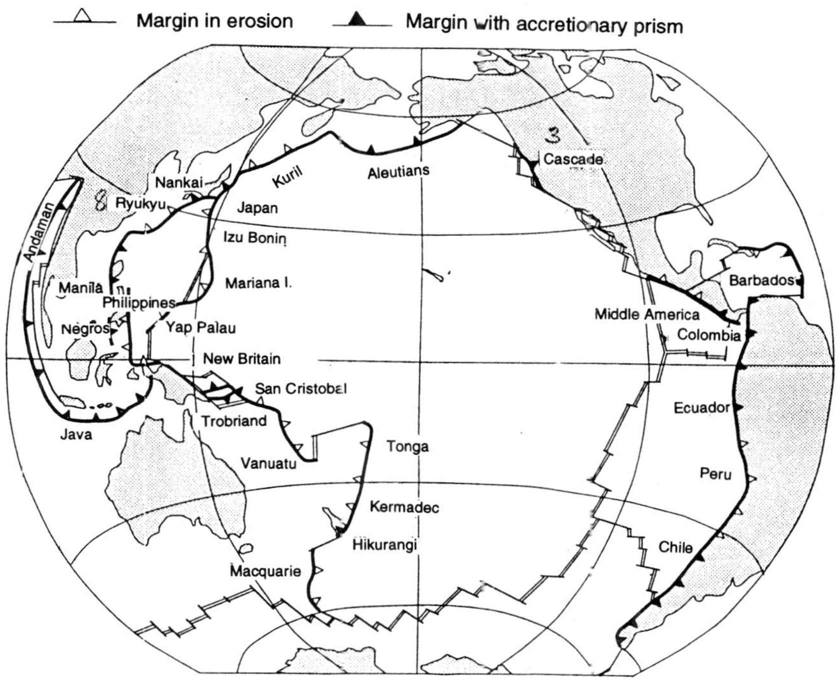

Fig. 111. Nature of convergent margins (Pacific, eastern Indian Ocean and Caribbeans); in this projection the Mediterranean margins (Aegean, Calabrian) and the Makran and South Sandwich margins could not be represented (after Von Huene and Scholl, 1991, *Rev. Geophys.*, 29: 279–316).

ophiolites (basalts, gabbros, ultramafics), i.e., slices of the oceanic crust, alkaline basalts derived from the common suboceanic volcanism, oceanic sediments (radiolarites, brown shales, pelagic carbonates) and sediments coming from the volcanic arc (turbidites, volcanodetrital sediments or greywackes, reefal limestones, etc.).

What is especially significant is the irregular association or succession of these facies, in contrast to the regular sequence of the overlying frontal basin when it exists.

Some authors (Scholl et al., 1977) have estimated, however, that this process of formation of an accretionary prism by 'offscraping' has not occurred systematically in all cases and that these prisms might have also been the result of a common tectonic folding of sedimentary formations deposited on the more or less thinned continental margins. The oceanic sediments themselves would be almost entirely absorbed in the zone of subduction (this is indispensable because a total 'offscraping', applied during 100 Ma to an oceanic crust carrying a layer of sediments hundreds of metres thick, should have given to the prisms an enormous volume which is obviously not observed in reality). These authors consider that the volume and proportion of oceanic sediments entering into the composition of accretionary prisms have been overestimated.

The fact that 44% of active margins are devoid of an accretionary prism has led several authors to distinguish *tectonic accretionary margins* from *tectonic erosion margins* (Figs. 110, 111). However, this erosion is a quasi-general phenomenon, as mentioned earlier, even though the nature of the margin in effect results from its efficacy. One can thus distinguish margins with typical prisms in which a large volume of material is accreted, margins with small prisms in which most of the material is subducted and margins without accretion in which all the material is subducted (Figs. 110, 111).

Factors determining the erosional nature or otherwise of the margin seem to be the rate of convergence (smaller rate for accretionary margins), volume of material arriving into the trench (a larger inflow of unconsolidated sediments favours formation of a prism), presence of asperities (leading to erosion of the base of the upper plate) and abundance of fluids (leading to hydraulic fracturing of the upper plate).

From the morphological point of view, the accretionary prism may appear either as a simple scarp in the flank of the volcanic arc, or as a free submarine ridge, sometimes emergent. Such a structure is called a *sedimentary arc.*

This is the case of the Mentawai Islands, facing Sumatra, the eastern branch of the Celebes (Fig. 112), and the Kenai peninsula with Kodiak Island in the eastern Aleutians (Fig. 109). The Barbados arc can also be included here (Figs. 17A, 17B).

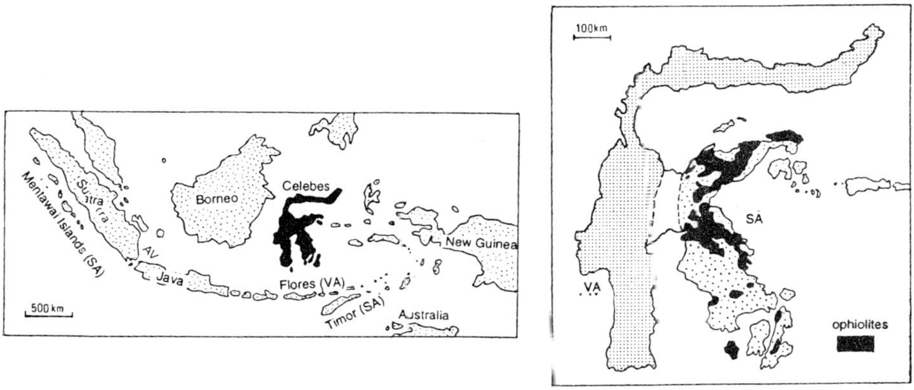

Fig. 112. Island arc of the Celebes and its Indonesian surroundings.
SA: sedimentary arc; VA: volcanic arc

The equivalents of fossil accretionary prisms are observed on Taiwan island (see p. 217), in the Makran region (southern Iran, p. 205) and on the southern parts of Hispaniola in the Greater Antilles.

Sometimes the Californian Coast Ranges, with their well-known 'Franciscan complex', are cited because these sedimentary belts pass in front of a present-day volcanic arc (Cascade Mountains) and a fossil one (Nevada Mountains). They show a sequence of parallel bands of increasingly younger age westwards, i.e., towards the ocean, conforming to the structure of accretionary prisms.

This interpretation is disputed, however. Some authors (see for example Roure et al., *BSGF* (1986) p. 931) consider that these Franciscan sequences rather represent the sedimentation of a marginal basin situated between the American continent and a volcanic island arc farther west with a continental basement. These two groups may have collided in the Middle Cretaceous and later may have been covered by molasse of Upper Cretaceous age (see Fig. 157).

The belts of an age gradually decreasing westwards represent only an appearance due to the outcropping, in the east, of the Franciscan series (Upper Jurassic-Lower Cretaceous) and superposed molasses of the coastal molassic basin, i.e., of a more and more recent age westwards.

We shall revert back to this question while discussing collision chains (p. 225). However, it seems very probable that the Californian Coast Range cannot constitute an example of fossil accretionary prism.

Volcanic Arc

A volcanic arc is sometimes also known as a *magmatic arc* because plutons, mainly granodioritic, exist at depth, which emerge onto the surface in older arcs that are folded, uplifted and eroded.

As mentioned earlier, the volcanic arc is one of the two main morphological elements of active margins. For a long time such an arc was described as the 'fire belt' of the Pacific (New Zealand,[3] Philippines, Japan, Kurils, Aleutians, North American Pacific Cordillera, Central America, Andes, Antarctic peninsula). The arcs of the Lesser Antilles or South Sandwich Islands in the Atlantic and those of Indonesia (Sumatra, Java, Bali, Flores, Western Celebes) in the Indian Ocean are also typical examples of volcanic arcs.

Volcanic arcs may occur in three ways:

— arcs situated on a continental margin proper, i.e., those developed on an emerged boundary of a continent (Andes);

— island arcs with continental basement, separated from the neighbouring continent by a marginal sea (Japan, New Zealand);

— arcs intermediate between the preceding two types, viz., those of the emerged active continental peninsulas (Kamchatka, Eastern Aleutians and Western Alaska) or those of a real but submerged continental margin (Indonesia).

These three types have in common a basement of continental rocks that is sometimes thinned out ('intermediate' type crust). It is this feature which distinguishes them from the intraoceanic volcanic arcs associated with subduction but in which no continental margin is involved.

The arcs of active margins are relatively narrow (width less than 50 km for half of them). Volcanism in this case appears only at a certain distance from the trench, never less than 100 km, because in order to trigger the volcanic activity the descending lithosphere should attain a certain depth (100 to 150 km in areas where its depth could be determined by seismic studies, Fig. 113).

For a slope of 45°, therefore, the subducted length ought to be 140 km. Taking the rate of subduction (5 to 10 cm/yr) into account, a lapse of 1.5 to 3 Ma is needed for volcanism to appear after commencement of the subduction process.

Contrarily, the time that passes between cessation of subduction and that of volcanism is less resolved, in general, due to lack of precise methods for determining the moment of this stoppage or the time during which the fragments of the subducted lithosphere might remain in the mantle. Figures

---

[3]The Tonga-Kermadec islands and the West Melanesian archipelago also form a part of this 'fire belt' geographically but represent intraoceanic arcs and not the ultimate member of evolution of a continental margin.

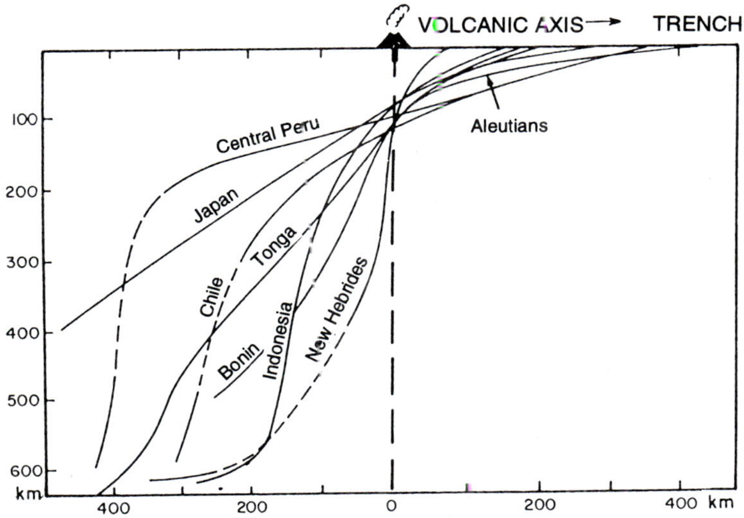

Fig. 113. Depth of subduction surface below the volcanic arcs of some active margins (after Maury, 1984).

of the order of 5 Ma have been proposed, but they would perhaps be greater in some specific cases—and, moreover, strongly disputed—as in the case of the Cascade Mountains in the USA. Though situated on a continental margin that is generally speaking active, this volcanic region is not in the boundary of a trench, however, and no subduction is seismically detected below this zone. Subduction, if any, appears to have ceased from about 20 Ma ago.

The difficulty in resolving this problem is also due to the fact that the active margins in the boundary of a continent are subject to an alternation of extension and compression regimes. In the latter case the crust is thickened and the corresponding 'orogenic' volcanism is not easy to distinguish from that of the earlier subduction.

Volcanism: The lavas of these arcs are, in general, saturated or supersaturated with silica and vary from basalts (especially common in intraoceanic arcs) to rhyolites through intermediate members of andesites and dacites. Depending on the contents of SiO_2 and K_2O, four types of volcanic sequences can be identified, which progressively pass on from one to another:

— 'arc tholeiites', low in potassium;

— calco-alkaline series (calco-alkaline basalts, andesites and dacites);

— potassic calco-alkaline sequence, whose dacitic members, the most common, are called latites;

— shoshonite series, whose extreme acidic members are liparites.

In general, high Ca and alumina, and low Ti are typical of these volcanic arc series.

These different types of lavas may coexist and, in this case, a certain zoning is generally observed in such a way that the K content tends to increase as the distance from the trench increases. The reason for this is still not clear. It is attributed to an increasing contamination of the magmas associated with a progressively increasing path in the lithosphere or to a gradual decrease in rate of partial melting due to stronger and stronger dehydration of the plunging plate.

The nature of the mantle above the subduction surface appears to play a predominant role in fixing the geochemical characteristics of the lavas. Further, the influence of the crust above the subduction surface, i.e., the basement of the volcanic arc, is evident in the sense that the thicker the continental crust, the richer the lavas in Si and K. Island arcs yield andesites and dacites while the arcs of continental margins proper show rhyolites (whose origin may also be influenced by the anatexis of the thick continental crust) (Fig. 114).

While this association is evident, the reason for the same is debatable. There could have been contamination of magmas by the continental crust or else the latter, being thicker, may have readily facilitated the development of magmatic reservoirs in which differentiation of acidic magmas becomes possible.

The volcanoes of the arc vary markedly in eruption styles; however, due to the high fluid content of the magmas, most of the eruptions are of the Pelean to explosive type. It is among this type that the most dangerous

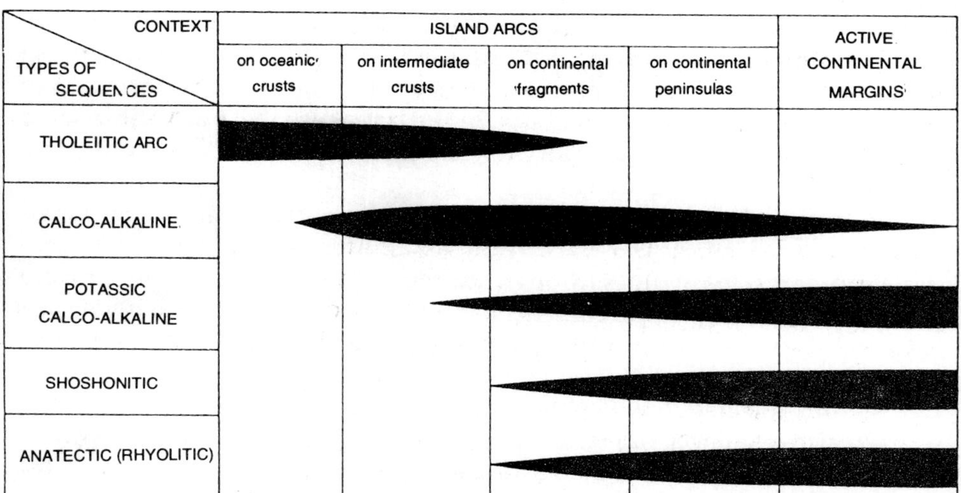

Fig. 114. Distribution of volcanic series in active margins and island arcs (after Maury, 1984).

volcanoes occur. Examples are: Mont. Pelée (in Martinique), Mt. Saint Helens (Cascade Mountains), Merapi and Krakatau (Indonesia), Santorin (Aegean Sea), etc. The products emitted are essentially pyroclastites ranging from classic tuffs to ignimbrites.

Origin of magmas: The origin of magmas in volcanic arcs is disputed. Any hypothesis on this aspect should satisfy the following constraints:

1. The depth of their formation is of the order of not less than 100 km, which corresponds to a pressure of 30 kb.

2. The magmatic liquid is derived only from peridotites of the mantle and from basic products of a metamorphosed oceanic crust (eclogites).

3. Water is imperative in the zone of partial melting to facilitate the melting of peridotitic or eclogitic material under a pressure of 30 kb and at temperatures of the order of 1000 to 1100°C that are compatible with the heat flows observed in the subduction zones.

Based on these considerations, a certain consensus is established, resulting in the following hypothesis for the process of generation of magmas in the arcs (Fig. 115):

1. The subducted oceanic crust undergoes metamorphism resulting in amphibolite and eclogite facies, the latter transformation being accompanied by dehydration. Thus an aqueous fluid is liberated which, being at high pressures, transports a certain quantity of silica and some other elements.

2. This fluid moves into the mantle above the descending lithosphere and induces partial melting of peridotites; this melting gives a relatively hydrated basic liquid, basaltic or andesitic, and perhaps even acidic liquids. All these fluids rise through the mantle and the overlying crust.

3. During this upward movement they undergo modifications due to three possible causes: interactions with the host rock (assimilation, transfer of fluids and transported elements) especially when the host rock is continental, differentiations by fractional crystallisation in magmatic reservoirs and eventual mixing with other magmas, both deep and shallow (anatexis of continental crust). These anatectic sialic magmas also certainly reach the surface independent of the preceding processes, thus giving rise to ignimbritic rhyolites.

At the end of this complex evolution, the magmas are stored in shallow reservoirs, from whence they feed the volcanoes through faults that are open or closed depending on the orogenic regime. They may also solidify at depth in the form of more or less delimited massifs.

However, this process, proposed by Ringwood (1974), has been contradicted by others. One interesting hypothesis (Westercamp, 1988) is based on the structure of the volcanic arc of the Lesser Antilles: in effect, the activity of this arc consists of two phases, one of Eocene-Oligocene (50–30 Ma) and the other of Miocene to the Present, separated by a period of quiescence of 8 Ma, during which the subduction process has not stopped. Moreover, the arc of the Lesser Antilles at places of maximum volcanic activity is

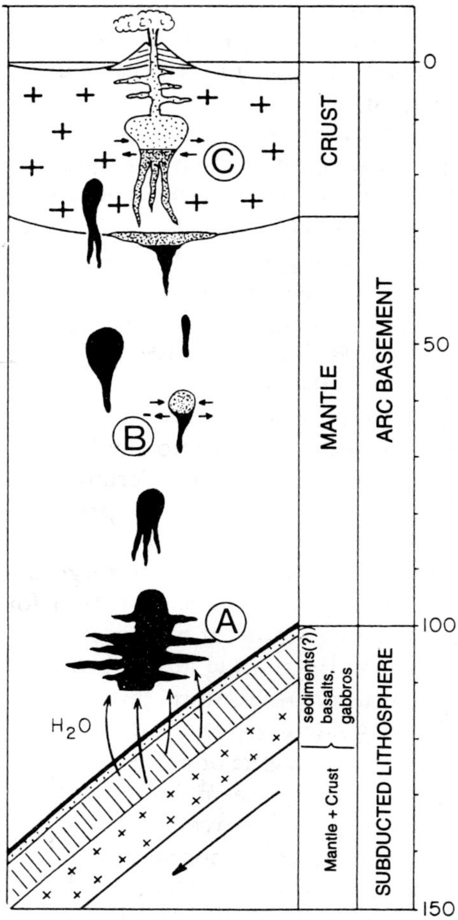

Fig. 115. Model of evolution of calco-alkaline magmas in subduction zones (after Maury, 1984).
A—By the effect of water, partial melting of the mantle above the lithosphere in subduction.
B—Diapiric ascent of magma into the mantle, accompanied by differentiations and magma-mantle interactions.
C—Arrival of magma into the crust; stay and differentiation in the magmatic reservoirs; possible crust-magma interactions.

170 km above the subduction zone, whereas the minerals of the plunging oceanic plate lose their water around 80 to 100 km depth (30 kb), i.e., at shallower depths.

To overcome these two difficulties, Westercamp proposed the *effect of subducted sediments in the generation of calco-alkaline magmas.* As a matter of fact, the most volcanic part of the arc corresponds to the zones in which the subducting Atlantic floor is covered by nearly 2 km of sediments (part of which vanished during formation of the accretionary prism). On the contrary, volcanism was weak and disappeared in regions where it had little or no sedimentary cover. The gap of 8 Ma might correspond to sinkage of that portion of the oceanic crust not covered by sediments (or whose sediments have been completely washed out). Further, various sets of geochemical data (trace-element and radiogenic isotope studies) of volcanic magmas suggest contributions of sedimentary origin.

Tectonic structure of volcanic arc: The structure of volcanic arcs reflects the fact that active margins are subject to both the processes of extension and compression. Hence, the following structures are observed:

— *extension structures* such as sedimentary basins, and submarine or lacustrine grabens, filled with predominant volcanodetrital sediments[4];

— *compression structures* in the form of folds and thrust faults. We shall revert to these in the chapter on 'Subduction Chains, p. 191.

Forearc Basin

A forearc basin corresponds to a hollow submarine region situated between a volcanic arc and a sedimentary arc when the latter bulges sufficiently. It is in general difficult to determine the nature of the crust below this basin because of the sedimentary cover. The little information available indicates a thin continental crust (hence, an 'intermediate' type crust), as well as an oceanic crust. This would imply that the subduction zone is located inside the oceanic crust, at a certain distance from the continental margin.

The morphology of these basins is highly variable depending on the quantity of sediments they receive and the upwelling of the sedimentary arc. The major morphologies are as follows (Fig. 116):

1. *Sedimentary arc is less bulging*, remains at great depth and receives little sediments. It thus forms a sort of terrace below a thick layer of water at the foot of the volcanic arc ('terraced forearc'). Examples: West Aleutian arc, Luzon arc (Philippines), Kermadec arc in the case of intraoceanic arcs (Fig. 116A).

2. *Sedimentary arc is more bulging* and almost reaches the surface. But, depending on the volume of sediments received, two types can be identified:

— Basin not fully filled, resulting in a hollow relief below a large thickness of water. Examples: basin between Sumatra and Mentawai Islands, and basin of the Lesser Antilles between Barbados and St. Vincent-Grenada for an intraoceanic arc (Fig. 116B).

— Basin almost completely filled, even though its surface forms a sort of continental plateau separating the volcanic arc from the sedimentary arc; the

[4]Greywackes, according to Anglo-Saxon authors.

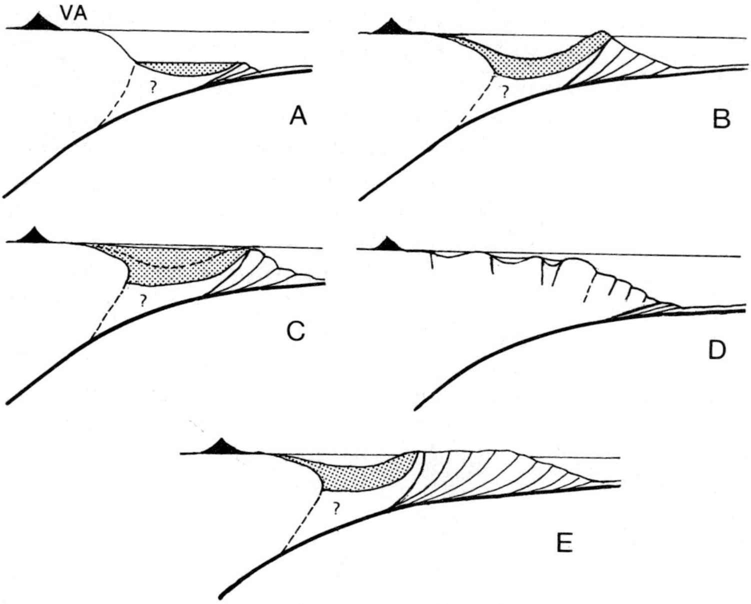

Fig. 116. Different types of frontal basins of volcanic arcs (VA) (after Dickinson and Seely, 1979).
A—Unbulged sedimentary arc, deep and unsubsided frontal basin (arcs of western Aleutians, Luzon, Kermadec).
B—Sedimentary arc reaching the surface but basin devoid of sedimentation and hence of deep water (basin between Sumatra and Mentawai Islands, between Barbados and St. Vincent in the Lesser Antilles).
C—Sedimentary arc reaching the surface and basin practically full, exhibiting a false continental plateau relief (submarine plateau off the coasts of Guatemala and Nicaragua).
D—Continental plateau devoid of forearc basin, which may be confused with type C (Peru).
E—Largely emergent sedimentary arc, shallow frontal basin (Cook Bay = Cook Inlet, Alaska).
The nature of the crust below the forearc basin is by and large not known.

latter remains visible by a thin crest or is totally masked by the sedimentation of the frontal basin ('shelved forearc'). Example: the submarine plateau separating the Central American trench from the Guatemala-Nicaragua trench (Fig. 116C).

Such a disposition should not be confused with the true continental plateaus that are more or less faulted, which sometimes extend the basement of the volcanic arc below the sea (example: North Peruvian coast, Fig. 116D).

3. *Sedimentary arc largely emergent*: The forearc basin then forms a strait or a creek of variable depth ('ridged forearc'). Example: Cook Bay between the eastern edge of the Aleutians and Kenai peninsula and Kodiak Island (Fig. 116E).

The sedimentation of the forearc basins mainly consists of detrital products, sometimes turbiditic, derived from the volcanic arc or its granito-gneissic basement, or from the sedimentary arc if it is sufficiently thick. The inference of their existence is based on the data of some wells drilled in this region and especially on the study of emergent fossil frontal basins. Cook Bay, at the eastern edge of the Aleutians (volcanic arc), provides a good example of such basins because part of the basin is exposed in the northern sector of Kenai peninsula. Detrital and volcanodetrital sediments ranging from abyssal Mesozoic facies to shallow Neogene facies are observed there.

The sedimentary arc may even be completely emergent as in Makran (southern Iran and Pakistan) (Fig. 143).

Structures at the Rear of a Volcanic Arc

These depend on the tectonic regime to which the margin is subjected:

In the East Asiatic type of extension margins a new sedimentary basin develops, known as a *back-arc basin*. It may present varied stages of evolution:

1. *The basin is still in granito-gneissic crust*. This is the case of shallow offshore platforms which join the volcanic arc of Sumatra-Java to the South Asiatic coast (Malaysia-Indochina), or the volcanic arc of Kamchatka to eastern Siberia, or the volcanic arc of the eastern Aleutians to the western edge of Alaska, or finally, the (extinct) volcanic arc of New Guinea to Australia.

The sedimentary deposits of these basins are the same as those of any partly subsident continental plateau, wherein they are represented by detrital products and platform carbonates including reefal limestones.

2. Often the incessant extensional process, mentioned earlier, may cause a thinning and rupture of the continental crust of the back-arc basin. *The basin is hence in a neo-oceanic crust*; this is a *marginal sea* (see p. 34). In this case the volcanic arc and its granito-gneissic basement are separated from the continent and form an island arc. This stage of evolution is very common on the Asiatic margin of the Pacific Ocean. It appears to be associated with a steep slope of the subduction plane and hence with the presence of an older oceanic crust that was thick and heavy, and rapidly and readily sank into the asthenosphere (in contrast with the active margins of the East Pacific where the younger and lighter oceanic crust subsides along a very gentle slope: island arcs are totally absent).

In the West American type of compression margins, there exists no back-arc basin but a suite of intracontinental type folded belts, just attached to the volcanic arc, which itself is folded. Some of these belts result from the folding of a sedimentary basin (sub-Andean belts of Peru, for example). It is hence probable that such margins have undergone dominant extensional stages of evolution.

This question shall be dealt with again when reviewing subduction chains.

Geophysical Data on Active Margins

Seismic data: Available information mainly concerns the subduction zone, marked by earthquakes up to a depth of 700 km. It can be inferred from this data that the inclination of these planes is highly variable, sometimes very small (and in this case volcanism does not appear), sometimes also bent (initially gentle and then steeper slope).

Focal mechanisms show a compression regime in the region of the accretionary prism and extension regime beyond it in the descending lithosphere. It may be compressive at the contact of the two participating lithospheres, however.

Gravity data: A strong negative free-air anomaly occurs over the oceanic basin and the sedimentary arc (when it exists); this anomaly is evidently associated with the thickness of sediments and the subduction which pushes the heavy material of the oceanic mantle deeper from the surface.

When subduction ceases, isostasy causes the sedimentary ridge to emerge, 'decoupling' (i.e., separating) the latter from the volcanic arc along a fault system.

A *positive anomaly* occurs over the volcanic arc, related to the presence of granodioritic magma reservoirs. If subduction stops, the volcanic arc will subside.

Thus an active margin is a region subjected to strong isostatic disequilibrium throughout the subduction process. The slowing down or cessation of the latter implies major vertical movements.

Geothermal data: The heat flow of sedimentary arcs is weak while that of volcanic arcs is high. The former is due to the presence of a cool oceanic lithosphere below the prism and the latter is evidently related to the uplift of magmatic masses.

PART THREE

COMPRESSIONAL STRUCTURES OF THE CONTINENTS

Compression results from the phenomenon of convergence of plates that are not balanced by the subduction process.

When a lithospheric plate carrying a continental crust is compressed, it undergoes deformation. Deformation obviously commences from the most fragile parts, wherein the tectonic inversion processes, mainly involving rejuvenation of pre-existing normal faults into reverse faults, are manifest. We have already seen an example of such processes (p. 85). If the deformation is excessive, a true mountain chain is produced.

In this part, we shall study the following topics:

1. Intracontinental chains
2. Continental-margin chains:
   - — subduction chains
   - — obduction chains
   - — collision chains.

The last chapter is devoted to those special structures which Anglo-Saxon geologists call 'collages', in which transcurrent movements play a major role.

CHAPTER 1

INTRACONTINENTAL CHAINS

These are structures which appear on the surface of the continental lithosphere subjected to compression. They would hence be contemporary to major global orogenic episodes. They can be grouped into two categories:

— *Independent chains*, i.e., isolated mountain chains in the middle of a platform region, such as the High Atlas or the Pyrenees. They may correspond to a simple upwelling or to a more complex folding of the platform, and most often are situated in a region of pre-existing structural disturbances, representing a specific fragile zone. The structural disturbance may be a fault system of crustal importance, a transform-fault zone or a graben, etc.

— *Annexes of major chains of continental collisions*. In effect, once the core of major continental collision chains is formed and if the compression regime continues, the deformation incorporates new terrains that are external to the chain being formed and plainly intracontinental (outer alpine zone, sub-Andean zone, for example). These new structures will thus be contiguous with the preceding ones and always folded with a time lag. Hence they will be presented later along with the major chains in question. However, some of the annexed structures that present sufficient morphological identity are mentioned here.

The intracontinental chains range from those resulting from a simple uplift of the underlying basement (these are the 'basement folds' of earlier authors) to those in which the basement is compressed, without upwelling, by internal cleavage of the crust or intracontinental subduction. In the latter case the sedimentary cover has to adapt itself to the shortening by folding over itself. The chain thus becomes a simple train of folds inside which the underlying basement no longer appears.

Earlier authors called these two extreme cases 'basement chains' and 'cover chains'. Unfortunately, this terminology is often difficult to apply because a given chain may be produced by the uplift of its basement, the latter remaining fixed to its sedimentary cover (Lebanon, Eastern High Atlas), or the chain may be basement or cover type at different points (Pyrenees). Finally, the concept of basement and cover is not distinct. For example, the Pyrenean Palaeozoic which is little or unmetamorphosed starting from the Upper Ordovician or Silurian, should it be classified in the basement group or in the cover?

These terms of basement or cover chains have only a descriptive value but no genetic significance. It is more important to differentiate the chains resulting from an upwelling of their basement and those emerging from a crustal cleavage or an intracontinental subduction.

CHAINS RESULTING FROM AN UPLIFT OF THEIR BASEMENT

The uplift of the basement ought *a priori* to represent either an overthickening of the crust or lithosphere, or a true arched structure of the latter following a corresponding upwelling of the asthenosphere below. Geophysical data generally indicate neither the one nor the other. In fact, because of the enormous energy these mechanisms require on the lithospheric scale, it is more probable that a tangential cleavage is produced inside the crust and only the layer above the cleavage is likely to assume a domal or duplex structure. The appearance of these structures is obviously facilitated by pre-existing weak zones in the granito-gneissic crust, such as a thin-out of the latter (subsident basins) or a system of faults that may function as ramps.

SIMPLE CASES

The surface of the continental crust forms a dome, eventually affected by reverse faults with double vergence, which uplift its cover.

Moroccan Anti-Atlas Chain (Figs. 117, 124)

This chain is merely a simple and large upwelling of Palaeozoic material with a Precambrian core, devoid of its Mesozoic and Tertiary sedimentary cover, which was uplifted for the first time during Variscan orogeny and again at the end of the Tertiary and Quaternary up to 2500 m. These latter movements have been recorded and hence dated by the Neogene deformations of the near-by basins.

Uplift and folding were accompanied by a certain amount of fracturing, creating a passage to Plio-Quaternary volcanism. The largest volcanic structure is that of Djebel Siroua, in western Ouarzazate, a stratovolcano showing end-Miocene (10 to 6 Ma) trachyandesites covered by Plio-Quaternary basalts and phonolites. The composition of this volcanic material is invariably alkaline.

Rocky Mountains, Wyoming and Colorado, USA (Fig. 118)

There are several ranges, sometimes reaching a height of 4000 m, oriented N-S (Laramie mountains, Front Range) or E-W (Uinta Mountains). All of them show a Precambrian core, largely uplifted along with some remnants

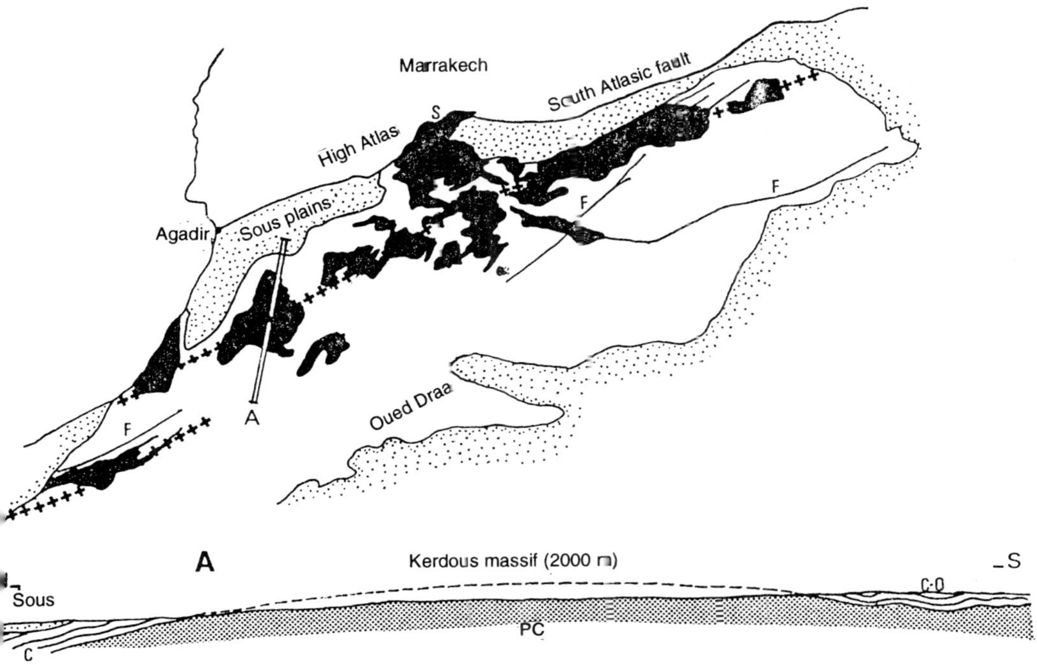

Fig. 117. Structural map of the Anti-Atlas mountains (after Choubert and Faure-Muret, 1971, simplified).
Dotted: Neogene; white : Palaeozoic; black: Precambrian; F: major faults; alignment of crosses: axis of uplift; S: Djebel Siroua
A—Cross-section; N—Neogene; O—Ordovician; C—Cambrian; PC—Precambrian.

of the primary and secondary cover: these formations are of shallow-water facies, plainly continental.

The significance of these mountains lies in the occurrence of reverse faults with opposite inclinations on the opposite flanks of their 'axial zone'. The tectonics of this chain is of Laramian orogeny (Palaeogene) and because of these reverse faults involves a more intensive compression than in the preceding examples.

In fact, the geophysical profiles of COCORP (Consortium for Continental Reflection Profiling) have revealed that the crust below these mountains was affected by very distinct reverse faults, some of which are extended to the surface by the overthrust faults mentioned earlier. It is not known whether they are connected at depth to the major flat cleavages In any case, they indicate a significant crustal shortening whose local variations reflect Precambrian discontinuities.

It should be added that the folding and uplift of these chains are accompanied by secondary fracturing, producing a passage to common alkaline volcanism.

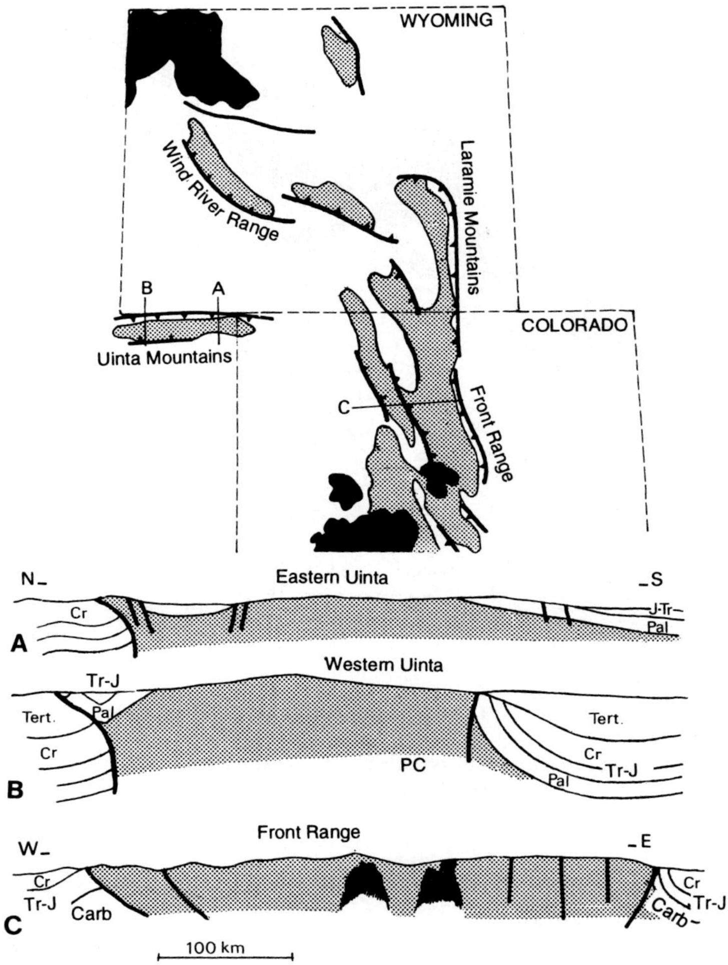

Fig. 118. Rocky Mountains of USA.
The Precambrian basement is shown in grey; Tertiary volcanism and magmatism in black.
Carb: Carboniferous; Cr: Cretaceous; Pal: Palaeozoic; Tert: Tertiary; Tr-J: Triassic-Jurassic.

Lebanon, Anti-Lebanon and Palmyra Mountain Ranges

The Lebanon mountains (3096 m) and their annex, the Anti-Lebanons (2814 m), separated by the Bekaa syncline, are two basement folds with a Jurassic core (Fig. 119) affecting the Arabian platform on either side of the Jordanian fault, which takes a SW-NE trend and is known as the Yammuna (or Yamuneh) fault in this region.

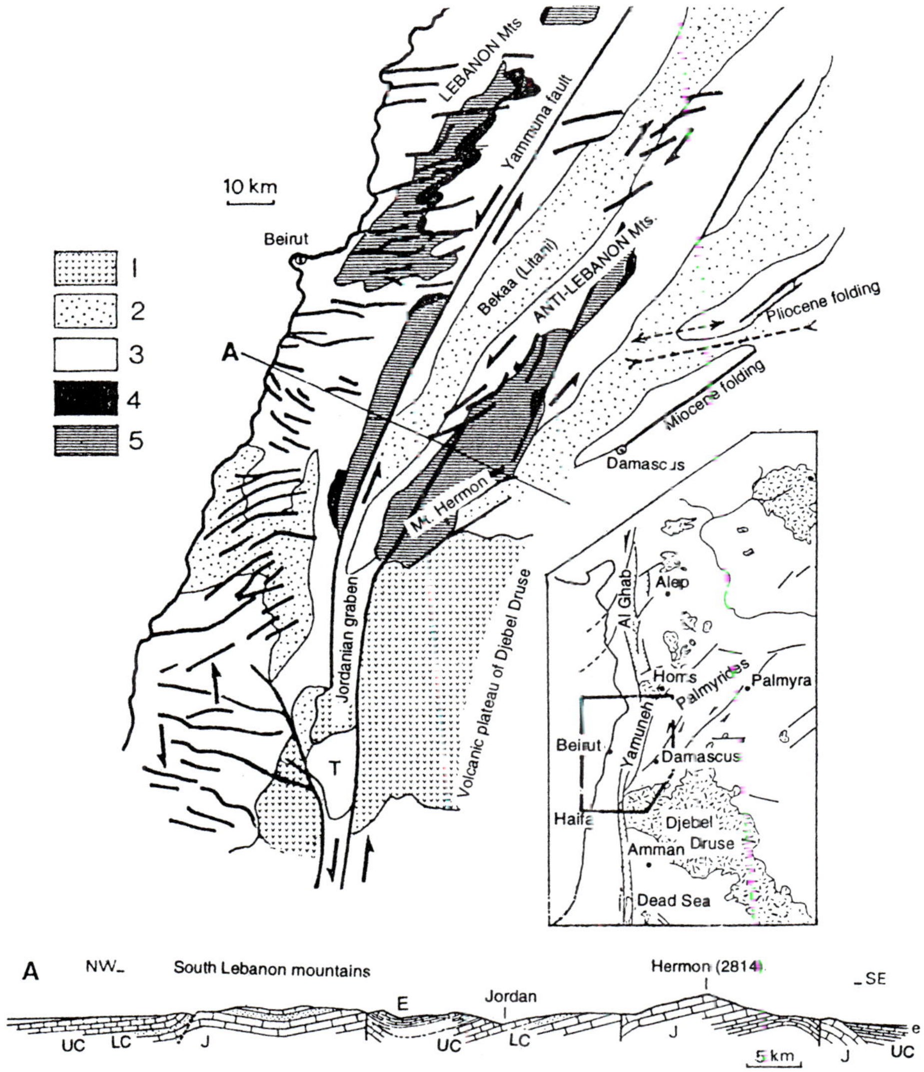

Fig. 119. Structure of Lebanon and Anti-Lebanon mountains.
The Lebanon mountains are a box-fold anticline, with a Jurassic core, whose uplift was not sufficient to expose the basement. The anticline is bounded on the east by the Yammuna (= Yamuneh) sinistral strike-slip fault, which is a major fault of the system and continuation of the Jordanian fault. The numerous minor faults of the western slope are late faults, subsequent to the formation of the massif.
Beyond the subsided syncline of Bekaa occurs the *Anti-Lebanon* anticline, also with Jurassic core (Mt. Hermon), intersected by major inclined faults.
In NE Damascus, the Figure shows the first anticlines of the Palmyras, in which the superposition of the two tectonic phases appears distinctly.
1. Recent volcanism; 2. Upper Cretaceous-Eocene; 3. Lower Cretaceous; 4. Lower Cretaceous basalts; 5. Jurassic.
Abbreviations in the cross-section: E: Eocene; J: Jurassic; LC: Lower Cretaceous; UC: Upper Cretaceous.

The Anti-Lebanon mountains are further continued in eastern Damascus, by the Palmyra Mountains, only of Cretaceous-Palaeogene formations, which extend over 400 km NE, and gradually attenuate before disappearing below the Neogene and alluvial deposits of the Euphrates.

While the Lebanon and Anti-Lebanon mountains are simple anticlines (Fig. 119), the Palmyras are more complex and consist of two superposed fold systems: (i) small NE-SW folds, parallel to the Lebanons and Anti-Lebanons (with which they are contemporary), of Miocene age because covered by a discordant Pliocene formation; (ii) major folds, roughly E-W, less accentuated, of Pliocene and synsedimentary age (which facilitates their dating).

This superposition represents two periods of activity (Fig. 120):

— *Miocene activity,* corresponding to a compressive stress field (σ_1 oriented NW-SE), already mentioned in the case of the Gulf of Suez and the Red Sea, and which then prevailed in all the peri-Mediterranean chains. This phase built up the first system of folds in which the Lebanon and Anti-Lebanon mountains also appeared. The Jordanian fault acts as a sinistral strike-slip fault but in a compression regime.

— Another, *Plio-Quaternary* phase, corresponding to a generalised extension period and to the opening of the Red Sea, thus representing a new strike-slip activity of the Jordanian fault, accompanied by the development of pull-apart basins.

In the sector with which we are concerned, σ_1 practically becomes N-S and gives approximately E-W folds in the Palmyra mountains (the Djebel Bilas anticline, for example, Fig. 120) superposed on the earlier folds. The Lebanon and Anti-Lebanon mountains are not modified but are affected by a complex system of minor strike-slip faults whose trends (N 120 and N 60) are compatible with the nearly N-S direction of σ_1.

The folding of all these chains is accompanied by eruption of normal alkaline basalts[1] distributed into two geographic groups, in the N and S respectively of the folded region. The most important is the southern one (Djebel Druse), in which the points of eruption are distinctly aligned along the NW-SE direction.

Radiometric dating and stratigraphic evidence show that volcanism starts in the Lower Miocene along this NW-SE direction, which is the same as that of the tension fractures parallel to σ_1. However, they are intensified in the Plio-Quaternary during which the sinistral strike-slip faulting of the Jordanian fault, operating in a transtension regime, opens up the preceding fractures.

The fact that the chains described above exist only up to the region of the Yamuneh fault and gradually attenuate away from it, as also the absence of volcanism in this sector, clearly indicate that it is the change of direction

[1]These should not be confused with the basalts of Upper Jurassic-Lower Cretaceous ages intercalated in the stratigraphic sequence.

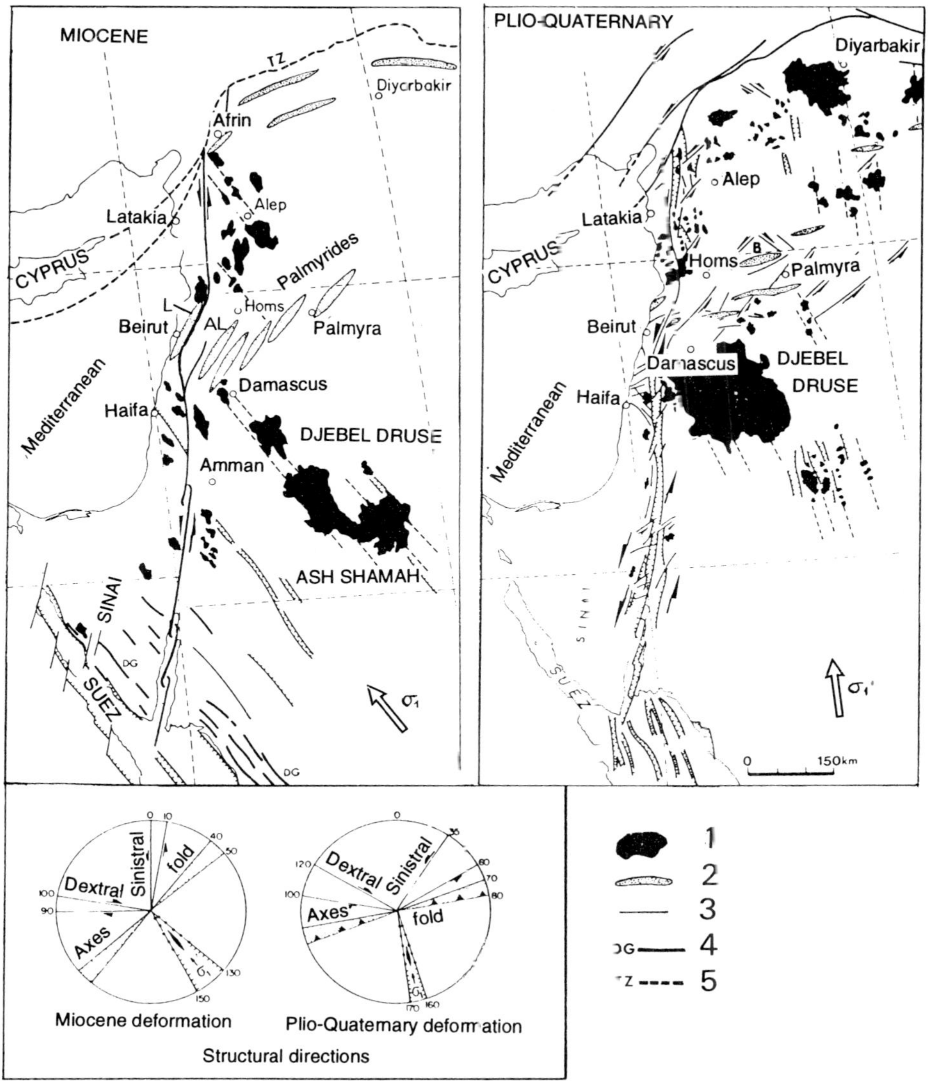

Fig. 120. Structural framework of the Lebanon (L), Anti- Lebanon (AL) and Palmyrides folding (after Giannerini et al., 1988, simplified and slightly modified).
1. volcanism; 2. major folds; 3. faults; 4. gabbroic dykes of the Red Sea and Gulf of Suez borders; 5. boundary of the alpine chains of Taurus and Zagros.
B. Djebel Bilas (NW of Palmyra).

of the Jordanian fault that produces them, causing a relative blockage for the movement of the Arabian plate northwards, and hence a local compression resulting in the folds in question. The latter obviously do not compensate the displacement of the Arabian plate, estimated to be 60 km for the Miocene

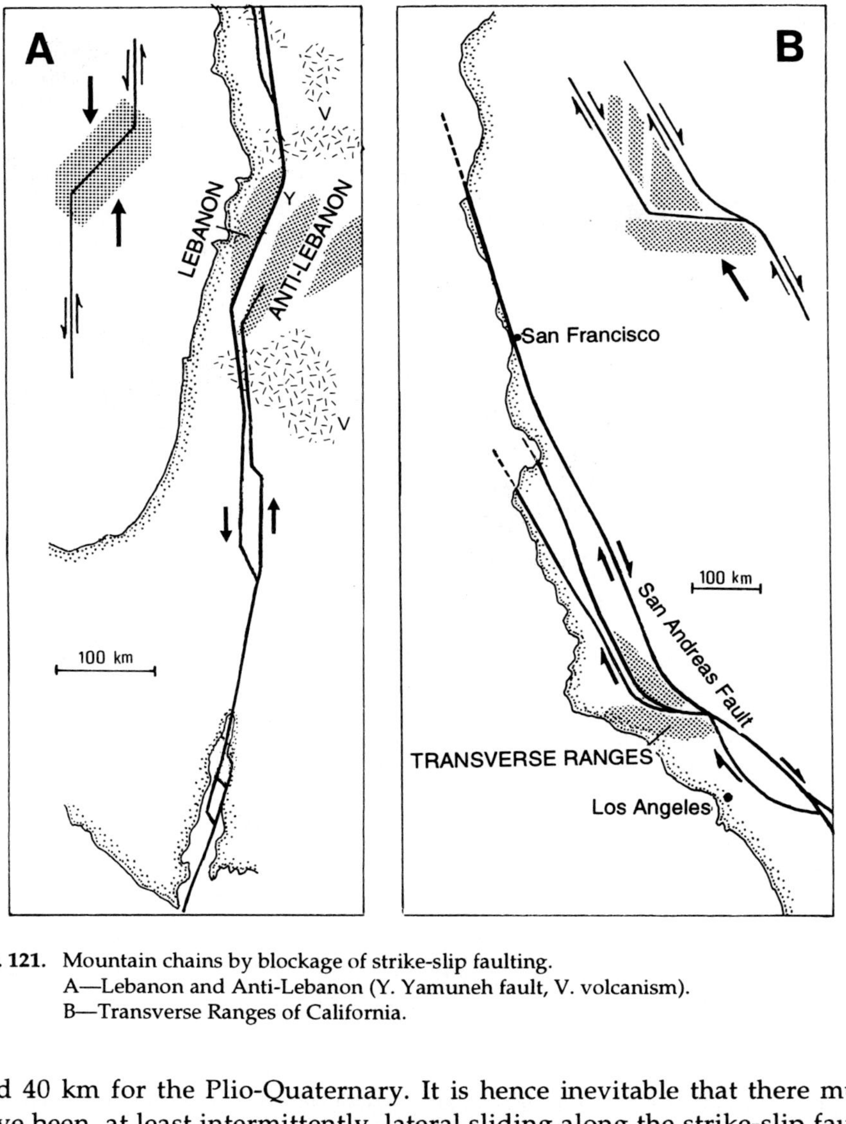

Fig. 121. Mountain chains by blockage of strike-slip faulting.
A—Lebanon and Anti-Lebanon (Y. Yamuneh fault, V. volcanism).
B—Transverse Ranges of California.

and 40 km for the Plio-Quaternary. It is hence inevitable that there must have been, at least intermittently, lateral sliding along the strike-slip faults since the corresponding consumption of energy was less than that required for the continuation of folding and its ultimate evolution into overthrusting.

Transverse Ranges of Southern California

A similar case of development of chains by blockage of strike-slip faulting is that of the Transverse Ranges of southern California, north of Los Angeles, where the San Andreas fault takes an E-W trend over a length of about a hundred kilometres ('Big Bend') (Figs. 121, 122). The Transverse Ranges arose subsequent to the origin of this fault and are evidently related to it.

This example is important inasmuch as the intermittent nature of strike-slip faulting is demonstrated here. In fact, the last seismic activity in the E-W segment dates back to 1857 and geodetic measurements have shown that, between 1959 and 1973, the displacements here have been smaller compared to those produced in the NW-SE branches (3 to 6 cm/yr). Thus the 'Big Bend' remains almost blocked and is subject to compression.

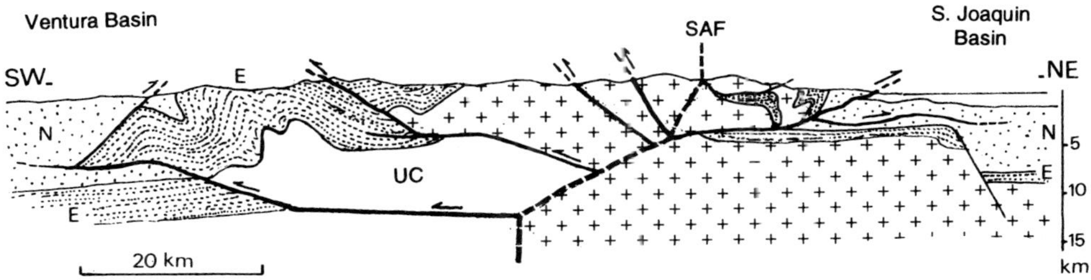

Fig. 122. Cross-section of the Transverse Ranges (California) (after Namson and Davis, 1988, simplified).
The cross-section shows a suite of folds and overthrusts formed since the Pliocene. Seismic data and the models of equilibrated sections imply the existence of a major cleavage plane at a depth of about 10 to 12 km, connected to the surface by ramps, as also a shortening of 50 km, i.e., a rate of roughly 20 mm per annum.
E: Eocene; N: Neogene; UC: Upper Cretaceous;
Crosses: diverse crystalline rocks of Mesozoic age (granites and ophiolites). SAF. San Andreas Fault.

High Atlas Mountains, Morocco

These represent somewhat more complex basement chains in which juxtapositions are observed of reverse faults with opposite inclinations (and hence a strong crustal shortening) and older structures that control the development and orientation of the chain.

This range, 600 km in length and 50 to 150 km in width, culminates as Djebel Tubkhal (4165 m) and consists, from W to E, of the Coastal High Atlas, Marrakech Atlas and Eastern High Atlas mountains (Fig. 123).

Structural Evolution: As early as the Permo-Triassic period, the African craton was subjected to extension and broke in the region of the future High Atlas mountains, into a series of grabens or half-grabens oriented along the faults inherited from Variscan orogeny (WSW-ENE), along with the eruption of tholeiitic basaltic products, dated as 200 to 195 Ma (Upper Triassic).

During the Jurassic the Atlantic ocean opened up between Morocco and Mexico, pushing the African plate eastwards. In this framework of extensional dynamics, a trough developed at the place of the future High Atlas.

The sedimentary filling of this trough was completed in the Middle Jurassic. This basin was bounded in the SW by the High-Atlas 'horst' of Marrakech, beyond which reappeared a subsidence zone, the future maritime High Atlas (Agadir-Essaouira region) (Fig. 123).

The Jurassic trough of the Eastern High Atlas was very subsident (3000 to 8000 m of calcareous-marly sediments with cephalopods, fringed by reefal formations). Subsidence was associated with a system of active (syn-sedimentary) faults, often strike-slip faults, that could give rise to pull-apart basins in the Eastern High Atlas region during the Middle Jurassic. The opening of these small basins was also accompanied by the origin of local anticlinal ridges (transverse folds) and alkaline magmatism (sills of dolerites, monzodiorites and leucogabbros marking the fracture network).

Extension and subsidence slowed down in the Upper Jurassic and Lower Cretaceous which witnessed a quasi-total emergence of the future High Atlas, once again accompanied by eruptions of alkaline basalts. This regime continued in the Upper Cretaceous in which only classic marine transgressions occurred.

Farther west, at the other edge of the chain, the future maritime High Atlas basin was merely a common, passive-margin basin, in which marine sedimentation continued even after the Dogger epoch. During the Tertiary the movement of Africa northwards modified the structural framework. This led to folding of the High Atlas in the course of a series of movements, extending from the end of the Eocene to the Pliocene, and reactivated the earlier fractures into dextral strike-slip faults. The whole of the High Atlas has in fact become a corridor of strike-slip faulting.

Uplift-folding was accompanied by intense fracturing and weak alkaline volcanism dated as 8 to 0.5 Ma, like that of the Anti-Atlas.

Present structure (Fig. 124): This varies from point to point.

The Marrakech High Atlas, which always occurred as a high zone during Mesozoic evolution and where the sedimentary sequence remained thin, is a typical basement fold with an 'axial zone' in which basement formations outcrop. This is a rigid block, bounded by reverse faults with a dual inclination, all representing the reactivation of Variscan faults. In the south the *south-atlasic fault* follows almost continuously in the form of a major reverse or vertical fault. In the north the *north-atlasic* fault is contrarily a swarm of reverse or vertical fractures, mutually continuing along SW-NE or E-W directions.

The eastern High Atlas and maritime High Atlas are actually cover chains characterised by a thick sequence of sediments and more supple style of folding: vast flat-bottom synclines filled with Upper Liassic and Aalenian marls, pointed anticlines with Lower Liassic to Triassic core. The folds are concentric, without schistosity; they obviously represent superstructure tectonics.

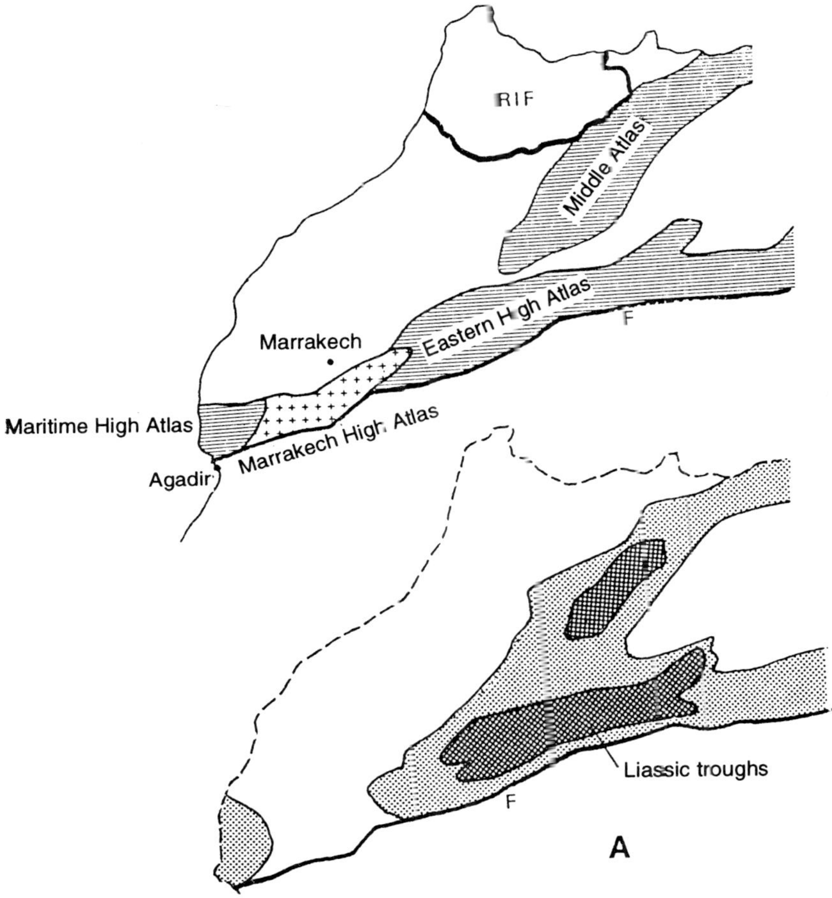

Fig. 123. Location and subdivisions of the Moroccan High Atlas (crosses: ancient basement; horizontal hatching: Mesozoic sedimentary cover).
A. Liassic palaeogeography (marine basins in grey): it predetermined the disposition of the chain.

The fracturing remains intense and it is interesting to note that the normal faults along which the Triassic basin was subsident, are reactivated as reverse faults or stretch-thrusts: these structures in fact correspond to facies changes on their limbs. This group always presents the characteristic dual inclination structure (Fig. 124) and the dual SW-NE and E-W orientations which facilitate the break-up of the chain into elongated diamond-shaped compartments.

Below the chain the crust shows no significant thickening. In fact, it is due to its break-up into pinchout segments, one sliding relative to another, that the atlasic basement is shortened.

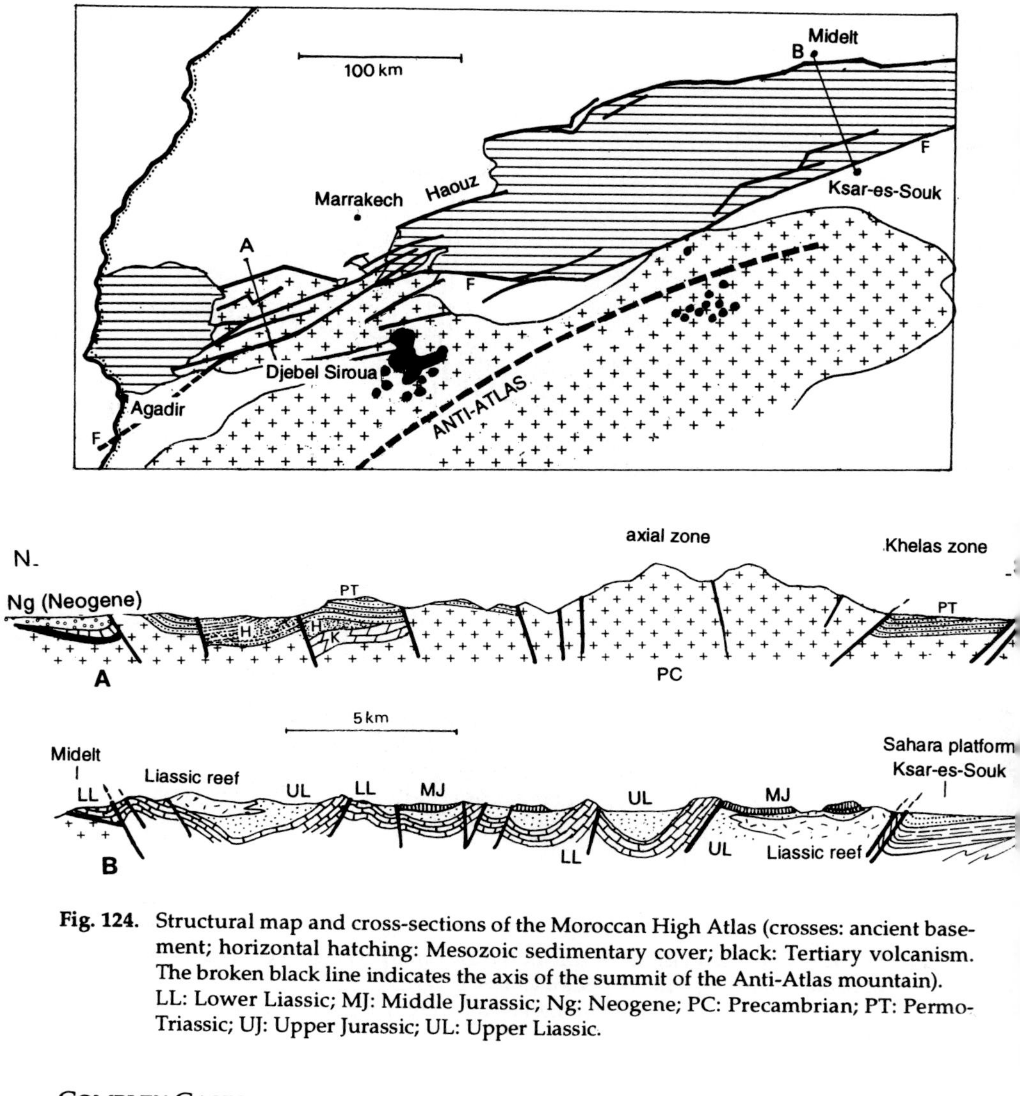

Fig. 124. Structural map and cross-sections of the Moroccan High Atlas (crosses: ancient basement; horizontal hatching: Mesozoic sedimentary cover; black: Tertiary volcanism. The broken black line indicates the axis of the summit of the Anti-Atlas mountain). LL: Lower Liassic; MJ: Middle Jurassic; Ng: Neogene; PC: Precambrian; PT: Permo-Triassic; UJ: Upper Jurassic; UL: Upper Liassic.

Complex Cases

Pyrenees (Figs. 125 to 128)

This is a 400 km long and 50 to 100 km wide mountain chain, with a 3400 m high summit, forming the boundary between the Iberian craton and the European craton.

At first glance it very much resembles the High Atlas with an 'axial zone' in which mainly the Archaean basement outcrops, affected by reverse faults of dual inclinations: in the north it is represented by the North

Pyrenean fault, an older Variscan structure reactivated; in the south several faults of very small dips give the southern slope of the axial zone a tangential sliced structure, at least for the surface crust (Fig. 125).

All these structures, as well as the axial zone itself, pitch westwards below their Mesozoic cover, starting from the Pau meridian.

The chain is much less symmetric than the High Atlas, however, because it is complicated on the French slope by a special zone known as the *North Pyrenean zone,* which exhibits special characteristics. Its basement, granito-gneissic, is very much imbricated ('satellite massifs' of earlier authors). Some minuscule layers of peridotites (lherzolites, named precisely here, after Lherz lake) are associated with the granito-gneissic slices. A frequent syn-sedimentary volcanism exists here, though weak in volume.

These three characteristics clearly indicate the thinning and fracturing of the basement in the neighbourhood of the old Variscan fault, viz., the North Pyrenean fault, which forms the southern boundary of the North Pyrenean zone. The latter thus corresponds to a weak zone in the crust, which is found to be the cause of most basement chains.

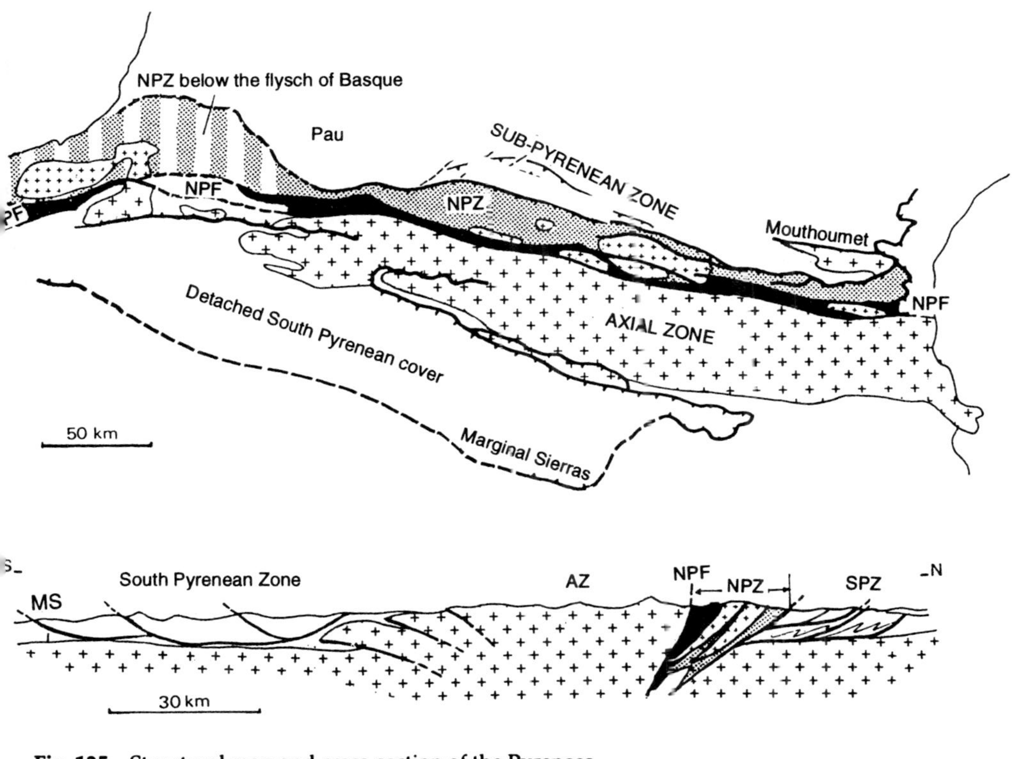

Fig. 125. Structural map and cross-section of the Pyrenees.
AZ: Axial Zone; MS: Marginal Sierras; NPF: North Pyrenean Fault; NPZ: North Pyrenean Zone; SPZ: Sub-Pyrenean Zone.

It may be added that the formations of the North Pyrenean zone confined to the axial part are metamorphosed into high temperature facies, with progressively increasing pressure facies accompanied by schistosity (sericitic schists).

Another interesting feature of the Pyrenees is that the North Pyrenean zone represents an ancient transform corridor which brushed past the rupture, i.e., the source of the crustal fracture, with the result that the completed mountain chain could almost be considered a collision chain. This is shown by the evolution of the Pyrenean region since the Triassic.

Structural evolution: The structural evolution of the Pyrenees is inseparable from that of the Aquitaine basin and the Gulf of Gascogne. We saw with regard to the latter (p. 131) that the position of the Iberian craton relative to the European craton varied during the Upper Cretaceous, which implies the existence of a highly deformable zone between the two cratons, remaining nevertheless in the continental crust since no ophiolites are known in the Pyrenees.

From the Triassic to the Jurassic there was a shallow-water subsidence basin here, a probable consequence of the disintegration of the Pangaea. This basin was certainly controlled by numerous faults, of which the 'Arcachon-Toulouse flexure' (Fig. 61) and, farther south, the North Pyrenean fault are the most important.

In the Aptian and the Albian, following the gradual opening of the North Atlantic, extension and subsidence took place suddenly. The future North Pyrenean zone appeared then as a region of maximum subsidence, directly bounded in the south by the fault of the same name ('slaty flysch' deposits of Pyrenean geologists, surrounded in the north and the south by Urgonian reefs). Moreover, this Aptian-Albian extension is accompanied by synsedimentary volcanism (alkaline basalts) and thermal metamorphism, especially well marked near the North Pyrenean fault. It may be concluded from this that by this stage a veritable rift had developed at the place of the Triassic-Jurassic subsident basin, a rift whose axis must have been at the contact of the North Pyrenean fault (Fig. 126).

At the end of the Albian the sinistral strike-slip faulting of the Iberian craton and opening of the Gulf of Gascogne commenced. These movements had two consequences: (1) Slip faulting probably occurred along the transform structures marked by small pull-apart basins, which can explain the appearance of lherzolite thrust outliers from the upper mantle.

The opening of the Gulf of Gascogne induced contraction of the eastern edge of the subsident zone, primarily of the North Pyrenean rift. The future axial zone was uplifted and dragged the North Pyrenean zone along with it. The subsidence axis hence moved towards the northern part of the ancient Triassic basin where a new trough known as the sub-Pyrenean trough, the future sub-Pyrenean zone, formed, representing a passage zone from the Pyrenees to the Aquitaine basin.

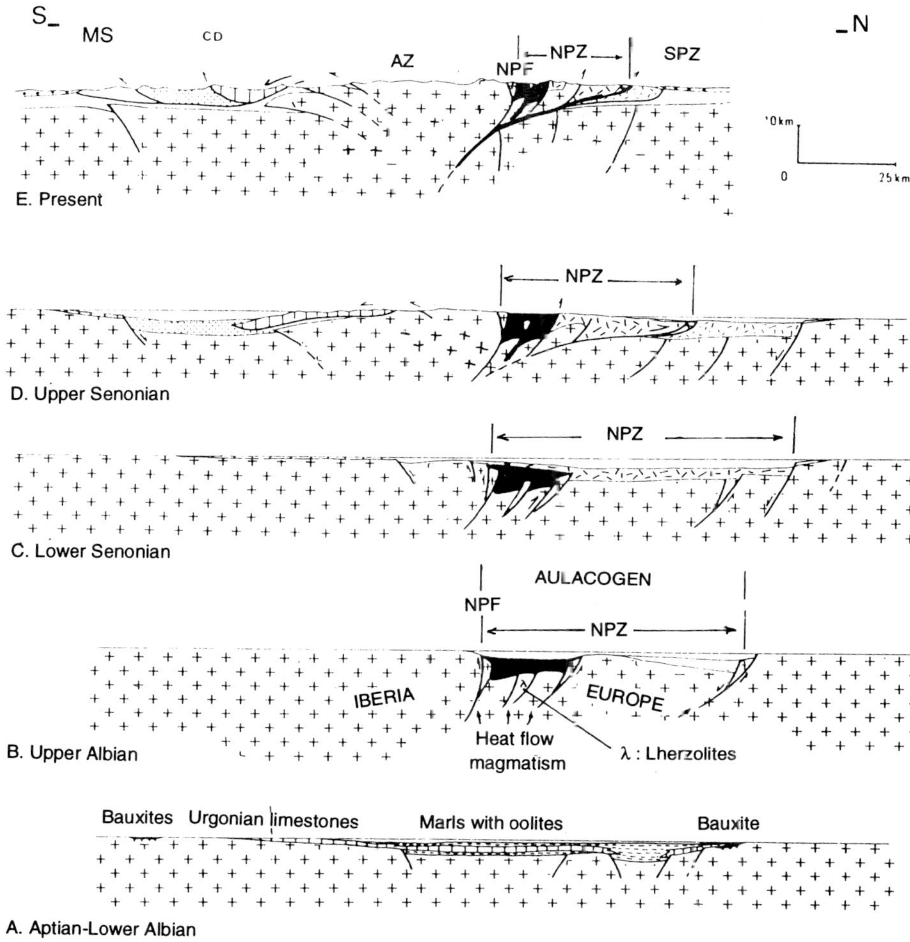

Fig. 126. Structural evolution of the Pyrenean region since the Aptian (after Souquet and Debroas, 1980, simplified).
Abbreviations same as in Fig. 125.

In the Campanian, following the opening of the South Atlantic, the Iberian craton starts its migration northwards (Fig. 127). The process of closure commences in the eastern Pyrenees where a low dynamic metamorphism, no longer thermal, is developed, associated with a slaty cleavage.

In the Eocene the process extends to the entire chain and reaches maximum in the Middle Eocene. At this moment the older North Pyrenean region becomes a swarm of folds in the sedimentary cover and a swarm of imbricated structures ('satellite massifs') in the basement. All these structures have a northern vergence and are partly overthrust by the axial zone. The southern slope of the latter is split, at least on the surface, into large

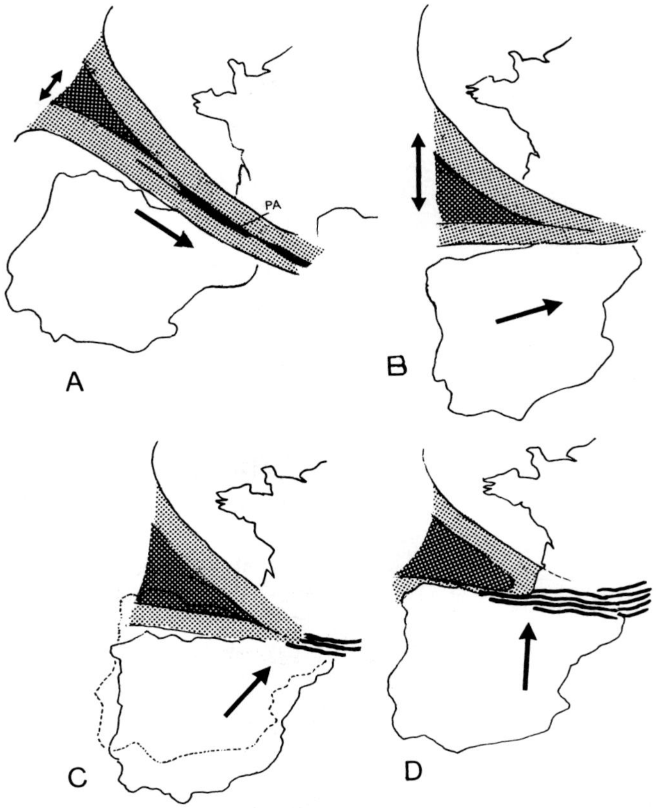

Fig. 127. Palaeogeographic framework of the origin of the Pyrenees (in light grey: continental talus; dark grey: oceanic crust. The outlines of Spain and France are indicated only as geographic markers).
A—From the Albian to the Campanian. The Spanish block slides counter-clockwise relative to France, resulting in the origin of small pull-apart (PA) basins considered to be the cause of lherzolite fragments. This sliding is accompanied by a displacement of the pole of rotation leading to the opening of the Gulf of Gascogne.
B—During the Campanian. The Gulf of Gascogne is entirely opened up. Spain initiates its movement NE.
C—End of the Cretaceous. The Spanish plate moves NE. Folding commences in the eastern Pyrenees.
D—Eocene. The movement of Spain once again changes towards the north. Folding extends to the entire chain.

sheets with a southern vergence; this feature gives the chain its dual inclination structure. These phenomena are contemporary to the subduction of the oceanic crust of the Gulf of Gascogne below the Cantabrian margin (see p. 131).

Recent geophysical profiles have revealed the following interesting facts (Fig. 128):

1. The Iberian crust is thicker than the European crust and its lower part is layered, corresponding probably to the granulitic facies which outcrops in the small 'satellite massifs' of the North Pyrenean zone.

2. The general structures of the chain are fan-shaped but slightly dissymmetric, with reflectors dipping northwards predominant.

3. The whole structure appears as though the European lithosphere was indented into the Iberian lithosphere. But opinions differ on the magnitude of this indentation because of the difficulty in interpreting the profile below the North Pyrenean zone, which is understandable since the more the structures are straightened, the weaker they become as reflectors.

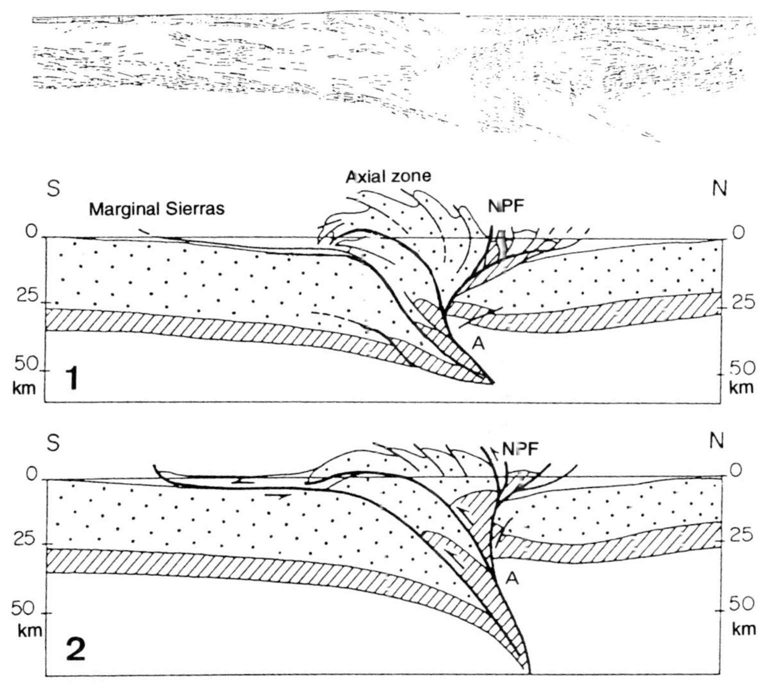

Fig. 128. Two interpretations of the ECORS profile of Pyrenees (after Mattauer, 1990).
In model 1, generally accepted, the North Pyrenean Fault (NPF), nearly vertical on the surface, is offset at a depth of about 10 km by a major thrust northwards. The indentation of the European basement is strong.
In model 2, the North Pyrenean Fault retains its vertical disposition over a depth of 20 to 30 km. Indentation of the European basement is weak to absent.
The only visible element in the profile of this sector of the North Pyrenean Fault is a strong reflector situated below the axial zone in the 'white zone'. It represents either the layered lower crust of the North Pyrenean massif (model 1), or the lower crust of the axial zone uplifted by slicing (model 2).

Whatever the manner and the magnitude of the indentation, the significant element revealed by this profile is the beginning of the underthrust of the Iberian crust below the European crust, implying an uplift of the upper mantle below the latter. Perhaps this might be the inheritance from a mantle diapir of Middle Cretaceous age, right below the North Pyrenean 'rift'. Such an uplift might facilitate the indentation mentioned above.

In conclusion, the Pyrenees are developed over a weak zone that originated in the Triassic, which itself was modelled on the Variscan structures including the North Pyrenean fault. This weak zone became a rift with local occurrence of crustal fractures. The two rims of the rift longitudinally slid relative to each other before colliding but the system remained essentially intracontinental. Nevertheless, it is an evolution preceding that of collision chains: to reach this stage, it would suffice that the crustal fracture be a little more open and there be enough space of oceanic crust.

But the crustal cleavage of the Pyrenees is large and introduces the second group of intracontinental chains.

CHAINS RESULTING FROM INTRACRUSTAL CLEAVAGE

Catalan and Iberian Chains (Fig. 129)

Between the Ebre basin and the Betica Cordillera a series of low-elevation (2400 m maximum) chains, of NE-SW (Catalanides chain) and NW-SE (Iberian chain) orientation are found. The Mesozoic sedimentation there reaches a thickness of 2000 m and starts with Triassic evaporites that obviously represent a quasi-general decollement zone. This presents folds of large radii of curvature bounded by overthrust and strike-slip faults, implying Mesozoic cover and Palaeozoic basement.

The age of these chains has been established mainly in the Catalan sector because of Tertiary marine deposits. They are contemporary to the Pyrenees (Eocene). However, during the Lower Miocene an extension phase developed which gave rise to Neogene basins superimposed on the older compression structures.

The following remarks can be made regarding the *interpretation of the geodynamics* of these chains:

1. The orientation of strike-slip faults (sinistral in the Catalan chain and dextral in the Iberian chain), as also the general direction of overthrusts, implies a roughly N-S compression, similar to the one responsible for the Pyrenean folding.

2. The existence of fragments of the basement included in the overthrusts shows that the basement has been as much compressed as its

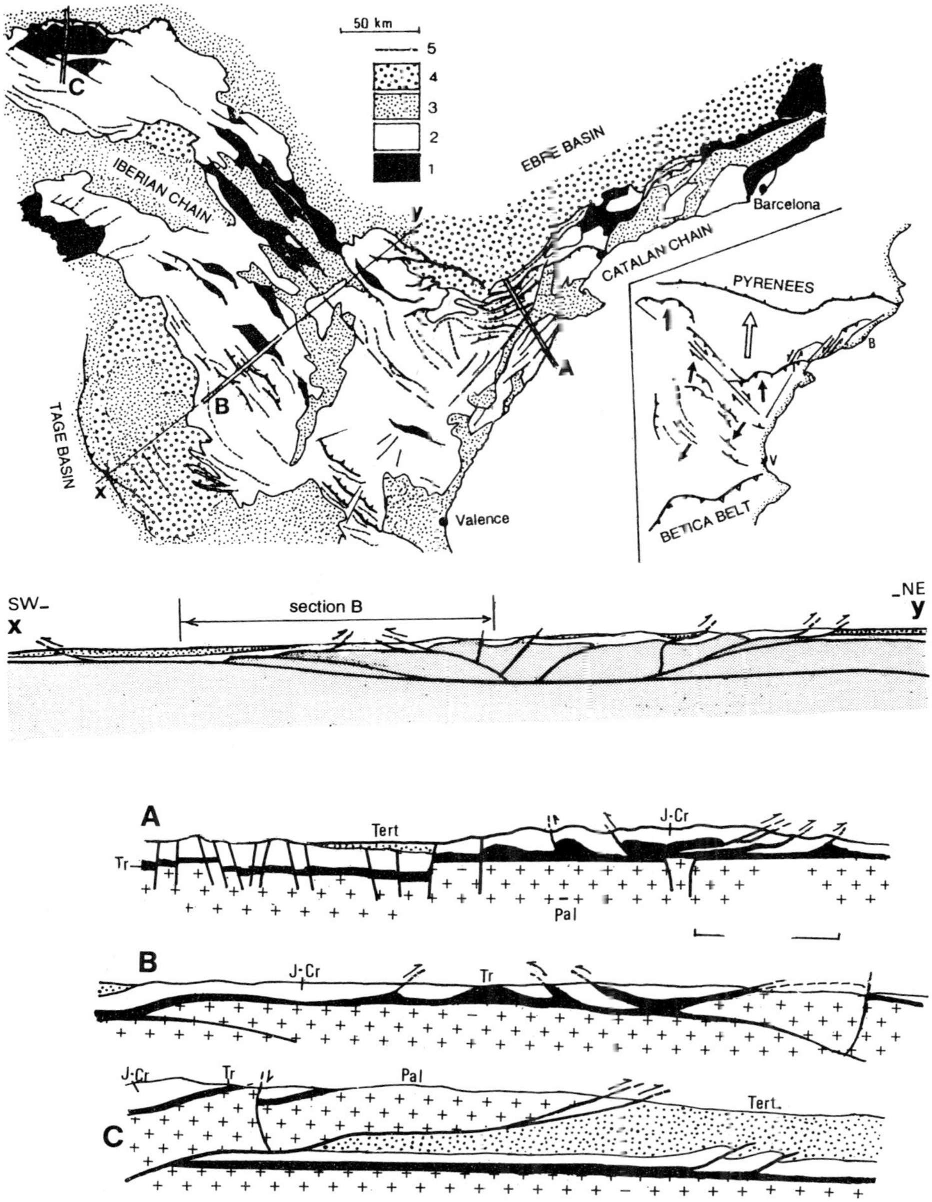

Fig. 129. The Iberian and Catalan chains (after Guimera and Alvaro, 1990, simplified).
Legend for the map: 1. Triassic evaporites; 2. Mesozoic; 3. Palaeogene; 4. Neogene; 5. folds.
Legend for the cross-sections:
J-Cr: Jurassic-Cretaceous; Pal: Palaeozoic; Tert: Tertiary; Tr: Triassic.
White arrow: general direction of compression; black arrows: vergence of structures.

sedimentary cover. It has not been uplifted en bloc or domed up into a basement fold; the cleavages corresponding to these fragments have remained very flat. Further, as the crust has a normal thickness, it may be concluded that a subhorizontal intracrustal cleavage plane exists, probably connected to the Pyrenean contraction zone, which explains the exact contemporaneity of the respective deformations. This cleavage plane might be situated around a depth of 7–11 km, where a low-velocity zone is known to occur.

Reverse faults that are more or less concave are developed on this cleavage plane which act as ramps and end as the thrust faults observed on the surface.

There are several types of chains intermediate between this type of chain and those in which the intracrustal cleavage mentioned above ends as a true *intracontinental subduction zone* (Type A subduction of the authors).

In this case, if a favourable decollement surface exists, the sedimentary cover does not follow the horizontal translational movement of its basement and its subduction. It is detached from the subsiding basement and is passively folded over itself, giving rise to a sequence of folds or overthrust fragments.

These chains appear in general as annexes of larger orogenic chains, as illustrated by the two examples chosen below, viz., the Jura Mountains and the Rocky Mountains of Canada.

Jura Mountains (Fig. 130)

This small mountain chain, of almost entirely Mesozoic deposits, forms an arc supported in the north by the German Black Forest and in the south by the uplifted Variscan block of Crémieu concealed below a thin Jurassic cover.

The stratigraphic sequence starts with a Triassic germanotype facies, including the Keuper formations rich in evaporites (1300 m of rock salt and gypsum, indicated from drilling i.e., the same thickness as the overburden: the latter literally floats on an enormous plastic rock cushion). The Jurassic and Cretaceous formations consist of limestones on the French slope and marls on the Switzerland slope. Their average thickness also increases from west to east (from 1000 to 2000 m), implying a basement tilted eastwards as early as the Mesozoic (Fig. 130). The facies everywhere are those of a typical platform.

Tectonic history. At the beginning of the Tertiary, the platform emerged and was subjected to erosion and an extensional tectonic regime contemporary to the formation of European grabens. It is these grabens that the Bresse basin belongs to, forming the boundary of the Jura in the west, whereas the Swiss molassic basin in the east has a different origin (see p. 99).

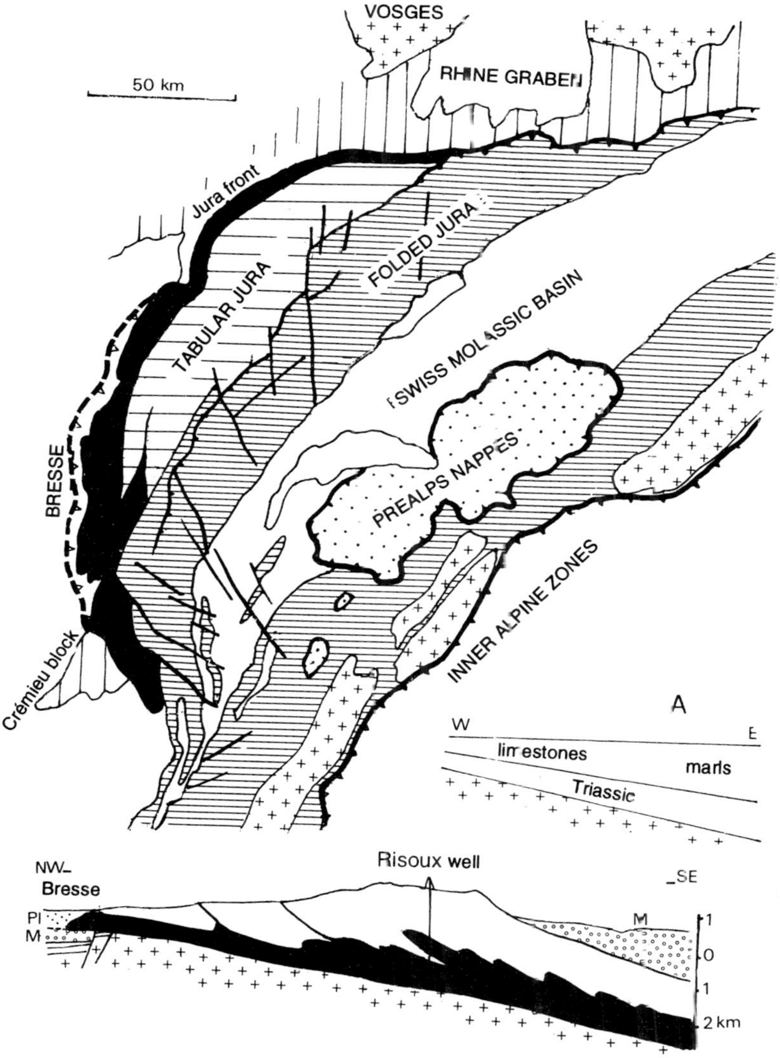

Fig. 130. Structural map and cross-section of the Jura (after Chauve, 1980, simplified). A—Model of Jura sedimentary cover at the end of Mesozoic sedimentation. The thickness of sediments is larger in the east than in the west, implying that the top of the basement was already inclined towards the east.
On the section, in black, the Triassic formation, mainly consisting of Keuper evaporites.

At the end of the Miocene a unique phase of folding took place. The western part of the Jura, with a thin limestone cover, presents a certain rigidity and acquires a box-fold style (tabular Jura), whereas the eastern part, with a more marly and thicker cover, gives a series of regular folds dipping westwards (folded Jura). At the same time, the cover detaches from the basement, moves westwards and thus overthrusts onto the lacustrine Miocene formations of the Bresse basin. The thrust front itself was ultimately buried below the Pliocene formations, also lacustrine, and hence has not been observed anywhere except in drill holes. The extent of overthrusting is of the order of 3 to 5 km. The style of this overthrust part of the Jura is still debated: are the successive blocks separated by normal faults or reverse faults (Fig. 131) or, in other words, was this front produced in an extension regime or in a compression regime? Field and drilling data provide no answer.

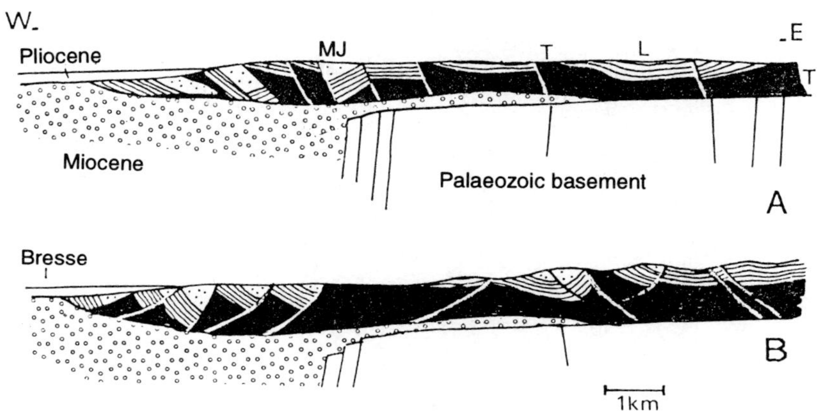

Fig. 131. Overthrust of the Jura front on Bresse basin.
A—Conventional interpretation: blocks separated by thrust faults (compression regime) (after Chauve et al., 1988, simplified).
B—Interpretation as blocks separated by normal faults (extension regime) (after Mugnier and Vialon, 1983, simplified).

Folding of the Jura is accompanied by uplift of the inner Jura where the folding, being more intensive, has considerably thickened the cover. The topographic surface of the Jura is thus inclined westwards, but geophysical data show that the slope of the basement at depth is always oriented eastwards on the one hand, and that the top of the basement has remained relatively flat on the other. The 'alpine' deformation hence appears to have affected only the cover.

Mechanism of folding: It was long thought that the Jura Mountains were produced due to gravitational sliding of the sedimentary cover detached from the basement farther east. This suggests its frontal overthrusting taking place in a non-deformed lacustrine basin.

However, the basement is and has always been inclined towards Switzerland, as already mentioned, and could not lead to a cover under gravitational sliding westwards. Moreover, it is not very clear from which basement the cover in question would be coming since east of the Jura the outer alpine basement exhibits a proper cover everywhere.

The simplest explanation of course is that of underthrust of the Jura basement eastwards below that part of the Alps where a crustal overthickening effectively exists. This implies that the basement of the external alpine chains (or subalpine chains) has followed the movement or, in other words, that it was combined with that of the Jura because the subalpine crust shows the same normal thickness of about 30 km (Fig. 132). This unity of behaviour of subalpine and Jura chains is quite normal since in the southern part of the Swiss molassic basin the folds of the Jura are linked with the outermost subalpine folds.

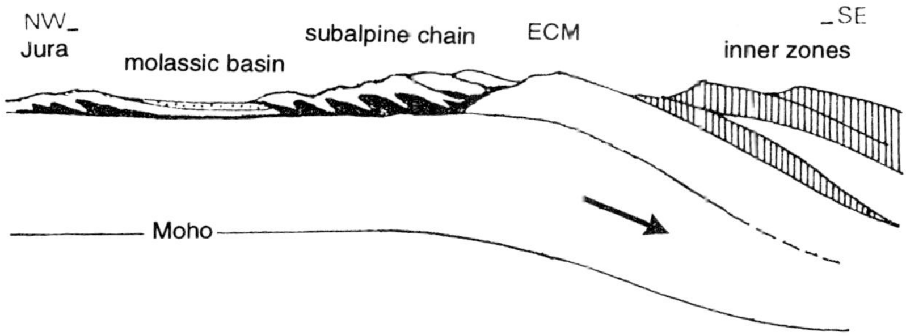

Fig. 132. Mechanism of folding of the Jura by underthrust of the basement of the Jura and subalpine chains below the inner alpine zones.
The underthrust may also be the cause of fragmentation of the external crystalline massifs (ECM).

The Jura thus appears to be an annex of the Alps and the mechanism of its folding cannot be separated from that of the external Alps, of which it is an exact contemporary.

Rocky Mountains, Canada (Fig. 133)

This cover chain also appears to be an annex of a major chain, viz., the North American Pacific Cordillera.

Structural evolution: These mountains result from the folding of a strongly subsident Mesozoic basin, itself succeeding, after the interruption of less-intensive Variscan movements, into a Palaeozoic or even Proterozoic basin (the Belt or Beltian sequence, thickness 10,000 m!). It is these Proterozoic deposits that form the apparent basement of the chain.

All the formations are detrital, monotonous, of shallow-water origin and fed from the disintegration of a Precambrian relief (the 'Beltian geanticline' of earlier authors), which separates them from the Pacific coastal basins.

The gaps, discordances and presence of some fissural alkaline basalts clearly indicate a continental platform.

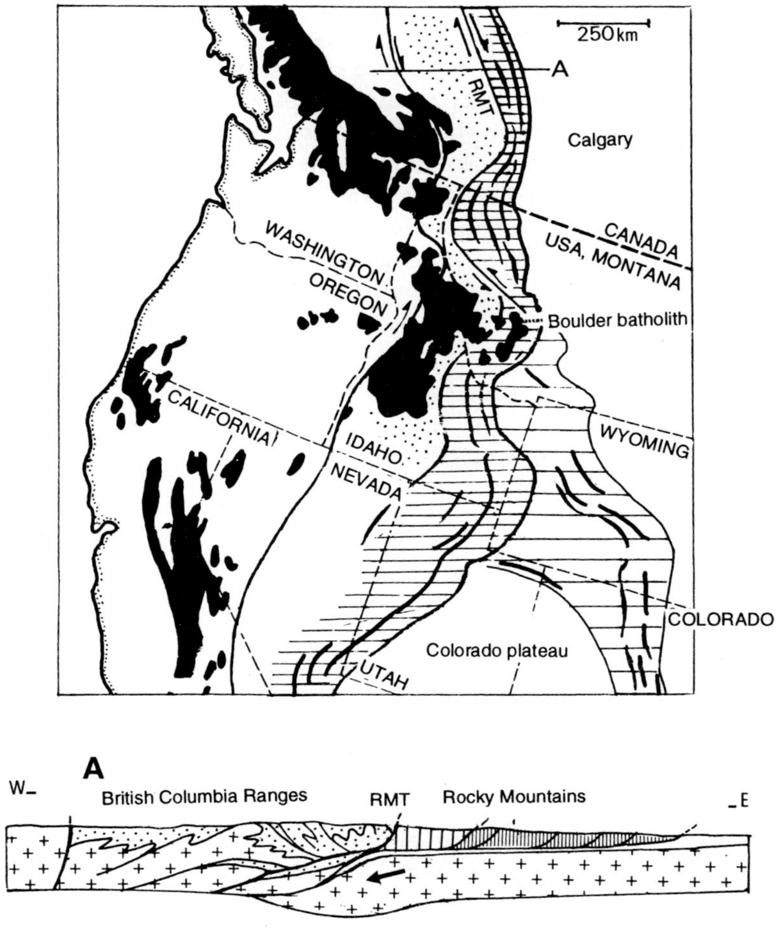

Fig. 133. Structural position of the Canadian Rocky Mountains (dense hatching). They extend into the USA as the Western Rockies (Eastern Rockies—widely-spaced hatching—are basement folds; see Fig. 118).
In black: granodioritic batholiths of the Pacific Cordillera. Dotted: metamorphic zones of British Columbia Ranges (collision chain). RMT. Rocky Mountain Trench (late strike-slip fault in the western boundary of the Rockies).

Evolution of the basin was stopped at the transition of the Cretaceous and Tertiary by Laramian orogeny, which extended up to the Eocene. The resultant chain is dissymmetric as follows (Fig. 134):

— the western half, the most uplifted, hence the most eroded, mainly consists of older material (Mountain Belt);

— the eastern half, less elevated and less eroded, is of Cretaceous material (Foothills Belt).

The whole chain has a dominant eastward inclination structure and a tectonic style like that of the very flat reverse faults affecting the two preceding zones.

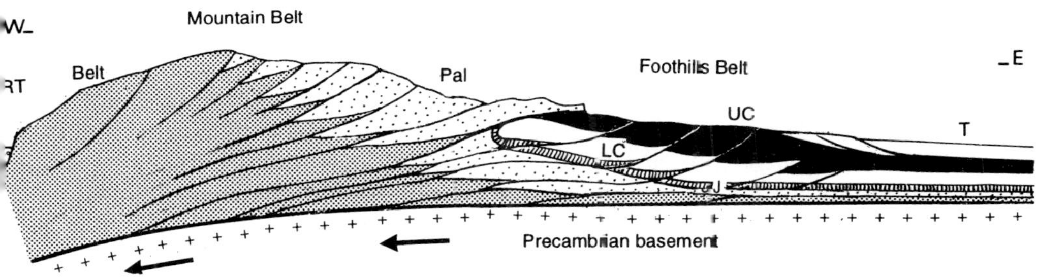

Fig. 134. Purely schematic section of the Canadian Rocky Mountains, showing piling up of layers of the cover in front of an underthrust zone. The chain resembles an accretionary prism.
RT: Rocky Trench (later strike-slip fault).
J: Jurassic; Pal: Palaeozoic; Belt: Beltian (Infracambrian); T: Tertiary; UC, LC: Upper and Lower Cretaceous.

Tangential tectonics of the Rockies is one of the most renowned structural phenomena of the world. It has become particularly well known because of the large number of petroleum wells drilled in this region, which enable us to trace the overthrusts in the subsurface. One can observe a pile up of sheets, concave upwards (Fig. 135), whose thrust contacts represent a succession of flats and ramps in which folds that are more or less intense but without any schistosity are associated .

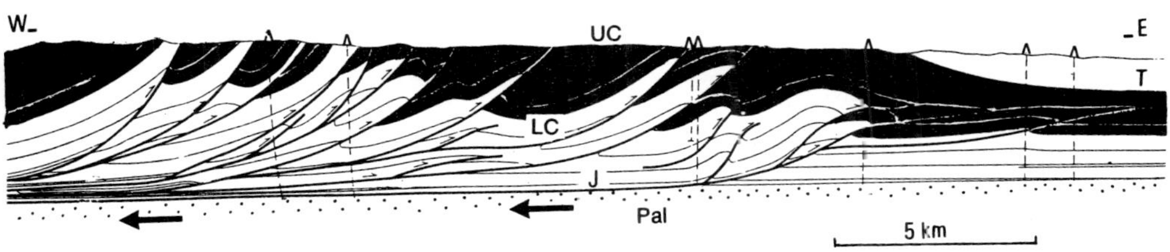

Fig. 135. Detailed cross-section in the 'Foothills Belt', west of Calgary (Canadian Rockies).
J: Jurassic; Pal: Palaeozoic; T: Tertiary; UC, LC: Upper and Lower Cretaceous;
Boreholes are indicated by arrows.

This very special style is generally explained by a kind of underthrusting of older Precambrian (pre-Beltian) basement westwards, in the direction of the zone of crustal fragmentation above the Pacific Cordillera, essentially of Jurassic age.

The mechanism of emplacement of these superposed slabs must have commenced by a sliding of one layer on the other, parallel to the stratification in the deeper parts of the sedimentary sequence. It evidently proceeded on par with a total decollement of the stratified cover, including the Beltian, relative to the crystalline basement.

These Laramian structures were subsequently slightly altered. In the Neogene primarily an uplift of the chain was produced with no modification of earlier tangential structures. This uplift was concomitantly an isostatic response to the crustal overthickening of the Pacific Cordillera and an effect of a new state of compressive stress. These new stresses—oblique to the Laramian structures—have reactivated the overthrusts into strike-slip faults, as in the case of the western boundary of the Canadian Rockies (Purcell Fault and its northern extension, the Rocky Mountain Trench, Fig. 134).

A *Neogene magmatism* also occurs in the most fragmented parts in the form of granitic batholiths, some of which are large in size (batholiths of Boulder and Idaho, Fig. 133). They are probably related to the crustal overthickening below the Mountain Belt.

CHAPTER 2

SUBDUCTION CHAINS

Subduction chains develop vertically above a subduction zone when a compression regime prevails there. In this case it is generally observed that the subduction zone underlying the chain has a very small slope.

The slope of the subduction zone depends on several factors, of which the most important appears to be the thickness and density of the subducted lithosphere and hence its age (older lithospheres are thicker, colder and denser) and, secondly, the more or less rapid convergence of the plates involved (factor related to global tectonics), and even the existence of 'floats' (Nazca ridge, for example, appears to lighten the plate that carries it).

Depending on the interferences of these different factors, subductions may be of steeper slope (greater than or equal to 30°), giving extension structures and magmatism at the surface of the thrusting plate, or of gentler slope (1 to 10°), resulting in compression structures and arrest of magmatism on the surface of the upper plate. The reason for this might be found in the strong 'coupling' of the two plates if the subduction zone that separates them is flat, as also in the absence of upper mantle in the upper plate.

Physical modelling results (Chemenda, 1993) suggest that a subducting lighter lithosphere rests 'glued' to the upper plate where it induces a compression regime. A dense lithosphere causes an extension regime and the appearance of a marginal basin. An intermediate lithosphere, slightly denser than that of the upper plate, leads to an alternating regime: the upper plate undergoes quite a long period of extension and, when the extremity of the subducted lithosphere is detached, rearrangement of the system induces a compression in the upper plate. This favours a succession of relatively short compressive episodes intercepting a generally extensional evolution.

When there is a distinct obliquity in the convergence of the two plates involved, a compression regime causes formation of major strike-slip faults almost parallel to the volcanic arc (Sumatra, Philippines, New Zealand).

We shall consider the example of the *Andes of Peru and Bolivia*, which are more interesting because both the dispositions of the slope of subduction zone mentioned above are encountered there side by side.

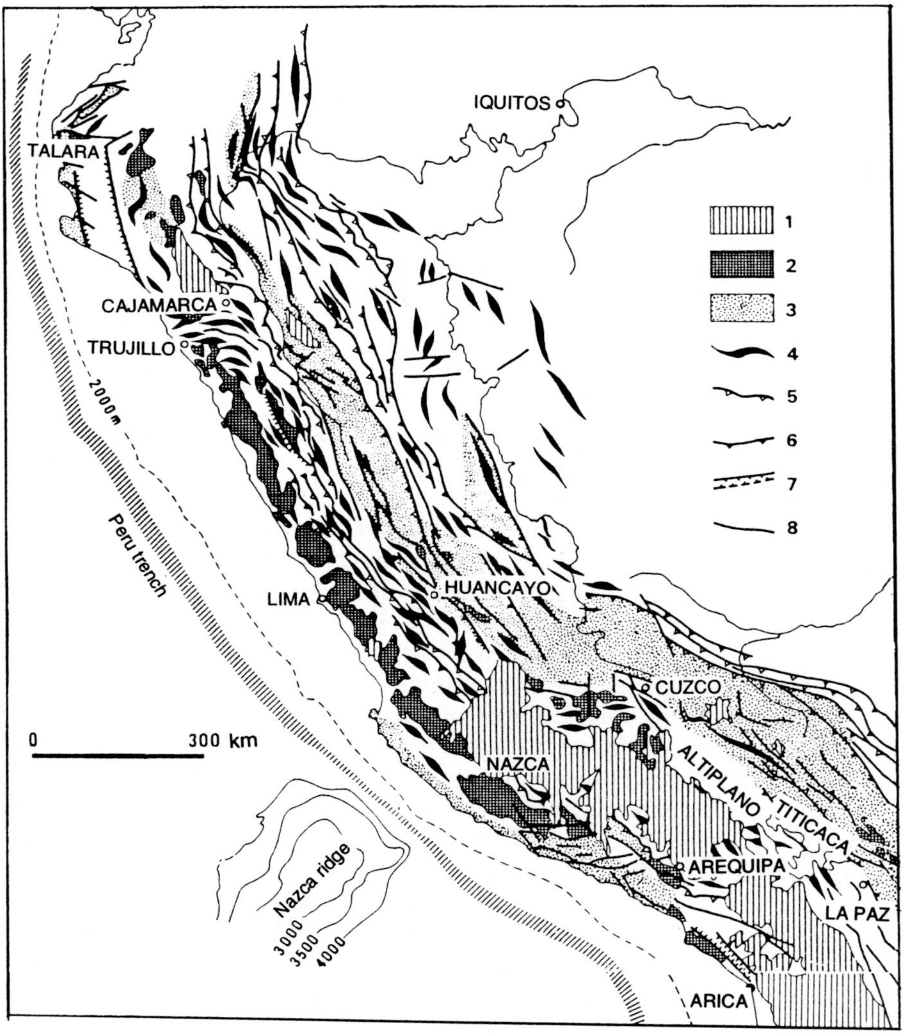

Fig. 136. Structural map of the Andes of Peru and northern Bolivia (after Mégard, 1987). 1. Plio-Quaternary volcanism; 2. granodioritic batholiths; 3. Pre-Mesozoic basement; 4. Major folds; 5. overthrusts; 6. reverse faults; 7. normal faults; 8. strike-slip faults.

Andes of Northern and Central Peru (Figs. 136 to 140)

In this region the oceanic plate of Nazca is subducted below the South American plate along a very low angle (10 to 15°), and even tends to vanish 250 km farther east, below the continent, where the oceanic lithosphere becomes almost horizontal (Fig. 137A).

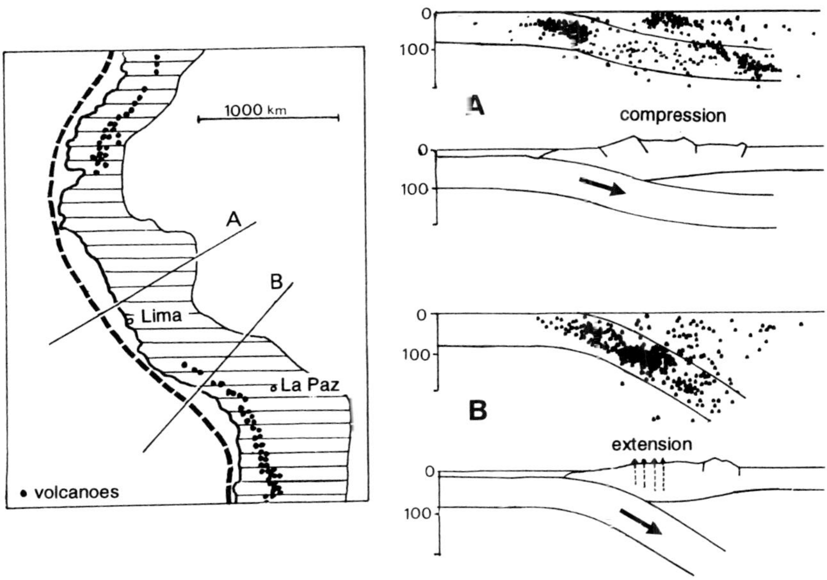

Fig. 137. Distribution of earthquakes and variations in the slope of the Peruvian subduction plane (after Mégard and Philip, 1976).
A—Northern Peru: low-dip plane, surficial compression regime, absence of volcanism.
B—Southern Peru: plane with steeper dip, surficial extension regime, volcanism present.

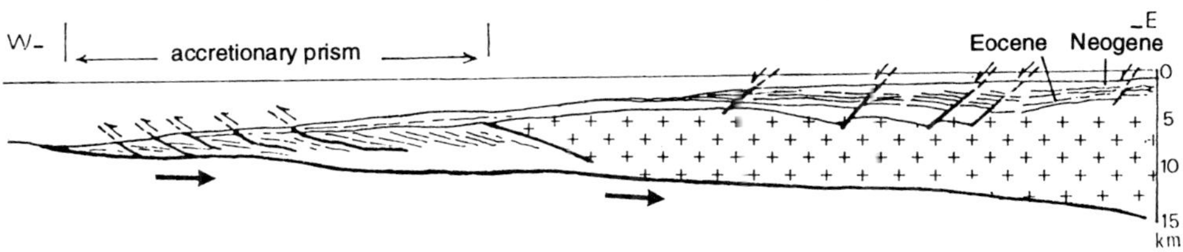

Fig. 138. Structural interpretation of the Peruvian margin around 9.5° S latitude obtained from seismic data (after Von Huene et al. (1989) *J. Geophys. Res.* 94: 1703–1714).

In this part of the mountain chain the following structural elements are observed from west to east, i.e., on moving away from the trench (Figs. 138, 139):

1. *A probable accretionary prism*, deeply buried and covered by solifluction flows. It is known only from seismic profiles (Fig. 138).

2. *A submerged continental plateau*, primarily of complex structure and revealing juxtaposed blocks, with a continental crust encountered in boreholes. It is difficult to say, however, whether these blocks are tilted or overthrust.

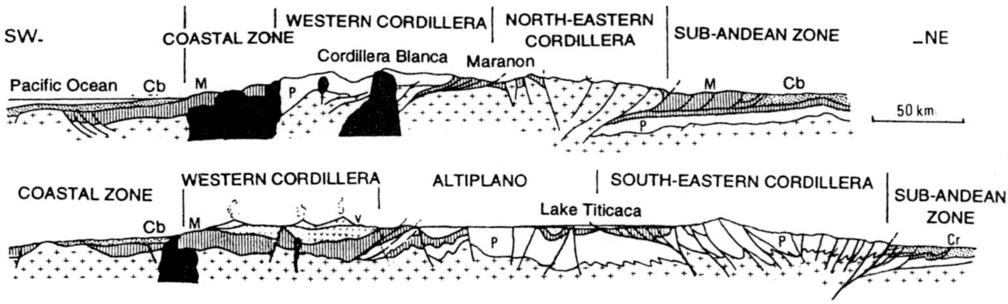

Fig. 139. Cross-sections of the Andes of northern (top) and southern (bottom) Peru (after Mégard, 1987, simplified and slightly modified).
Cb: Upper-Cretaceous-Palaeocene red beds; M: Mesozoic; P: Palaeozoic. In black: granodioritic batholiths of 100 to 32 Ma; v: Plio-Quaternary volcanism.

3. *The coastal zone,* where outcrops of the Precambrian and Archaean basement are observed below a thin Mesozoic sequence. This sector has been a zone of positive movement since the Palaeozoic.

4. *The Western Cordillera,* an old oceanic arc, presently inactive since the Palaeogene, with the exception of some eruptions of acidic tuffs and rare flows of common basalts. The Cordillera is made up of a thick sequence of Jurassic and mainly Cretaceous volcanodetrital sediments, associated with calco-alkaline volcanic products (flows and tuffs). The entire Cordillera is folded into a more or less regular succession of anticlines and synclines oriented NW-SE, with straight axial planes, accompanied by nearly vertical cleavage. These folds are locally disturbed by nearly E-W virgations (termed 'deflections' here), similar to lateral ramps, in which the influence of older structures can sometimes be seen.

This tectonic style is complicated only on the eastern slope of the Cordillera where one can observe a swarm of folds and subhorizontal imbrications, the 'Marañon thrust slices', of eastern vergence (antithetic faults).

However, there is no metamorphism and cleavage is moderate, indicating that deformation took place at a shallow depth.

This group is intersected by granodioritic *intrusive plutons* of Upper Cretaceous to Neogene age, which were brought to the surface by the recent uplift of the mountain chain. The two principal plutons are the 'coastal batholith', more than 1000 km long, and the Cordillera Blanca which carries the highest peaks of the Western Cordillera (Huascaran, 6770 m). This uplift is still continuing as indicated by the escarpments of the active faults that offset the recent moraines at the foot of the Cordillera Blanca and which are the zones of earthquake foci (for example, the Quiches fault; 3.5 m throw in 1946), as also the deformation of the piedmont terraces and cones on the coast. The earthquake focal mechanisms indicate a compression regime.

5. *The Eastern Cordillera* is poorly distinguished from the preceding one. The Mesozoic sedimentary sequence here is thinner even though the ancient basement, Variscan to Precambrian, is largely exposed, with a very distinct fan-shaped structure (Fig. 139).

The uplift of the entire group also occurs along active faults (like that of Huaytapallana, with 1.6 m vertical throw and 0.7 m horizontal throw during the earthquake of 1969).

6. *The sub-Andean zone*, already occupied by the Amazon forest, shows a thick sedimentary sequence of Mesozoic and Tertiary formations, non-volcanodetrital, subsequently folded, with an eastern vergence. Geophysical profiles show that it is detached and overthrust eastwards on the Amazon shield, along a subhorizontal plane situated at about 9 km depth, the plane in which all the subsidiary faults are rooted.

Andes of Southern Peru and Northern Bolivia

In this zone the Nazca plate is subducted below the South American plate along a steeper angle (about 30)°. The subducted part of the continent is similar, as in the previous case, but the blocks are typically tilted.

The *Western Cordillera* has roughly the same stratigraphic sequence and the same style of folding as farther north but is covered by a thick discordant layer of calco-alkaline volcanic formations of Neogene to Quaternary age, which are non-deformed and carry the active or recently extinct volcanoes (Misti volcano, 5842 m, for example). The volcanic arc thus formed is aligned parallel to the trench, situated nearly 250 km from the latter, and shows a characteristic zoning of the active margins in such a way that the shoshonitic sequence appears there east of the calco-alkaline products.

Eastward, the Western Cordillera overthrusts on the *Altiplano*, a vast plateau of mean elevation close to 4000 m, on which Lake Titicaca is located. This new morphostructural element, in-between the two Cordilleras, the Western and the Eastern, shows a quite complicated folded and overthrust structure, broken during the Neogene into a mosaic of horsts and grabens. The latter have been filled with thick lacustrine to continental detrital products, separated by numerous unconformities, which indicate that the extension regime was interrupted by several brief compression phases.

This rapid filling of the grabens might explain the general planation of the relief but the fact that the plateau has retained a relatively regular topography in spite of its high elevation may be interpreted as the result of a recent uplift since erosion has not yet reached it. In fact, stratigraphic evidence has shown that the uplift took place immediately after the Quechua 3 phase (Late Neogene).

The Eastern Cordillera of southern Peru and northern Bolivia differs from that of northern Peru by its dual-inclination structure (Fig. 139) and the magnitude of Palaeozoic material there which forms high peaks (Ancohuma, 7114 m and Illimani, 6889 m).

In general, volcanism and the subsident structures indicate that this Andes segment is at present in an extension phase, which appears to be clearly related to the slope of the subduction plane. The focal mechanisms of local earthquakes confirm an extension regime.

The separation between these two segments of the Peruvian Andes corresponds to the arrival in the trench of the 'Nazca ridge', a submarine fossil structure of unknown origin, which appears to play the role of a 'float' in the middle of the Nazca plate.

Orogenic evolution

From the end of the Permian (i.e., after Variscan folding) *up to the Middle Cretaceous*, the palaeogeography remains simple (Fig. 140).

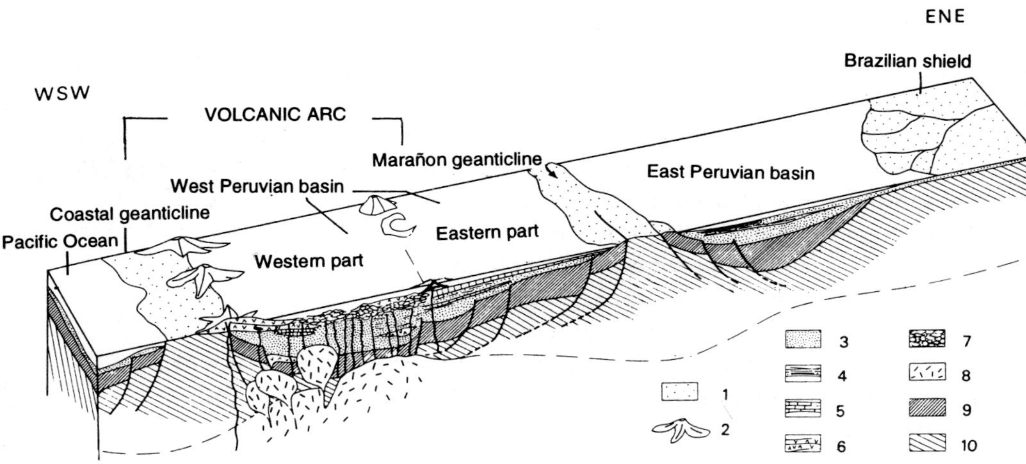

Fig. 140. Block diagram of the Peruvian region from the Neocomian to the Albian (after Mégard, 1987, simplified).
1. emergent zones; 2. volcanoes; 3. sandstones; 4. shales; 5. limestones; 6. volcanic flows and sills; 7. pillow lavas; 8. granodiorites and gabbros; 9. Liassic and Triassic beds; 10. Palaeozoic and Precambrian.

In the boundary of the ocean, a *magmatic arc* appears, which is produced simultaneously on shoals (the geanticline of Paracas, the future coastal zone) and in subsidence zones where volcanic activity is often submarine (future Western Cordillera). This *Western Peruvian subsident basin* comprises two parts:

— *The Western part*, of high volcanic activity and a thin crust. During the Middle Albian, for example, there were volcanic rocks (pillow basalts and gabbros) nearly 10 km thick, indicating a strong uplift of the upper mantle. The heat flow likewise must have been high (low thermal metamorphism).

— *The Eastern part*, of platform type, without volcanic products. Thinning of sediments eastwards suggests the existence in this direction of a high zone, known as the 'Marañon geanticline'.

Beyond these occurs the *East Peruvian basin*, which again becomes subsident, since there are sediments locally almost 10,000 m thick fed exclusively by the disintegration products of the Brazilian shield.

It is thus evident that an extension regime dominated above the subduction zone during this initial period of development of Andean orogeny.

From the Middle Cretaceous onwards commences a complex period in which compression and extension regimes alternate.

A deformation episode appears to exist in the Upper Albian but as yet remains uncertain since the Mesozoic volcanodetrital and volcanic structures are not sufficiently established.

The following phases are clearly identified, even though some are only of local significance:

— Santonian phase, known as 'Peruvian';
— Middle-Upper Eocene phase, known as 'Incaic';
— Neogene phase, known as 'Quechua'.

Of these, the *Incaic phase* is the principal one. It is chronologically well documented because it produces the folding of Upper Cretaceous formations of the Western Cordillera and these folds are covered, with unconformity, by conglomerates and volcanic rocks dated as 40 Ma. This phase is hence Lower to Middle Eocene and must have produced the swarm of folds and overthrusts of Marañon, with eastern vergence, in northern Peru, and the Huancane folds and thrusts, with western vergence, in southern Peru, in the boundary of Altiplano and the Eastern Cordillera (Fig. 139).

The *Quechua phase* is divided into three episodes, of which the latest, Quechua 3, at the transition of the Miocene and the Pliocene, is the most important because it is common to the whole Andean axis where it gives overthrusts with eastern vergence in the sub-Andean zone.

It is during this Upper Cretaceous-Tertiary orogenic period that the *granitic batholiths of the Western Cordillera* were emplaced: first of all, the coastal batholith (1600 km x 65 km), between 100 and 32 Ma, parallel to the trench and unaffected by surface geology (see Fig. 136), then farther west and later, the batholith of the Blanca Cordillera (150 km x 15 km), between 35 and 7 Ma.

It is also during this orogenic period that an intensive *calco-alkaline volcanism*—andesitic to rhyolitic—was manifest; the episodes of this volcanism correspond to the major emplacements of granitic magma, i.e., Upper Eocene-Lower Oligocene (40–30 Ma) and Mio-Pliocene (18 to 2 Ma).

In conclusion, the most important fact is the *constancy of the magmatic arc* which appears in the Lower Jurassic and continues up to the present epoch, even though its activity is intermittent. This arc is built up on a continental crust, sometimes very thin (Northern Peru, in the Albian, Fig. 140). Further, it occupied the same place from the Lower Jurassic to the Eocene and then became displaced eastwards, remaining everywhere at the same distance from the trench. This feature excludes any explanation of the origin of the Peruvian Andes by collision with any volcanic arc[1]. In this case, the magmatic arc should have been displaced westwards, following the movement of the oceanic trench.

The Peruvian Andes thus constitute a good example of the origin of a continental margin orogeny exclusively associated with the subduction of an oceanic lithosphere below the margin in question. Yet, how to explain the fact that up to the Albian the margin functioned in an extension regime and subsequently in both extension and compression?

The present tendency is to associate these variations of regime with those of the slope of the subduction plane.

In the Mesozoic, up to the Albian, a steeply dipping subduction plane prevailed, giving an active magmatic arc, a prolonged extension regime, and hence subsident basins, with crustal thinning that is sometimes intensive, not only along the volcanic arc but far away in the interior of the continent.

From the Albian to the Present, evolution consisted of successive compression and extension periods (conforming well with the results of modelling mentioned earlier), the former corresponding to flattening of the subduction plane. The major compressive phases, which affected the western margin of South America over thousands of km and, moreover, are contemporary to the classic episodes affecting all the mountain chains of the world, are evidently related to the equally major events of plate tectonics, notably an acceleration in the rate of sea-floor spreading, which increased the velocity of approach of the plates[2]. Such is the case of the Middle Cretaceous, Upper Eocene (Incaic phase) and Upper Miocene-Pliocene (Quechua 3 phase). The minor phases, of local significance, correspond to the sectors of 'flat' subduction within a heterogeneous subduction zone, i.e., consisting of successive segments with variable regimes, such as the present disposition.

It may be noted, however, that even the major episodes result only in moderate tectonic structures, quite often without cleavage or metamorphism, except in some local cases such as the Marañon or Huancane thrust zones.

---

[1]Its absence could readily be explained by a subsequent displacement, parallel to the South American coast, as we shall see in the case of the Colombian Andes.

[2]By calculating the rates of sea-floor spreading and the convergence along the western coast of South America from the magnetic anomalies of the East Pacific and South Atlantic, Frutos (*Tectonophysics* (1981) p. 72) showed that it is the *beginning* of the periods of high rates of spreading that is contemporay to compressive movements. It is, in effect, the moment when the coupling of the two plates is the strongest. Subsequently, isostasy tends to re-establish the earlier equilibrium.

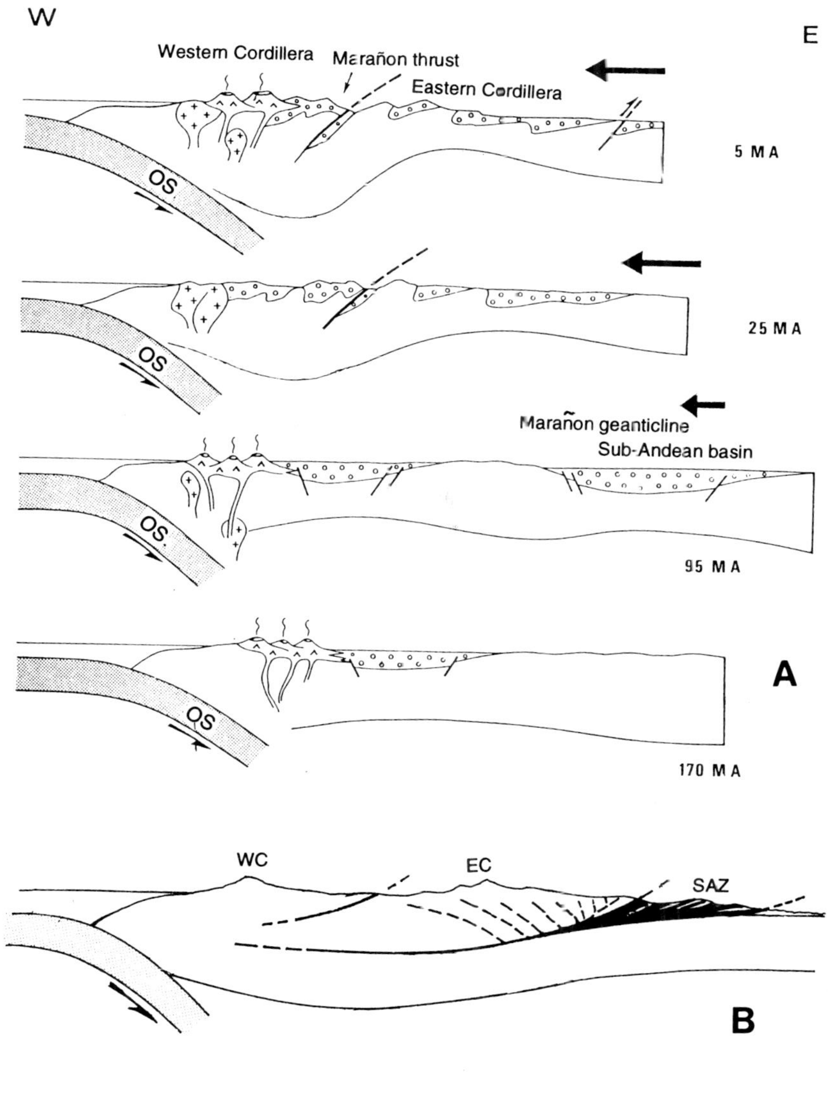

Fig. 141. Geodynamic evolution of the Andes under the assumption of an underthrust of the Brazilian shield below the Western Cordillera.
A—after Bourgois and Janjou, 1981, modified.
OS: oceanic subduction.
B—after Roeder, 1988, simplified.
WC, EC: Western and Eastern Cordillera. SAZ. Sub-Andean zone.

The magnitude of shortening, calculated on structures visible on the surface, is of the order of 100 to 150 km and since this shortening occurs inside the continental crust, it can explain the origin of a very thick *crustal root* that may attain a thickness of 75 km below Altiplano. This root is evidently the cause of the high elevation of this segment of the Andes.

The mechanisms that led to this crustal thickening are still disputed but two factors must have combined to produce it:

1. The quantity of granodioritic magma emerging from the mantle and the different materials carried by the plunging oceanic lithosphere (oceanic sediments and eventual fragments of continental crust extracted from the margin). This mechanism must have played only a minor role since the maximum thickness of the sub-Andean crust does not occur in the region of plutons of the Western Cordillera but farther east (see Fig. 141).

2. The compressive mechanism itself, which cut the crust into slices, subsequently superposed one above the other. However, views about their disposition vary considerably. Some authors emphasise the cleavage touching the surface in the Marañon thrusts (Fig. 137A); others, the overthrust of the sub-Andean zone (Fig. 137B). In effect, these two structures, respectively associated with the Incaic (Eocene) and Quechua 3 (Neogene) phases, correspond to the events of uplift represented by the arrival of large masses of detrital products.

CHAPTER 3

OBDUCTION CHAINS

In the case of subduction phenomena, the denser oceanic lithosphere sinks below the lighter continental lithosphere. However, in some archipelagos of the SW Pacific (such as New Guinea, New Caledonia) or in Oman (Arabian peninsula), reverse dispositions are observed, viz., large slabs of ophiolites overlying a gneissic basement and its sedimentary cover. In other words, a part of the oceanic lithosphere is overthrust on a continental margin, an *a priori* abnormal situation in view of the respective densities of these two compartments.

These obducted slabs never have the thickness of the normal oceanic lithosphere (40 to 60 km): they thus represent either abnormal or young lithosphere, or are the result of the cleavage of a normal lithosphere, implying the existence of decollement zones (of serpentinised rocks, for example). As a matter of fact, mylonites of serpentinised ultrabasites are always observed at the base of the obducted masses. The transition from such a cleavage plane to a thrust plane would be produced by a ramp, which may be any fault in the ocean bottom.

The term *obduction* was introduced to describe these structures that are reverse to subduction. According to the American geologist, Coleman, who coined the term in 1971, this disposition may be due to *severe compressive tectonics* that renders the density differences immaterial. It might in effect be the collision of two lithospheric plates, the boundary of one passing above the other, irrespective of their mutual densities (Fig. 142A). When the compression ceases, isostasy is re-established by the appearance of a classic subduction plane, which separates the obducted slab.

It can thus be seen that Coleman considered obduction and subduction as completely different processes. But many authors, on the contrary, consider the two processes interrelated or, more exactly, that obduction is only a phenomenon associated with subduction in some cases of convergence. This is the *blocked-subduction theory* (introduced in 1971 by Davis and Smith for New Guinea and by Dewey and Bird for Newfoundland).

This theory can be summarised by Figure 142B.

The mechanism starts with an intraoceanic subduction in the proximity of a continental margin A. The latter follows the movement and sinks below B, with the effect that obduction is produced. But isostasy promptly stops

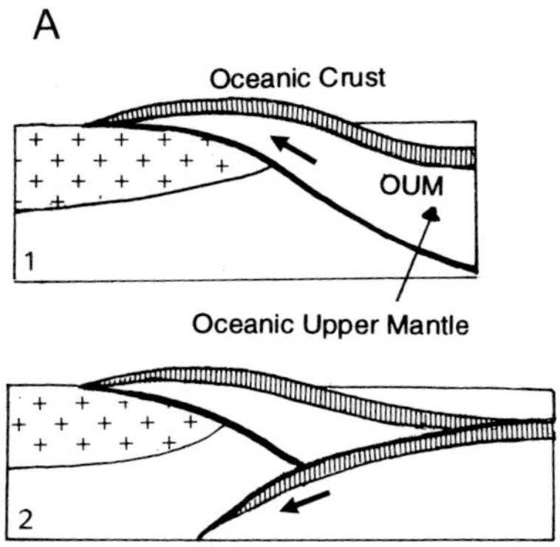

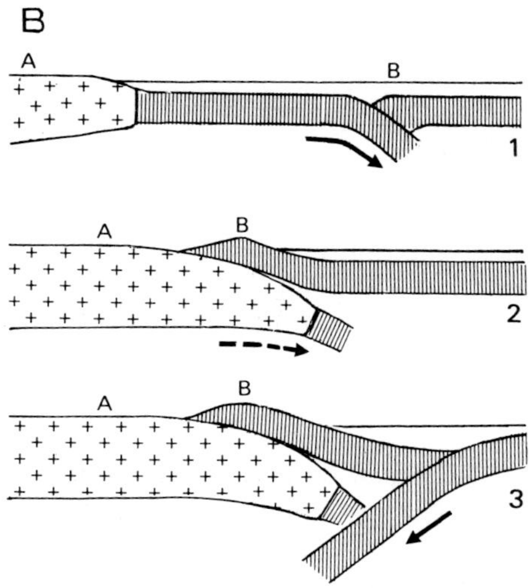

Fig. 142. Interpretative model of obduction.
A—according to Coleman;
B—blocked-subduction theory.

the system because of the relatively lighter mass of A. Equilibrium is re-established; a new subduction plane might appear but the disposition of the system implies that it is generated in a direction reverse to the preceding one (a phenomenon known as 'flip'). The obducted slab thereafter becomes fossilised.

Obduction often constitutes the first stage of folding. It is hence followed by other compression phases such that the original structures are totally masked and even the original oceanic region disappears. It is not possible to determine the mechanism of emplacement, which remains entirely hypothetical.

However, a small number of cases exist in which the obduction structures remain intact after the emplacement of an oceanic complex and in such cases one can actually speak of *obduction chains*.

Physical modelling studies (Chemenda, 1993) provide a new interpretation which takes into account structural features that were neglected in the two preceding models, viz., the existence, at the rear of the ophiolitic nappe and underneath, of units of continental-margin nature below blueschist or eclogite facies (see Fig. 143). In this model, subduction of the oceanic lithosphere involves the subduction of the thinnest part (base of the margin) of the continental margin up to such a level that the latter finds itself in the conditions of blueschist or eclogite facies. A cleavage is produced in the midst of the continental material which rapidly ascends by piercing the oceanic material above and isolating an ophiolitic fragment (see Fig. 144). It thus results in a configuration comprising an overthrusting shear below the metamorphic material and a normal shear in ridges that are active simultaneously.

Oman Mountains (Fig. 143)

The Oman mountain chain (500 km by 100 km approximately) occurs in the eastern part of the Arabian peninsula, rising to peaks of a little more than

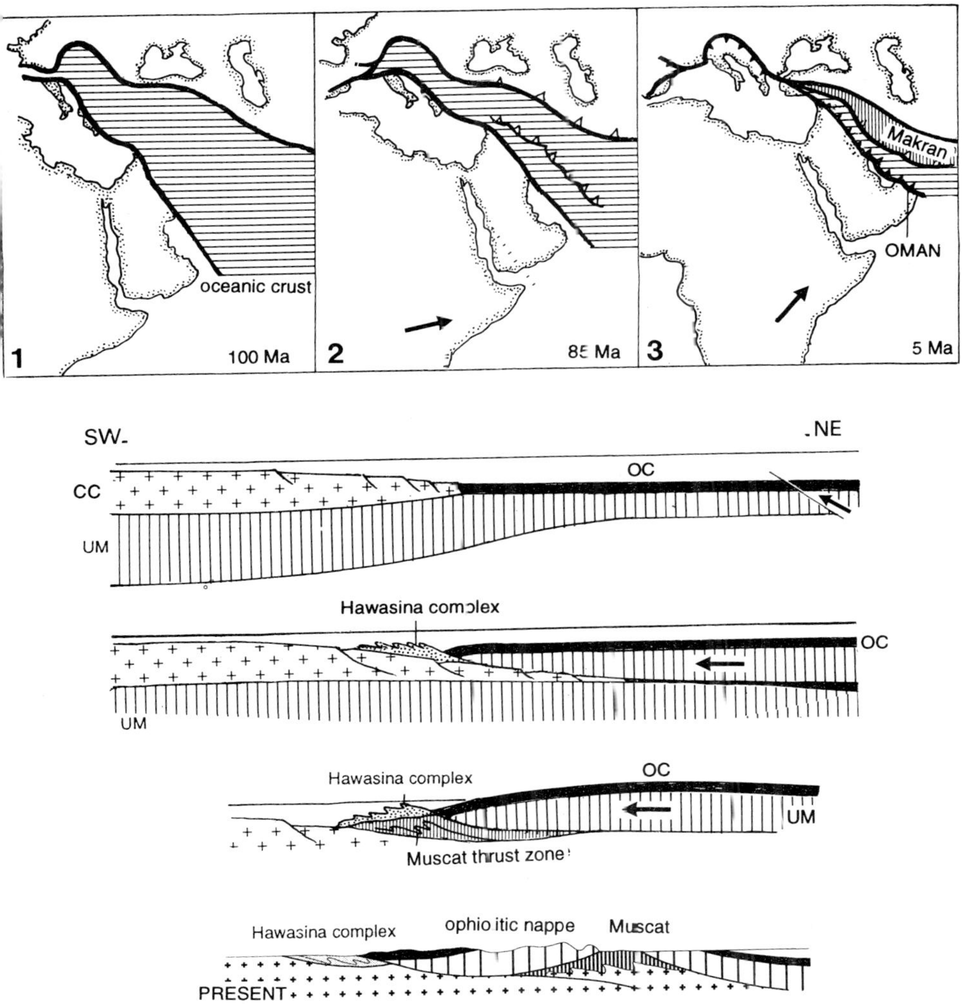

Fig. 143. Obduction chain of Oman (after Michard, 1987, simplified). CC: continental crust; OC; oceanic crust; UM: upper mantle.

3000 m and consisting of an enormous ophiolitic nappe lying flat on the Arabian platform.

The autochthon is first covered by a group of allochthons of oceanic origin, known as the *Hawasina nappes,* consisting of oceanic basalts associated with deep sediments (radiolarites) and platform limestones. This is followed by the oceanic nappe proper, consisting of peridotites and basaltic products, sometimes topped by a sedimentary pellicle still in place.

Counting the width of the Arabian continental plateau (300 km) and the extent of thrust zone (100 km), the total range of thrusting can be considered to be 400 km. The thickness of the nappe is of the order of 15 km, with an 8-km thick peridotite layer, which certainly represents the upper mantle for a major part as indicated by its internal structure.

Stratigraphic data enable us to estimate the rate of displacement of this nappe. In fact, the most recent sediments carried by the nappe are of the base of the Upper Cretaceous (about 95 Ma) and the most recent of the overthrust material of late Upper Cretaceous (75 Ma). This represents a period of 20 Ma for the displacement of 400 km, i.e., a rate of 2 cm/yr (mean value of the rate of sea-floor spreading).

Displacement of the ophiolitic material occurring above the peridotites has produced a foliation of the latter and a stretch lineation oriented SW-NE. This obviously marks the direction of displacement of the nappe. The mineralogy of these basal peridotites indicates that their deformation took place at a temperature of 800–1000°C (hot lithosphere), which conforms with the minerals of metamorphism developed in the outliers of sediments or the basalts remaining at their base. In the front of the nappe these figures are smaller (400°C) and indicate a greenschist facies.

What does the Oman chain indicate regarding the mechanism of detachment of the ophiolitic slab?

The oldest ages of the ophiolitic rocks, measured in the lenses of metamorphic rocks of the nappe, range from 100 to 95 Ma, i.e., Middle Cretaceous. This is, within a few million years of accuracy, the age of the sedimentary pellicle locally covering the ophiolites. This feature shows that their detachment was produced *a little after they were emplaced at the sea bottom.* It thus represents a very young lithosphere, of age lower than or equal to 5 Ma, and still thin.

The Middle Cretaceous corresponds to a sudden change of regime of the major global plates. In the Tethys region, from which this crust has emerged, the tectonic regime changes from plate divergence, i.e., sea-floor spreading—and hence formation of oceanic lithosphere—to a convergence regime associated with the approach of the African and Eurasian plates towards each other. It is known that this approach takes place at the rate of 2 cm/yr, which is precisely the same as that of the nappe, as we have already seen.

Thus compressed, the oceanic Oman lithosphere broke probably in its thinner, and hence younger part. Then started an intraoceanic subduction. A tectonic accretionary prism was formed comprising oceanic sediments (radiolarites of Hawasina nappes). Then subduction affected the thinnermost part of the Arabian margin of which part of the sedimentary cover was incorporated into the accretionary prism. To summarise, the ophiolitic slab acted as a bulldozer, pushing the unmetamorphosed sediments in front of it and thus building the Hawasina nappes.

It is probable that this overthrust was facilitated by the fact that the Arabian platform, under the weight of the ophiolitic nappe, had to bend, as appears to be indicated by the facies (the limestones of the Middle Cretaceous platform are overthrust on Late Cretaceous turbidites with continental material).

However, the older sedimentary cover of the thinnest part of the margin was subducted quite deep as evidenced by the blueschist facies and eclogites observed in the Muscat nappes (11 kb, 450°C) (Goffe et al., 1988).

A moment arrived when this continental material, even though eclogitised, was no longer likely to be subducted. It moved along an overthrust shear zone at its base, its top showing a normal fault activity, and pierced through the overlying structure isolating the ophiolitic nappe (Fig. 144) (Chemenda et al., 1995; Chemenda and Mattauer, 1996).

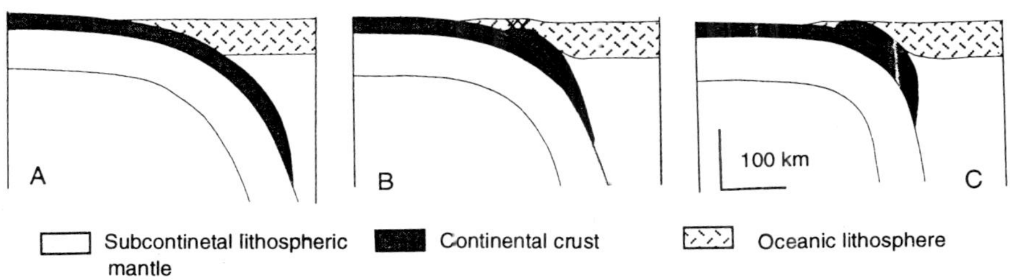

Fig. 144. A possible mechanism of obduction (after Chemenda, 1993).

The phenomenon was subsequently arrested due to lack of energy and it has become easier for the oceanic lithosphere of the Gulf of Oman to sink—the lithosphere has become denser and thicker—below the Iranian coast along a subduction zone of NE vergence. In fact, the accretionary prism of Makran (southern Iran) (see Fig. 143) is built up there, beginning precisely with the Lower Eocene, and always seems to be developed at the rate of 2 cm/yr, according to the calculations related to the opening of the Atlantic and the Gulf of Aden.

At this rate, and if the phenomenon continues, the Gulf of Oman will disappear in another 10 Ma. At that moment there will be a collision between the Arabian and Iranian plates and an alpine type of mountain chain will be developed.

In conclusion, though it may not be possible to generalise the conclusions drawn from the Gulf of Oman to all obduction structures, it appears that the obduction process occurs only in the case of young and thin lithospheres that are still light and are suddenly subjected to compression. A thin continental lithosphere may subduct under such lithospheres.

On the contrary, the lithosphere of major oceans, generally being thick and dense because of its age, would rather tend to be buried in the subduction zones. Hence original basins of obducted ophiolites should be searched for in the crust of marginal basins, crustal fractures or oceanic gulfs such as the Red Sea and the Gulf of Aden or the Gulf of California.

New Caledonia Chain (Fig. 145)

This island, 400 km long and 40 to 50 km in width, culminating to only 1650 m, mainly consists of a Mesozoic basement that supports a peridotite slab emplaced in the Upper Eocene.

The evolution of this chain comprises two main stages:

1. *From the Permian to the Eocene,* units of the substratum are formed in a complex structural framework, such as island arcs and back-arc basins in the Gondwana margin (NE of Australia) during the Upper Jurassic. The corresponding sedimentary and volcanic products form an accretionary prism which is subsequently disintegrated into several units starting from the Upper Cretaceous when the marginal basins of Tasmania, New Caledonia and Loyauté Islands opened up. In the Eocene the entry of these units into the Loyauté subduction zone led to their superposition, while brecciaceous flyschs and discordant olistostromes were deposited.

We shall not enter into the details of these basement units. Suffice it to say that the lower ones (*foreland units:* Noumea, Teremba, Koumac) show normal facies, sedimentary and volcanic, of arc basins of the Permian and Senonian. These are overlain, always in tectonic superposition, by two older, more or less fragmented, ophiolitic suites metamorphosed to high-pressure facies: one (Central Chain) in the Upper Cretaceous and the other (anticlinorium of Mont Panié) in the Eocene. The latter is the best known, with eclogites of 25 kb/550°C and glaucophanites which gave their ages as 51 and 37 Ma. Exhumation of the Mont Panié anticlinorium occurred in an extension regime.

The group of these basement units is separated from the peridotite nappes by the *Poya nappe,* or basalts nappe, comprising basalts and oceanic sediments of the upper Cretaceous-Palaeocene. This nappe may represent the ancient cover detached from peridotites.

2. *The peridotite nappes* are emplaced in the Upper Eocene (Fig. 145): This thrust sheet constitutes one-third of the surface of the island; it is mainly represented in the southern part of the island (the Southern Massif), where it attains maximum thickness (3 to 4 km); numerous klippes are scattered in the rest of the island. The nappe consists, from the base to the top, of the following rocks:

— A sole of *mylonitic serpentinites.*

— A *major peridotitic mass* of massive harzburgite, with intercalations of banded harzburgite or pyroxenite horizons. Frequent occurrences of dunite are observed as irregular pockets. Lherzolites, sometimes feldspathic, are known to occur in the Tiebaghi Massif (northern part of the island) on the surface of harzburgites.

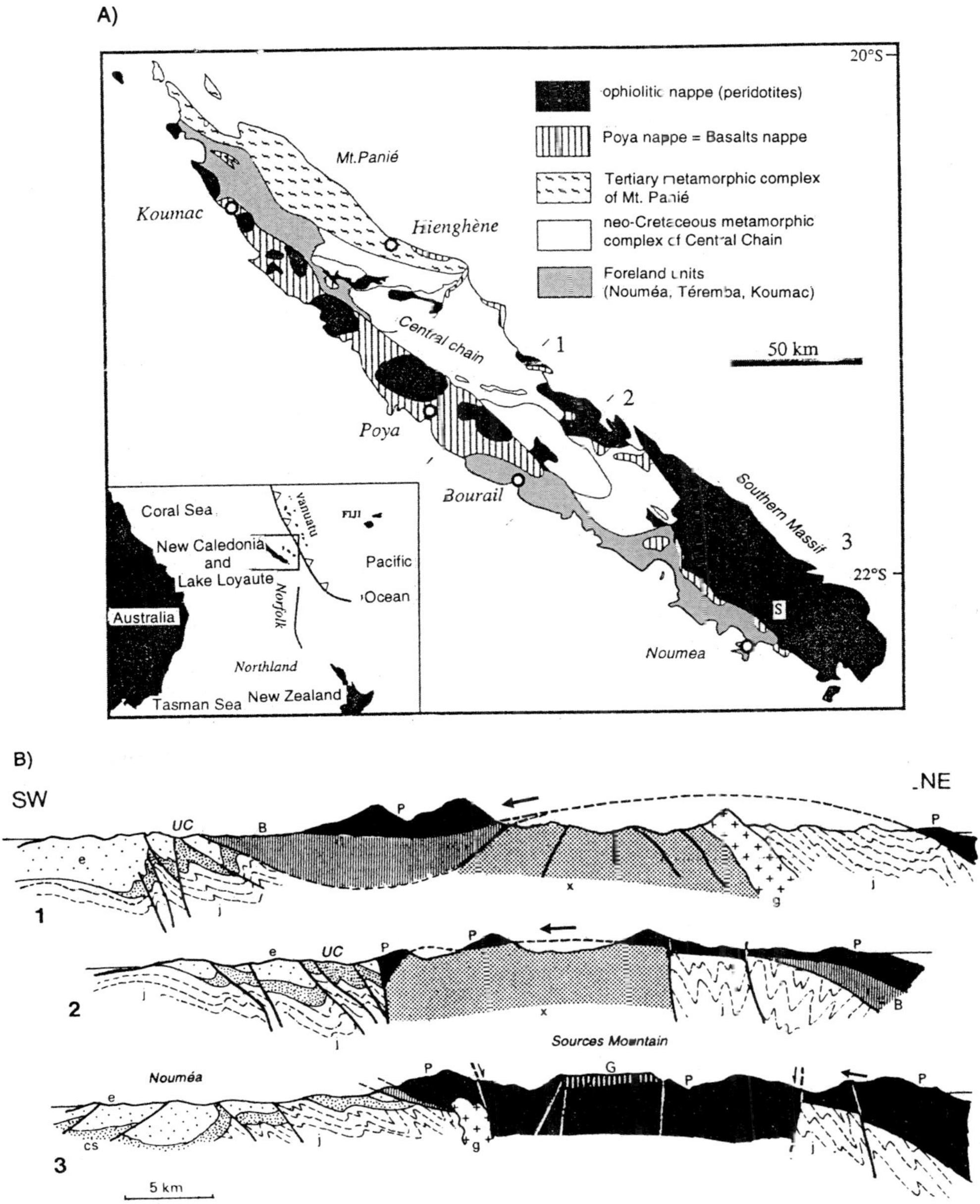

Fig. 145. Major features of the geological structure of New Caledonia (after Paris, 1981).
Map: T. Tiebaghi Massif, S. Sources Massif.
Section: B. basalts, e. Eocene, g. gabbroic intrusion (Oligocene), G. gabbros, j-Jurassic-Triassic, P. peridotites, UC. Upper Cretaceous, x. pre-Permian formations

All these rocks have textures of tectonites and represent the upper mantle. They are intersected by veins of pyroxenites, gabbros and doleritic basalts and contain plagiogranite intrusions.

— Above the ultrabasites, but only in the minor Sources massif, near Nouméa, are found cumulates of *dunite* and *gabbro,* with a very small discordance (Fig. 145). This is the only known evidence of an oceanic crust *sensu stricto.* Overlying basalts, if they existed, are not known, unless they are separated from the underlying ultrabasites by an earlier decollement and have thus given birth to the *Poya basalts nappe.*

Either way, with or without basalts, the ultrabasic rocks of New Caledonia are considered a mass obducted after the subduction chain was subjected to an intensive compression. This obduction does not seem to have produced any specific structures except the sole of mylonitic serpentinites. The earlier magmatic structures, porphyroblastic to blastomylonitic, with a foliation parallel to the magmatic layering, were passively transported, undergoing only long wavelength undulations.

On the contrary, after its formation the slab was fragmented by numerous later faults into juxtaposed compartments, sometimes subsident and thus more readily preserved as klippes. These late structures are often strike-slip faults.

Age of formation: The ultrabasites are reworked in the end-Eocene. Moreover, in the Nouméa region, they are thrust over these Eocene formations or compress them in front of the nappe. The thrusting is hence of Upper Eocene and thus would be contemporary to the metamorphism in the northern part of the island and hence to exhumation of the high-pressure facies which occurred by tectonic denudation (Cluzel et al., 1995). Here we find a structure comprising a compression regime in front of ophiolites and an extension regime in their rear.

Direction of thrust: It is certainly from NE towards SW. A belt of high gravity anomalies occurs on the eastern coast, which indicates that the peridotites have come from this region. Further, the peridotite slab has produced folds and SSW-dipping slices at its front, in the southern part of the island.

New Guinea Chain

The geological conditions of the New Caledonia obduction are also observed in New Guinea, which is its northward continuation but shows some additional structures, notably a complete and much thicker ophiolitic sequence.

This island consists, at least in its eastern part, of three components (Fig. 146):

— A *Southern zone,* bordering the Australian continental platform, covered by shallow-water detrital deposits. This zone is bent below the

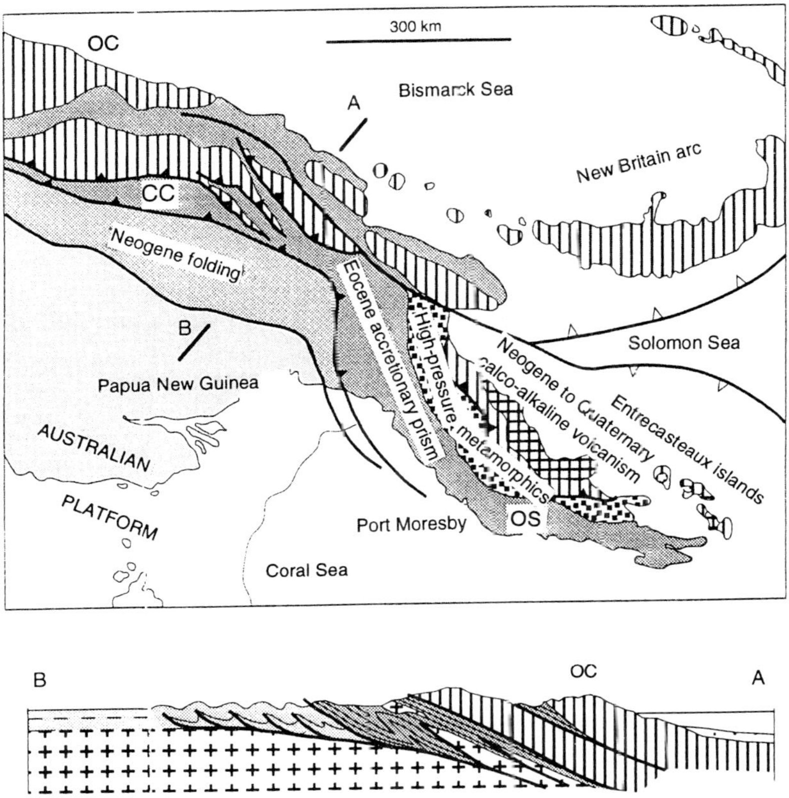

Fig. 146. Major features of the geological structure of the Eastern New Guinea chain. CC: Central chain; OC: oceanic crust; OS: Owen-Stanley chain.

chain giving rise to a flexural basin whose southern border is deformed (zone of Neogenic folds).

— A *Central zone,* mountainous, equivalent to the New Caledonia Central Massif, with a Mesozoic to Eocene sequence, metamorphosed into a high-pressure facies during the Upper Eocene.

— A *Northern zone,* which is exposed over a length of 400 km and a width of 40 km, on the NE flank of the Owen-Stanley mountain chain. It exhibits, from top to bottom, tholeiitic basalts, as massifs or pillow lavas, with a few dacites (4 to 6 km), gabbros of cumulate texture (4 km), and lastly peridotites (harzburgites), first as cumulates (about 0.5 km), then with a structure of tectonites, at times metamorphosed (4 to 8 km), which evidently represents the mantle.

The whole chain is traversed by tonalitic intrusions (diorites) dated as Eocene (50 to 55 Ma).

The age and the tectonic conditions of emplacement of this group are the same as those of New Caledonia. There also exists a relatively high-pressure metamorphism (glaucophane, chlorite, albite) below the overthrust peridotites. This is also interpreted as an effect of overloading of this layer, perhaps by pressure of fluids.

Another peculiarity of New Guinea is that the peridotite nappe is intersected at its eastern extremity by numerous volcanic flows of Neogene to the Present age that are calco-alkaline in the north and shoshonitic in the south. Their origin is still disputed but this volcanism might represent the existence of a late subduction with southern vergence, even though the oceanographic data does not exactly indicate the location of the starting point of the subduction zone. Perhaps it is an example of a 'flip' (see p. 202). This situation prompts us to try to relocate these obductions and their subsidiary phenomena in the general framework of crustal evolution of this part of the Pacific.

Structural evolution of SW Pacific

Based on the magnetic anomalies of the ocean bottom, it has been established that the Eocene was an epoch of major changes in the organisation and relative movement of plates. In fact, the Australian plate was separated from the Antarctic plate during this period and displaced northwards while the marginal basin of the Coral sea continued to open. These diverse movements led to the confrontation of the Australian margin *sensu lato* (New Guinea and New Caledonia) with the Pacific margin. This confrontation then took the form of a subduction with a SW dip, and later, at the moment of paroxysm of convergence, the form of an obduction with the same vergence. The angle of collision must have been quite oblique, which explains the numerous strike-slip faults developed later, perhaps parallel to the older transform faults emerging from the South Pacific ridge.

The reorganisation of the stress field produced on a global scale at the end of the Miocene, has given rise to a new structural framework. Around -12 Ma, a new subduction zone appeared that is offset northwards and has a Pacific, i.e., NE vergence, in the Solomon and Hebrides islands[1]. New Guinea and New Caledonia became residual arcs.

---

[1]The general origin of the Loyaute and New Caledonia islands might be related to the upwelling of the Australian plate before its plunging into the Hebrides subduction zone.

CHAPTER 4

COLLISION CHAINS

As their name indicates, these chains result from a phenomenon of collision between a continental margin and some other structure which could be:

— Either an oceanic crust, generally abnormal oceanic plateau type or, more frequently, an island arc. The resultant structure would be a *liminary chain;*

— Or another continental margin. The result would be an intercontinental collision chain or *collision chain sensu stricto* (ancient 'geosyncline' or 'biliminary' chain of some authors).

According to Cloos (1993), a collision can occur only when the objects transported by the subducted plate have sufficient dimensions. For a continental block, thickness of the crust must reach 15 to 20 km; this means that very thin continental crusts can enter into subduction, and we find that this is effectively the case. In the case of oceanic plates and volcanic margins the limiting thickness of the crust is of the order of 17 km; only thicker objects are hence likely to enter in collision. Volcanic islands of very large dimensions (height greater than 8 km) may escape subduction and engender local collisions. Most of the time they are likely to undergo subduction and may be dismantled giving rise to strong earthquakes. Volcanic arcs, having functioned for at least 20 Ma, possess a sufficiently thick crust to participate in collision.

In all cases where an object enters in collision, it results in 1—a change in the position of the subduction surface, 2—a modification in kinematics of plates, and 3—a formation of a mountain chain.

LIMINARY CHAINS

Andes of Colombia and Ecuador (Figs. 147 to 150)

These chains constitute an example of collision of an oceanic plateau in the Cretaceous. They represent the northward continuation of the Peruvian Andes and hence all the characteristic features of the latter are observed in them:

—A volcanic arc in the west. The Western Cordillera of Peru is continued by the Eastern Cordillera of Ecuador and the Central Cordillera of

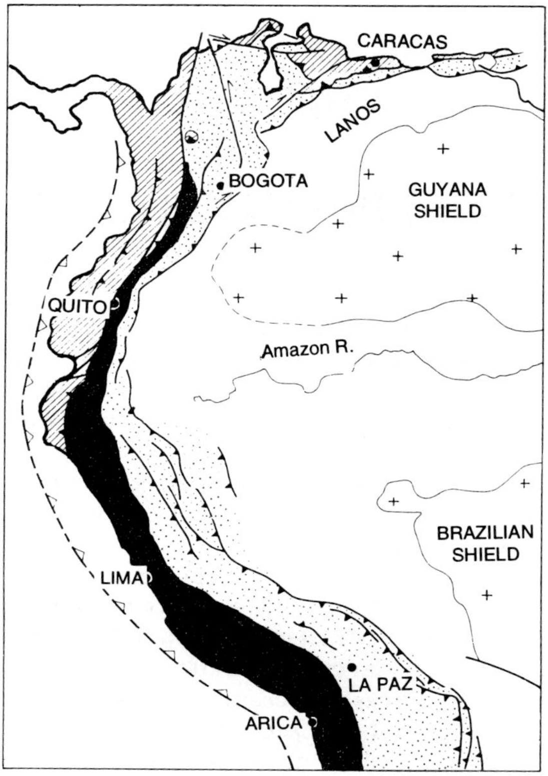

Fig. 147. North Andes (Peru, Ecuador, Colombia and Venezuela) (after Mégard, 1987, simplified and modified).
In black: volcanic arc.
Dotted: Eastern Cordillera and sub-Andean zone.
Hatched: oceanic material entered in collision with South American plate ('accreted').

Colombia. The volcanoes of this chain form the highest peaks (in Ecuador: Chimborazo 6272 m, whose peak constitutes the farthest point from the centre of the Earth due to its position below the equator and the form of the Earth; Cotopaxi 5897 m and Sangay 5230 m; Nevado del Huila 5760 m and Nevado del Ruiz 5400 m in Colombia). Calco-alkaline granodioritic plutons of several generations also exist in this belt: Jurassic (190–140 Ma), Upper Cretaceous-Palaeocene (110–60 Ma) and Neogene (15–10 Ma), the last offset westwards relative to the preceding ones, as we shall see later.

This magmatic suite occurs on a continental basement or host rock of Precambrian, Palaeozoic or Mesozoic age, often metamorphosed to a greenschist facies. The Mesozoic formation has exhibited K/Ar ages of 80 to 50 Ma (Upper Cretaceous to Palaeocene), contemporary to a 'Laramian' compression phase which has given the chain its present structure (Fig. 150).

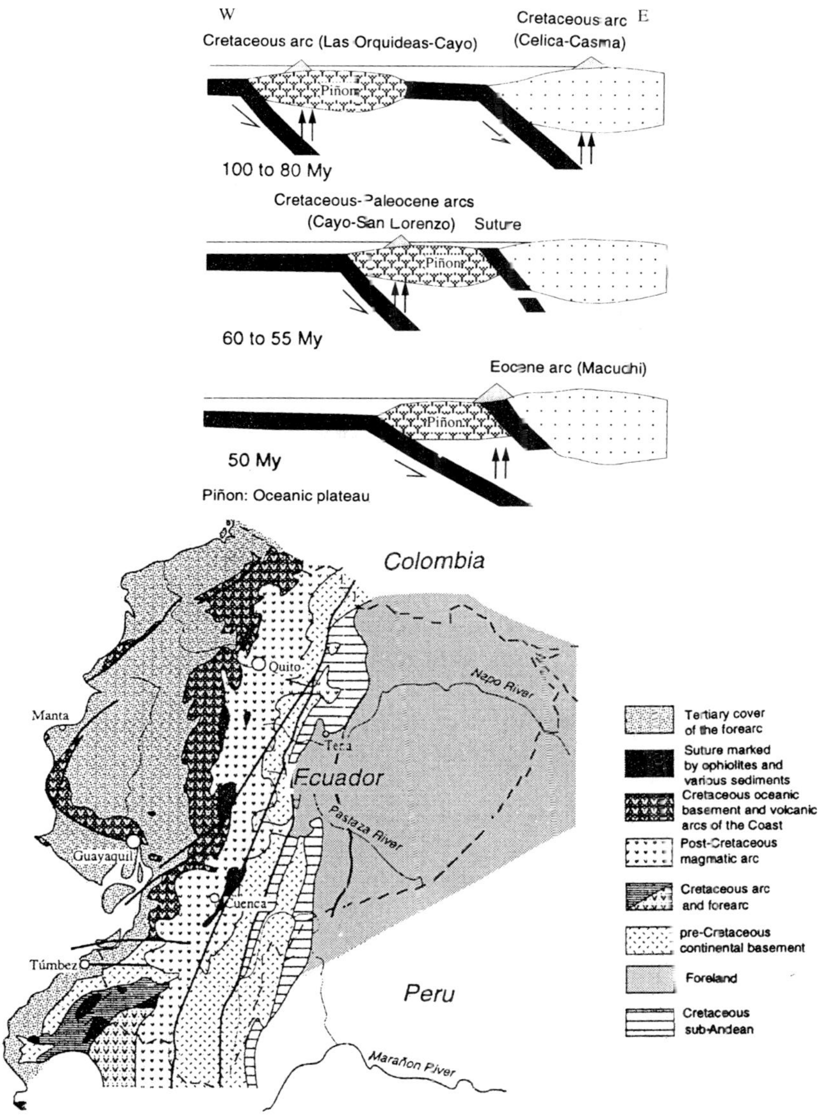

Fig. 148. Structural map of Ecuadorian Andes and interpretation of its evolution during the Upper Cretaceous

— *Intracontinental type chains* in the east, representing continuation of the Eastern Cordillera and the sub-Andean zone of Peru. These chains mark the 'sub-Andean zone' of Ecuador and Eastern Cordillera of Colombia. They correspond to migration of deformation over the flexured craton and to folding of the sedimentary cover of the back-arc basin. Sedimentary evolution shows a transition from a carbonate platform stage in the Jurassic to a subsidence basin in the Cretaceous (more than 10,000 m thick detrital sediments are deposited here). The Laramian orogeny is mainly manifest by the cessation of subsidence. The principal tectonic phase is of the Neogene period and is characterised by flat overthrusts with an eastward vergence (Ecuador) and a dual vergence (Colombia).

All these 'andean' elements (volcanic arc and chains derived from the folding of the back-arc basin) form the Ecuadorian and Colombian 'Oriente'. However, the new fact is that the chain is observed to be joined here, in the west, on the Pacific side, by oceanic material, including true ophiolites. This part of the Andes, with ophiolites, is called the Ecuadorian and Colombian 'Occidente'.

1. *'Ecuadorian Occidente'* (Western Cordillera and Coastal Zone) (Fig. 148). The largest part of the *Western Cordillera* of Ecuador is formed by a complex of intraoceanic island arcs and their basement. It mainly consists of marine pelagic and volcanoclastic sediments with Upper Cretaceous fauna. This group is covered by a flysch of Maastrichtian to Palaeocene age, which changes upwards and laterally into conglomerates with quarzitic gravels derived from the continental basement of the Eastern Cordillera. This indicates that the arcs and their basement are close to the older South American continental margin. In the Eocene a thick volcanic sequence of the arc developed comprising pillow basalts, basaltic andesites and calco-alkaline tuffs (Macuchi). This suite overthrusts the basement of the Cordillera eastwards. The contact itself is marked by very straight ultrabasite slabs (Fig. 148), a true oceanic suture zone. It has been rejuvenated by recent strike-slip faulting of the Guyaquil fault.

In the *coastal zone*, the oceanic material outcrops at some points in the form of massive pillow diabases, of the OFB type (see p. 43), gabbroic cumulates and harzburgites which represent an oceanic plateau type abnormal oceanic crust. The K/Ar ages of plagioclases in the basalts vary from 113 to 107 Ma. This complex is overlain by greywacke turbidites of Upper Cretaceous to Palaeocene age produced from erosion of intraoceanic island arcs (Cayo and San Lorenzo arcs). The suite is in tectonic contact with the Western Cordillera.

2. *'Colombian Occidente'* (The Western Cordillera and the coastal mountains known as 'Serrania de Baudo') (Figs. 149, 150).

The *Serrania de Baudo* is a continuation of the Ecuadorian coastal zone and subsides as an underthrust below the Western Cordillera. Stratigraphic data (no sediments later than Palaeocene) indicate that this phenomenon must have been produced in the Eocene but some authors consider it as Miocene; the magmatism of the Colombian Central Cordillera, which is suddenly displaced westwards in the Middle Eocene tends to confirm the Eocene age.

In the *Western Cordillera* the oceanic material forms a pile of nappes, of eastward vergence, in which the klippes extend onto the western slope of

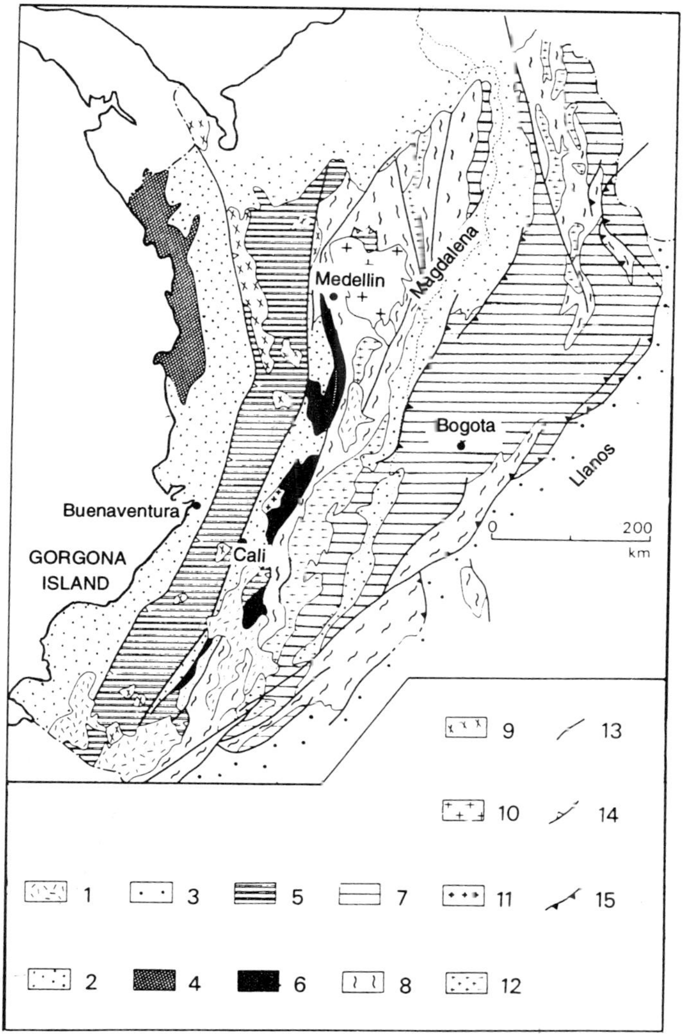

Fig. 149. Simplified geological map of Colombia (after Mégard, 1987).
1. Plio-Quaternary volcanism; 2. Plio-Quaternary sediments; 3. non-deformed Mesozoic and Tertiary cover of Llanos; 4. ophiolites of Serrania de Baudo; 5. ophiolites of Western Cordillera; 6. pre-Lower Cretaceous ophiolites; 7. Mesozoics of Central and Eastern Cordilleras; 8. Precambrian and Palaeozoic basement of the Colombian Oriente; 9. Eocene to Miocene granites; 10. Upper Cretaceous to Palaeocene granites; 11. Lower Cretaceous granites; 12 Pre-Cretaceous granites; 13. faults; 14. obduction; 15. intracontinental overthrusts.

the Central Cordillera. The easternmost units, also the oldest, are associated with glaucophane-bearing blueschists dated as Lower Cretaceous (125–104 Ma), the probable age of their formation. The most recent units were overthrust in the Upper Cretaceous-Palaeocene since their material is dated as 80 to 100 Ma and intruded by granodioritic plutons whose ages vary from 60 to 80 Ma.

Thus the evolution of the North Andean 'Occidente' is characterised by the adjunction of an oceanic material to an andean type chain.

In *Ecuador* this adjunction seems to result from the Palaeocene-Eocene accretion (= collision) of an oceanic plateau and the arcs it carries. The oceanic region situated between this plateau and the South American margin might actually be represented by the ophiolitic slices of ultrabasites of the suture (Fig. 148).

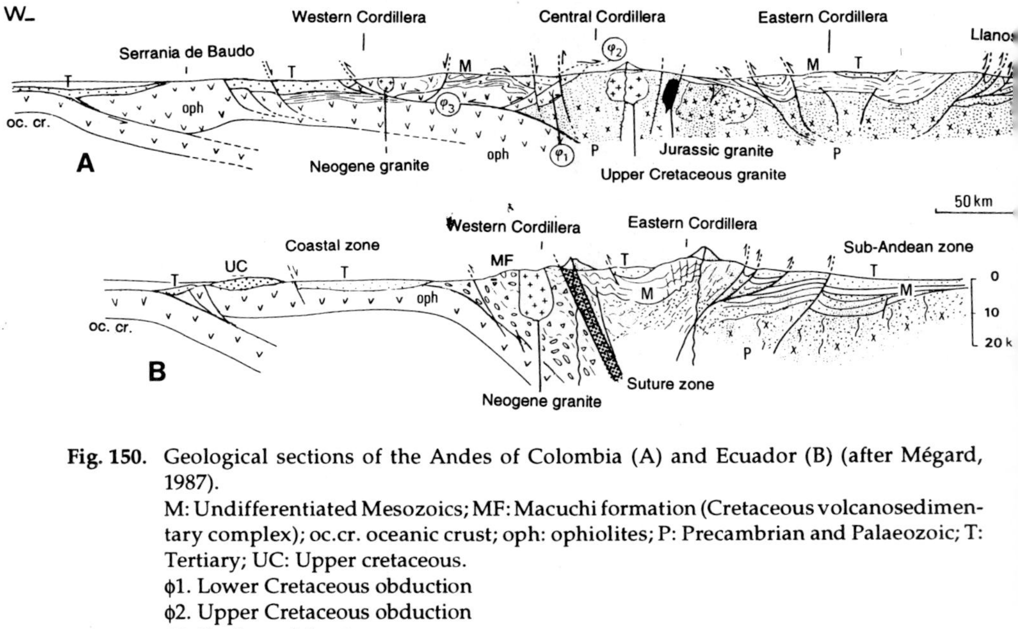

Fig. 150. Geological sections of the Andes of Colombia (A) and Ecuador (B) (after Mégard, 1987).
M: Undifferentiated Mesozoics; MF: Macuchi formation (Cretaceous volcanosedimentary complex); oc.cr. oceanic crust; oph: ophiolites; P: Precambrian and Palaeozoic; T: Tertiary; UC: Upper cretaceous.
ϕ1. Lower Cretaceous obduction
ϕ2. Upper Cretaceous obduction
ϕ3. Tertiary obduction

In *Colombia* the evolution is certainly more complex since two 'accretions' occurred in the Lower Cretaceous and Upper Cretaceous-Palaeocene, leading to the formation of thrust zones, of eastward vergence, on the American margin (obduction) followed by an underthrust, also of eastward vergence, under the American margin (subduction) of a disputed age (Eocene or Miocene).

It may be surprising that only a few island arcs responsible for accretion in Ecuador are traced in Colombia. Two interpretations are possible:

— either these arcs were subsequently displaced northwards by lateral strike-slip faults and their remnants are presently found in the Caribbean complex;

— or these arcs never existed and the ophiolitic nappes of the Colombian Western Cordillera resulted only from obduction of the oceanic material on the South American continent in the Cretaceous.

Taiwan Island (Figs. 151 to 154)

This island has resulted from the collision, presently in progress, of the Philippine arc with the Chinese (Euro-Asiatic) margin.

The Philippine archipelago is a volcanic island arc (known as the Luzon arc) which was separated from the Asian plate by the South China Sea in the course of its opening up (beginning of the Tertiary). The Manila trench is the point of departure of an eastward dipping subduction zone. It is associated with a significant accretionary prism.

Towards the north the South China Sea is found to be gradually closing. At the same time, the Manila trench becomes blurred, seismicity becomes diffuse and the volcanic arc loses all its activity. These changes result from the collision of the Luzon arc with the Asiatic continental margin. The part emerging from this collision system is Taiwan island.

The age of the collision can be inferred from conventional stratigraphic data (viz., the age of the most recent sediments of the Chinese margin overthrust by the island arc is the Miocene), from the cessation of volcanism at the northern extremity of Luzon arc (Miocene-Pliocene boundary) and lastly from the age of greenschist-facies metamorphism in the collision (5 Ma), which is also the transition between the Miocene and the Pliocene.

The orientation of regional stresses is given by the direction of movement of the Pacific plate, i.e., NW.

Taiwan Island proper consists of two parts:

1. *The Eastern Coastal Chain,* which is very narrow. It forms the northern extremity of the Luzon arc and is highly deformed. An intensive Miocene andesitic volcanism, associated with greywackes, is observed there.

This coastal chain is bounded on the west by a zone of slices backthrust eastwards, marking a major, seismically active strike-slip fault. Structural and focal mechanism analyses indicate very active compression NW or W, compounded by a sinistral transverse component.

2. *The western segment* forms the main part of the island and represents the deformed margin of the Euro-Asiatic plate. Without going into the details of its structure, it may be simply stated that the age of the formations and the age and intensity of deformation in general decrease westwards, which is in accord with a collision process produced from east to west.

The interior of this segment, and especially the *Central Chain* (which locally exceeds an elevation of 4000 m), comprises a group of thrust slices, part

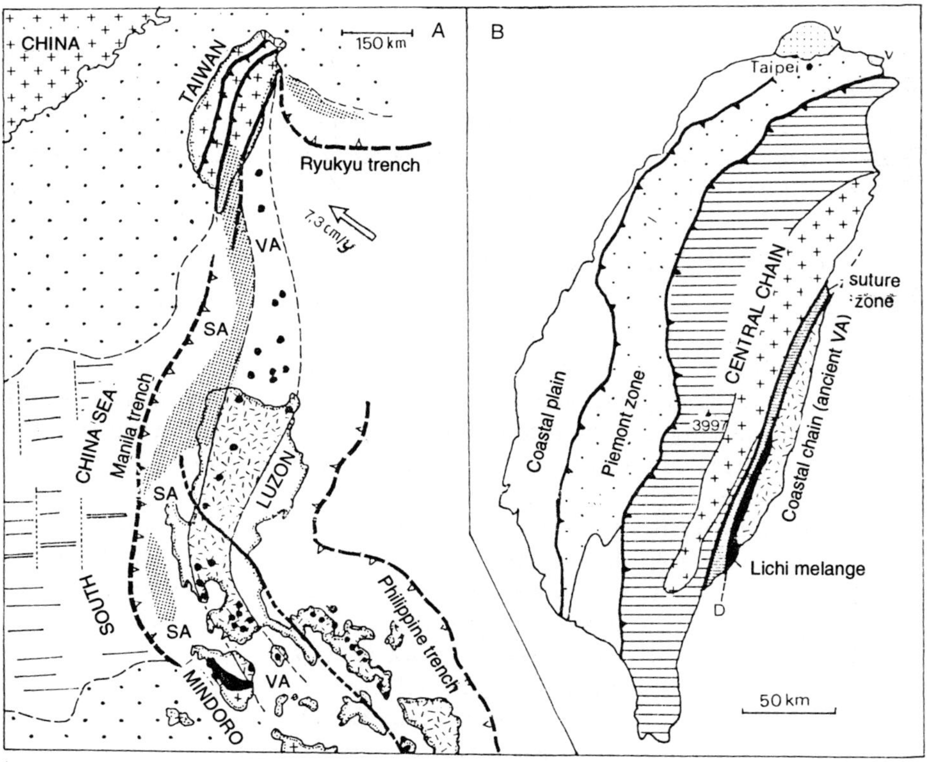

Fig. 151. Taiwan Island and its geological framework (after Pelletier, 1985, simplified).
A—Position of Taiwan relative to the Philippine arc and the South China Sea (SA: Sedimentary arc = accretionary prism; VA: volcanic arc; ophiolites of Mindoro Island in black).
B—Major constituent units of Taiwan (D: Strike-slip faulting of the suture zone; v: Quaternary andesitic volcanism in the northern edge of the island, associated with the Ryukyu active margin system). In the Central chain the pre-Tertiary basement and its sedimentary cover are distinguishable.

of which belongs to the basement and sedimentary cover of the Chinese margin; the other represents the emergent and folded part of the Manila accretionary prism, preceding the Luzon arc. It consists of turbidites, becoming more and more shallow-water facies up to the end of the Tertiary, which were not disturbed until the Quaternary.

The Western Coastal Plain is an especially important sector because deformation is still continuing there; the alluvial and marine terraces are tilted and deformed and seismicity intense.

Structuring of Taiwan island commenced in the Plio-Pleistocene and is still continuing, as mentioned above, and indicated by intensive seismicity. This is associated with a metamorphism which extends up to blueschist and

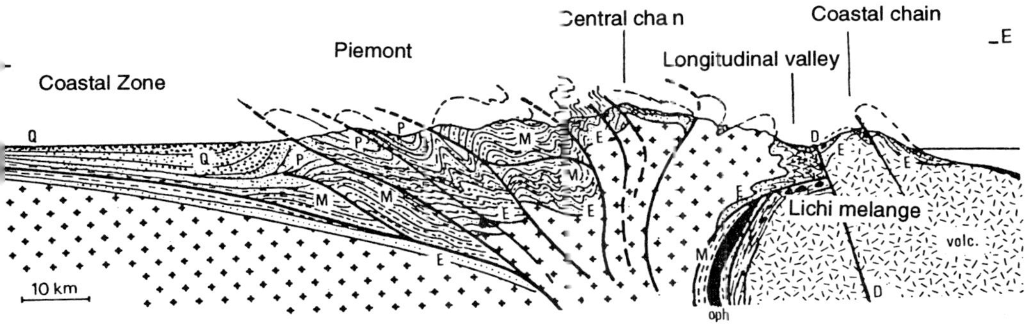

Fig. 152. Cross-section of Taiwan Island (after Pelletier, 1985).
D: strike-slip faulting of the longitudinal valley; E: Eocene; M: Miocene and Upper Oligocene; oph: ophiolites (suspected); P: Pliocene; Q: Quaternary; volc: volcanic complex representing the northern edge of the Luzon arc.

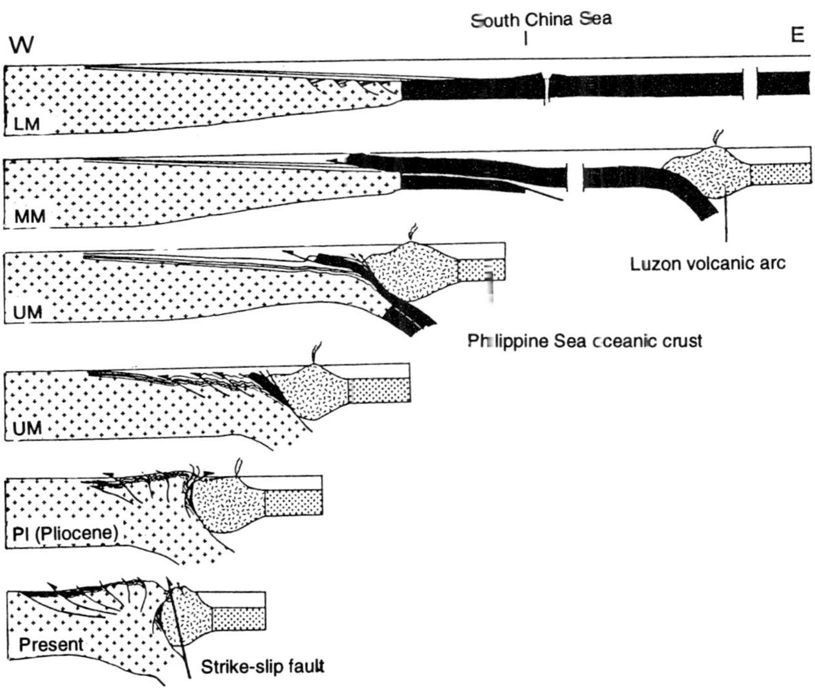

Fig. 153. Evolution of the Luzon-Taiwan system since the Lower Miocene (LM) (after Pelletier, 1985, simplified). MM: Middle Miocene; Pl: Pliocene; UM: Upper Miocene.

greenschist facies in the Central Chain, radiometrically dated as end of the Miocene (5 Ma).

The absence of ophiolites is noticed in the suture zone. These rocks may have existed but were destroyed by erosion because their traces have been

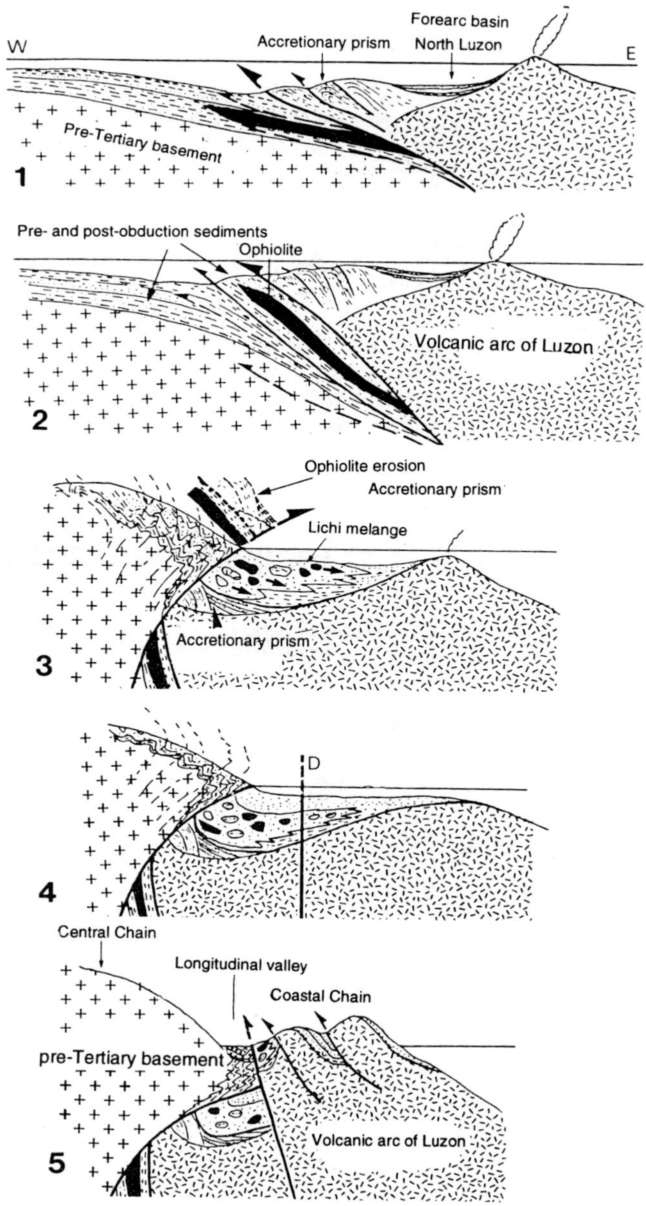

Fig. 154. Evolution of the suture zone during collision of Taiwan with the continent of China (after Pelletier, 1985, simplified).
1 and 2. Upper Miocene: commencement of collision; 3. Middle Pliocene: backthrusting and origin of the Lichi melange (with ophiolitic blocks); 4. End Pliocene-Lower Pleistocene: sinistral transverse faulting of the coastal chain; 5. Middle Pleistocene-Present: structuring of the coastal chain. D: strike-slip fault.

found in olistolites in the sediments that constitute the backthrust slices on the Eastern Coastal Chain. Nonetheless it is also probable that no major ophiolitic nappe ever existed.

Ophiolites are exposed only in the other extremity of the Philippine arc (Mindoro Island).

Finally, it may be noted that the vergence of the structures is oriented towards the continent, as in the case of the Colombian example. However, this is not a general trend. We find an example in the case of California where the vergence of collision structures is towards the ocean. For this reason, some authors oppose the 'Taiwan type' and 'California type' collisions. In fact, it seems that the general inclination of the slices is the same as that of the precollision subduction plane or, in other words, the structures are mainly of the synthetic type.

West American Pacific Chains

The mountain belt of West America is a heterogeneous region in which the following chains can be distinguished from east to west:

— *Rocky Mountains sensu lato*, a group of intracontinental chains resulting from a Palaeocene-Eocene (Laramian) deformation in the western margin of the North American platform. They are hence equivalent to the South American Eastern Cordilleras.

Several island arcs have been accreted with these mountains always on their western side:

— *Sierra Nevada-Klamath Mountains axis*, at the end of the Jurassic (Nevadan orogeny).

— *Coastal Chains (Coast Range)* in, Middle Cretaceous (Oregonian orogeny).

Sierra Nevada-Klamath Mountains axis (Fig. 155)

This range consists of Palaeozoic, Triassic and Jurassic oceanic sediments (radiolarites and greywackes), ophiolites and calco-alkaline volcanics, in other words an association characteristic of island arcs. The group was folded and sliced with an eastward vergence at the end of the Jurassic (because Tithonian and Lower Cretaceous molasse are observed here unconformably overlying the preceding formations—known as the Knoxville formation—and blueschists produced in this collision and dated 150 Ma). Thus there was a collision in this epoch between a block containing at least one island arc and the North American margin. This collision implies the disappearance, by subduction, of the oceanic region that separated them. However, an ophiolitic suture is not observed. It is assumed to be masked by subsequent overthrusts. Ophiolites are visible in the Nevadan complex; however, they are older or indicate a Mesozoic

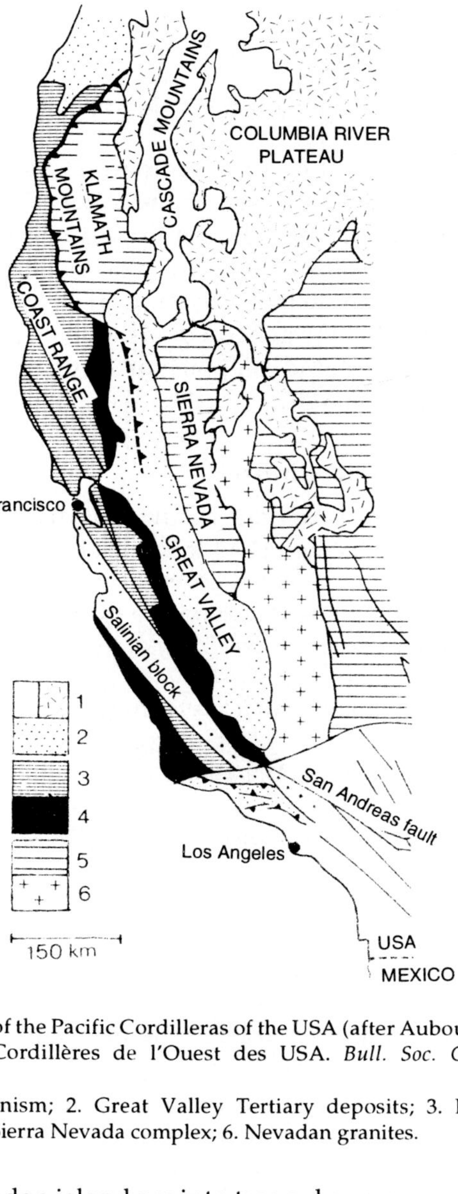

Fig. 155. Structural map of the Pacific Cordilleras of the USA (after Aubouin et al. (1986) Traits généraux des Cordillères de l'Ouest des USA. *Bull. Soc. Geol. France*, p. 741, simplified).
1. Recent volcanism; 2. Great Valley Tertiary deposits; 3. Franciscan complex; 4. ophiolites; 5. Sierra Nevada complex; 6. Nevadan granites.

break-up of the Nevadan island arc into two subarcs separated by a narrow marginal basin (Fig. 156).

The eastward vergence of all the Nevadan structures leads to the conclusion that the end-Jurassic subduction must have had a westward dip. The collision was followed by the intrusion, in the Nevadan arc, of a number of small granodioritic bodies (dated 145–130 Ma) derived from

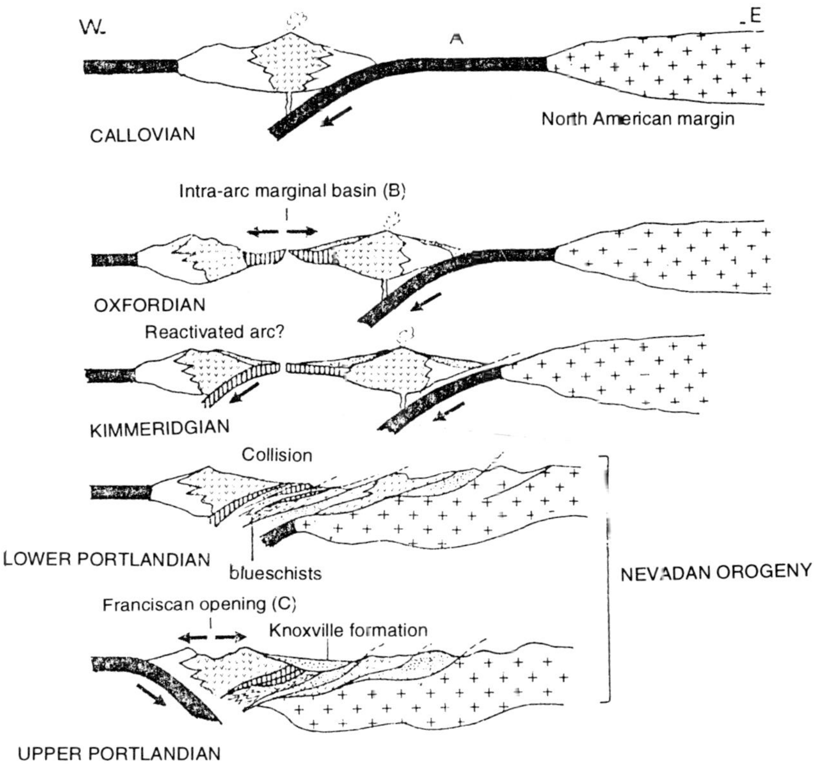

Fig. 156. Possible evolution of the Nevadan complex (Nevadan collision and orogeny) (after Lagabrielle et al., 1986).
A, B, C. Marker points (see text p. 227).

melting associated with crustal overthickening. These plutons were subsequently cut off from their roots by the Middle Cretaceous truncation planes (Fig. 157).

These late shear zones make the Nevadan structures difficult to identify. They have an eastward vergence in Sierra Nevada (but their western boundary is not seen because masked by the Cretaceous and Tertiary molasse of the Great Valley) and a westward vergence in the Klamath Mountains (but their eastern boundary is not observed because covered by the large and recent volcanic plateau of the Columbia River or the recent volcanism of the Cascade Mountains).

Nevertheless, it can be assumed that starting from the Middle Cretaceous, the Sierra Nevada-Klamath Mountains complex, incorporated into the American margin, had become a dual-vergence coastal chain, below which an eastward dipping Pacific subduction zone started to develop (Fig. 157). In fact, a calco-alkaline volcanism of this age exists in the Baja

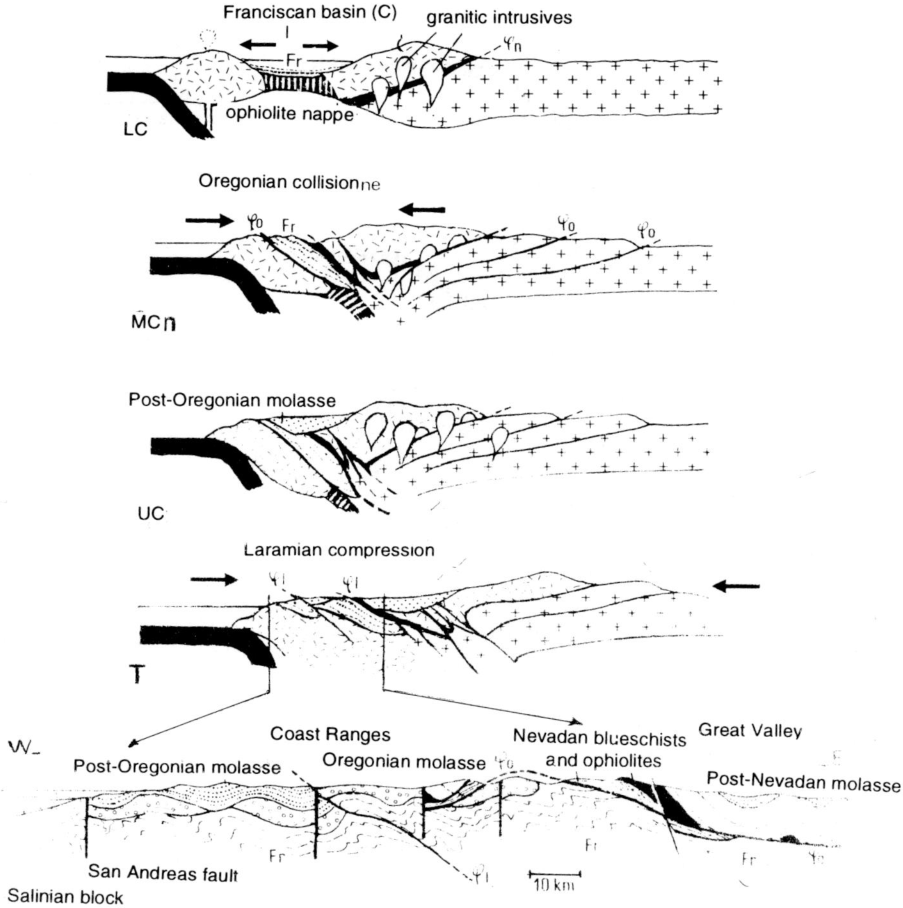

Fig. 157. Evolution of Californian ranges in the Cretaceous and beginning of the Tertiary. This evolution follows that shown in the preceding Figure.
LC, MC, UC: Lower, Middle and Upper Cretaceous; T: Tertiary; ϕn, ϕo, ϕl: Nevadan, Oregonian, Laramian thrusts.
Bottom: Present cross-section of the Coast Ranges north of San Francisco (after Roure et al., 1986). Fr. Franciscan complex

California and Sonora mountains. On the contrary, the same is absent in Oregon and California where, however, some intrusive plutons of 120 to 130 Ma are observed.

Laramian orogeny has been mainly active in the Rockies farther east. Here it is represented by some minor overthrusts and especially granitic intrusions, indicating a new crustal overthickening of the Nevadan arc. This, in any case, is a purely intracontinental deformation.

Coast Range Chains

These mountains consist of a pile of sedimentary units of westward vergence, hence dipping eastwards, which plunge below the Klamath Mountains and the Sierra Nevada range. These units are covered by molasse of a differing age and have been divided into two groups: the *Franciscan complex* in the west (name derived from San Francisco) and the *Great Valley series*, with an ophiolitic basement (Coast Range ophiolites), which is overthrust on the Franciscan complex (Coast Range thrust) (Fig. 157).

The Franciscan complex, very heterogeneous, exhibits the following units from top to bottom:

— *Blueschists* (of sedimentary or volcanic origin), produced from Nevadan orogeny but occurring here in Cretaceous structures of western vergence.

— *Detrital sequence (greywackes)*, less metamorphosed, palaeontologically dated Tithonian to Lower Cenomanian. Its metamorphism (light blueschists) is of Middle Cretaceous age (110–80 Ma). The base of the sequence shows pillow lavas and radiolarites: it thus represents sedimentary filling of an oceanic basin opened in the end-Jurassic (Fig. 156).

— An *olistostrome with serpentinite-clay matrix*, attributed to terminal Cenomanian-Coniacian. It contains blocks of all sizes, resulting from disintegration of preceding units at the time of their westward movement during Middle Cretaceous orogeny and its recurrences at the beginning of the Upper Cretaceous.

Post-Cenomanian molasse lie atop the three preceding suites, thus indicating the existence of a Middle Cretaceous orogeny (known as Oregonian). These molasse themselves are intersected by Laramian thrusts (Palaeocene-Eocene) or more recent strike-slip faults.

Ophiolites of the Great Valley: Ophiolites mark the eastern border of the Coast Range extending over a length of more than 600 km and overthrust westwards. They are unconformably overlain by a volcanodetrital molasse (Knoxville formation) dated by Upper Jurassic-Lower Cretaceous fossils (Fig. 156). This formation constitutes the beginning of the thick conglomerate sequence of the Great Valley, produced from the destruction of the Nevadan arc and of Cretaceous-Tertiary age. These ophiolites evidently represent the Nevadan material covered by an Upper Jurassic-Lower Cretaceous molasse, both subjected to Oregonian thrusting of westward vergence and later overlain by Cretaceous-Tertiary molasse of the Great Valley.

Interpretation: This region, exhaustively investigated by a number of authors, is a good example of emergence of ideas in terms of different theories. In effect, with the advent of the plate tectonics theory, the Californian ophiolites were considered the remnants of a Pacific oceanic crust that disappeared by the effect of an eastward-dipping subduction below North America. In this interpretation the Coast Range was regarded as a typical accretionary prism with tectonic 'melanges' and high-pressure metamorphism.

Further, the subduction-blueschist association was detected here for the first time. The molassic basin of the Great Valley itself represented a typical example of a 'forearc basin' (see p. 159).

This model appears too simplistic because it has subsequently been learned that the Franciscan sediments are not the cover of a Cretaceous Pacific crust, but the filling of a marginal type basin occurring between the fossil Nevadan arc and a new intraoceanic arc situated farther west (Fig. 157). The existence of the latter is justified by the presence of calco-alkaline detrital products in the westernmost Franciscan sediments.

The question arises: what happened to this arc and what was the nature of its basement—oceanic or continental? Geophysical data indicate that a continental crust exists in the Californian range, below the Franciscan complex, which might represent an autochthon relative to the latter. Perhaps this same crust reappeared on the coast, south of San Francisco, in the 'Salinian' block (Figs. 155, 157) which might represent the remnant of the missing arc, overthrust by the Franciscan complex. But it is also possible that the major part of this arc is no longer opposite to California and now occurs in the 'exotic blocks' attached to the Alaskan plate under the effect of major strike-slip faults of the San Andreas system (p. 287).

Whatever be the fate of that arc, at the present moment the Coast Ranges are considered the consequence of an Albian collision occurring at the end of subduction of the marginal basin for which they represent the sedimentary filling.

The Californian chains thus show two successive collisions, the second involving a part of the material already deformed by the first. This model is even more complex because the Nevadan arc, which enters into contact with the North American margin in the Kimmeridgian, was itself formed by the suturing of older arcs.

In conclusion, the history of the Pacific chains of North America, at least in the region of California, can be described as follows, starting with the Callovian (Figs. 156, 157):

In the *Callovian* an island arc, viz., the Nevadan arc, is built up above a westward-dipping subduction zone, probably bordering a microcontinent formed from the suturing of several arcs. It is separated from the North American plate by an oceanic region (point A of Fig. 156).

In the *Oxfordian* this island arc separates into two parts, thus causing the appearance of a marginal basin (point B of Fig. 156) between the two ridges; the eastern rise alone seems to have had volcanic activity.

In the *Kimmeridgian*, at the end of disappearance of oceanic region A, collision occurred between the eastern arc and the American margin. The corresponding ophiolitic suture is at present concealed. Oceanic region B, separating the two rises, is compressed, giving rise to ophiolitic slabs (associated with blueschists) on which the Knoxville molassic formation is deposited. The Nevadan arc is sutured to the North American continent into a structure with an eastward vergence.

A new orogenic cycle starts in the *Neocomian*. A Pacific subduction zone with eastern vergence is produced below the Nevadan arc. It opens there a new oceanic basin (point C of Fig. 156), viz., the Franciscan basin, associated with an island arc.

During the *Albian*, at the end of disappearance of basin C, the arc comes into collision with the older Nevadan arc, giving rise to overthrusts with westward vergence of the Coast Ranges. This is the Oregonian orogeny which, in fact, corresponds to a contraction of the structure formed by the Nevadan arc and the Coast Ranges. This structure acquires a dual inclination disposition.

A post-Oregonian orogeny (sometimes known by an obsolete term, 'subhercynian') in the *beginning of the Upper Cretaceous*, and the Laramian movements in the *beginning of the Tertiary*, represent new contractions which bring about changes to a structure that was earlier purely continental. These latter effects are further complicated by the disintegration due to longitudinal strike-slip movements of the San Andreas system.

INTERCONTINENTAL COLLISION CHAINS

These chains resulted from the collision of two continental margins previously separated by a region of oceanic crust. Most of the present chains of this type are found along the orogenic axis running from Gibraltar to Burma, or the Tethys axis, which was long interpreted as due to collision of the African and Indian blocks on the one side, and the Euro-Asiatic block, on the other, even before the lithospheric plates were defined (Argand, 1924).

As the origin of these chains implies the closure of an oceanic region, their evolution comprises two stages, viz., disappearance of this region and collision proper.

1. *Disappearance of the oceanic region*, inevitably pertaining to one of the plates involved, occurs by the subduction of one below the other. This stage leaves its traces behind: usual volcanic and metamorphic phenomena on the margins, ophiolitic nappes or thrust slices.

2. *Collision of the margins*: When the two margins actually come into contact, two cases are possible:

— The process of the plates approaching each other is blocked and transferred to another margin of the plate. The chain is then reduced to ophiolitic slices wedged between the least deformed continental margins (the 'coloured melange' zones of central Iran) (Fig. 158).

— The process of approaching is not stopped anywhere. Pressure is maintained in the contact zone and the two margins are shortened by undergoing cleavage into granito-gneissic slabs, most probably at the boundary of the surficial crust and the deep crust because their thickness rarely exceeds 15 km. Peridotite sheets are exceptional and indicate thinning due to other causes. These gneissic sheets are piled up one above the other, giving rise to a 'crustal root', i.e., a thickening of the continental crust that might reach 50 to 70 km. These roots, being relatively lighter, generally trigger isostatic uplift of the chain.

It has been observed that the intercontinental collision chains very often have an *arcuate form*. The group of peri-Mediterranean chains is a good example for the same with its seven successive arcs (Fig. 159).

The origin of these arcuate structures is probably due to several factors:

— either compression of a curved palaeogeographic region, in which case it is known as an *'inherited arc'*;

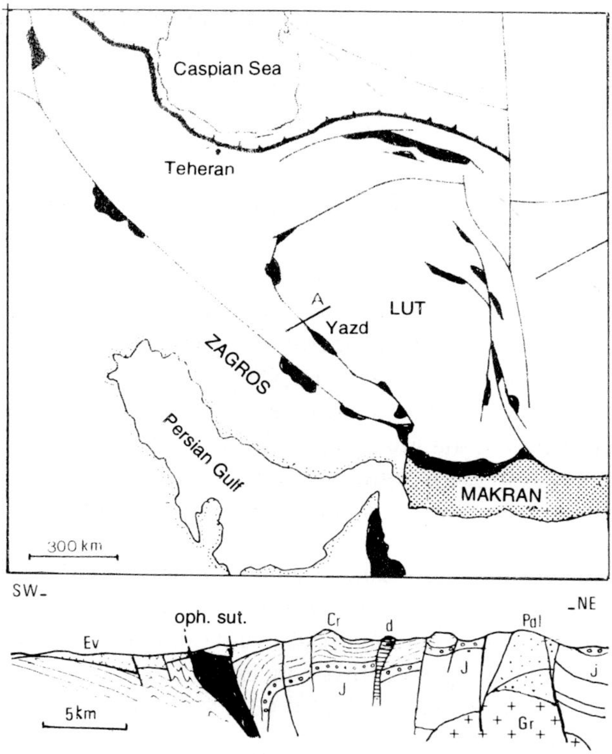

Fig. 158. Ophiolitic sutures of Iran (after Desmons (1982) *Geol. Alpine Grenoble*, 58: 23, simplified).
Cross-section taken from north-western Yazd (A).
Cr: Cretaceous; d: Quaternary dacite vein; Ev: Eocene volcanics; Gr: Mesozoic granite; J: Jurassic; oph. sut: ophiolitic suture; Pal: Palaeozoic.

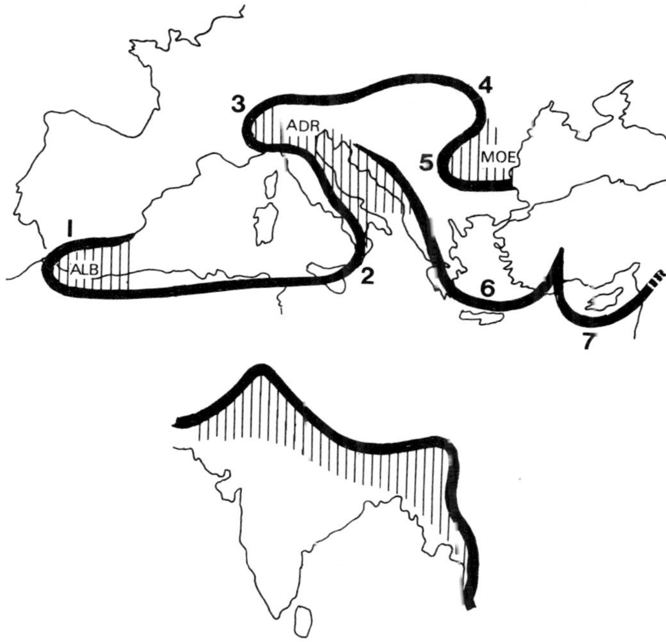

Fig. 159. Arc-shaped pattern of collision chains. Examples of Mediterranean and Himalayan belts.
1. Betica-Rifain arc (ALB. Alboran promontory); 2. Sicilian-Calabrian arc; 3. Alpine arc (ADR. Adriatic promontory); 4. Arc of North Carpathians; 5. Arc of Eastern Carpathians (MOE> (Moesian promontory); 6. Aegean arc; 7. Taurian arc.

— or moulding of a promontory of one or two of the margins present, which indents in the opposite margin. An arc due to such an impact, known as an *'induced arc'*, is thus produced, which may also be perfectly superposed on an inherited arc.

The alpine arc, for example, moulds the *'Adriatic'* (or 'insubric') promontory of the African plate; the arc of the Eastern Carpathians moulds the Moesian promontory of the Euro-Asiatic plate (Fig. 159). Also, sometimes the chain moulds the margin of the entire plate, as in the case of the Himalaya around the northern boundary of the Indian plate between its two 'virgations' of Assam and Kashmir (Fig. 159).

In summary, the intercontinental collision chains, often curved, show two major characteristics:

1. An *ophiolitic zone*. If this zone is narrow, it is called an ophiolitic suture (Iran, Himalaya): it enables us to mark the position of the ancient oceanic region that has disappeared. But this zone may be enlarged by the effect of stretching into a nappe of ophiolite-bearing units, which may even become allochthons and be separated from their zone of origin. Determination of the

position of the ancient oceanic region then becomes much more difficult (as in the case of the Alps).

2. *Piled up crustal slabs.* Very diverse modes may be particularly observed in this case. The crustal cleavage may in effect affect:

— only one margin (Nepalese Himalaya) (Fig. 160);

— two margins with opposite vergences (Western Himalaya-Pamir, Hellenides-Balkans pair, European Variscan chain);

— two margins but with the same vergence (one margin overthrusts the other). Such a structure is observed in the Alps, which is referred to as 'hypercollision'.

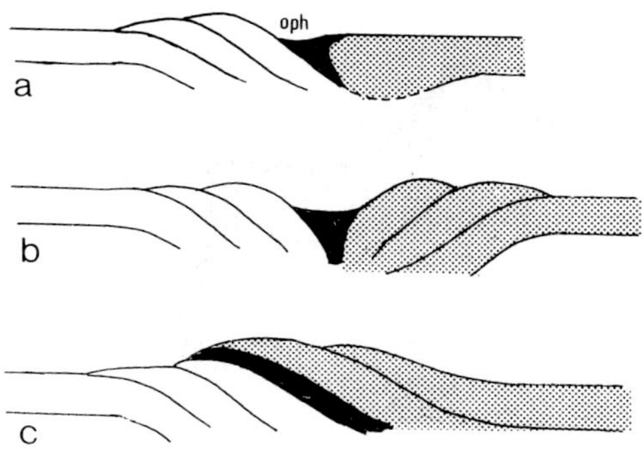

Fig. 160. Some types of collision chains.
a. Nepalese Himalaya; b. Western Himalaya-Pamir; c. Alps.
oph. ophiolitic zone

EXAMPLES

Central Himalayan Chain

Constituent units (Figs. 161, 162)

Along the central Nepal (Khatmandu) meridian, this chain exhibits, from south to north, the following clearly distinguishable units:

1. *The Siwalik Range* forms a series of chains consisting of Neogene to Quaternary molassic type sediments. They represent a group of slices overthrust southwards on the Quaternary formations of the Ganges basin along the 'frontal thrust' (MFT or MST, 'Main Frontal Thrust' or 'Main Siwalik Thrust').

2. *The Lesser Himalaya* are thrust over the Siwaliks along the 'Main Boundary Thrust' (MBT). They show essentially Precambrian, and in places younger sediments, which form a sequence of multideformed nappes. The main deformation of this group is characterised by an essentially horizontal regional cleavage and a stretch lineation nearly perpendicular to the strike of the range.

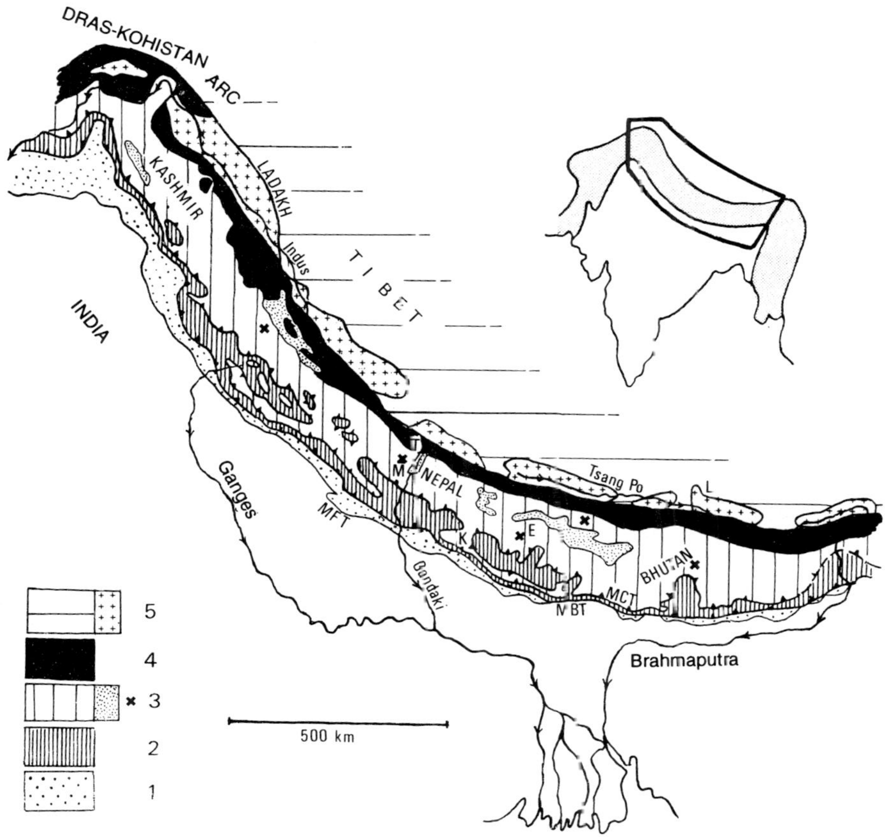

Fig. 161. Schematic structural map of the Himalaya.
1. Siwaliks; 2. Lesser Himalaya; 3. Greater Himalaya (crosses recent leucogranites; dotted: Plio-Quaternary basins); 4. Indus-Tsang Po ophiolitic suture; 5. Transhimalaya (small crosses: Ladakh batholith).
E: Everest; K: Khatmandu; L: Lhassa; M: Manaslu; T: Thakkola graben.

The intensity of metamorphism increases towards the top of the sequence (regressive metamorphism), reaching the amphibolite facies.

3. *The Greater Himalaya* are overthrust on the Lesser Himalaya along the 'Main Central Thrust' (MCT), continuously extending all along the chain. The thrust consists of a Precambrian to Cambrian basement forming a 5-km thick slab known as the 'Tibetan slab'. This is overlain by a more or less complete, 10,000 m thick sedimentary sequence ranging from the Ordovician to the Lower Eocene. Everest mountain (8848 m), at the top of which occur the Ordovician limestones, belongs to this sequence. The facies are those of a subsident continental platform.

Metamorphism decreases upwards in the sequence, from amphibolite facies at the base (6 to 8 kb, 650 to 750°C), and reaches varied horizons

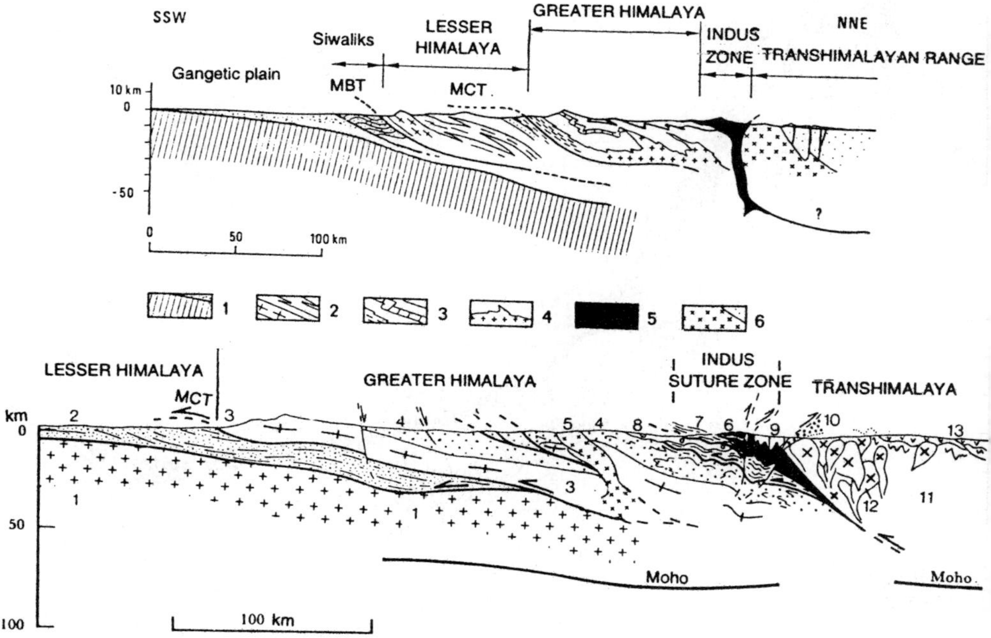

Fig. 162. Simplified cross-sections of the Himalayan chain.
Top: general section (after Bassoullet et al., 1984, simplified).
1. Indian crust and its sedimentary cover; 2. Lesser Himalaya (gneissic basement and its sedimentary cover); 3. Greater Himalaya (gneissic basement and its sedimentary cover); 4. Manaslu type leucogranites; 5. Indus-Tsang Po ophiolitic suture; 6. Transhimalayan range (Ladakh batholith, in small crosses).
Bottom: cross-section of the central zone (after Burg et al., 1987, in *Tect. Setting of Region. Metam., Phil. Trans. R. Soc.*, London, A 321: 67–86, simplified).
1. Indian basement; 2. gneisses and schists of Lesser Himalaya; 3. 'Tibetan slab' (metamorphic basement of Greater Himalaya); 4. Sedimentary cover of the Tibetan slab; 5. Recent leucogranites; 6. ophiolites; 7. oceanic sediments associated with ophiolites; 8. flysch, unconformable with exotic blocks; 9. Neo-Cretaceous to Palaeocene turbidites; 10. Eocene conglomerates; 11. Transhimalayan basement; 12. Ladakh batholith (Upper Cretaceous-Palaeocene); 13. Palaeogene calco-alkaline volcanism of Transhimalayan range.

depending on the region: Ordovician in Central Nepal, Upper Cretaceous in Kashmir.

The tectonic structures show a succession of nappes or folds inclined S or N, resulting from a polyphase deformation (Fig. 163).

In the Greater Himalaya, the existence of *inner crystalline massifs* has been recognised very recently (Guillot et al., 1995). The latter are represented in the western part of the chain by Tso Morari and Gurla Mandata domes.

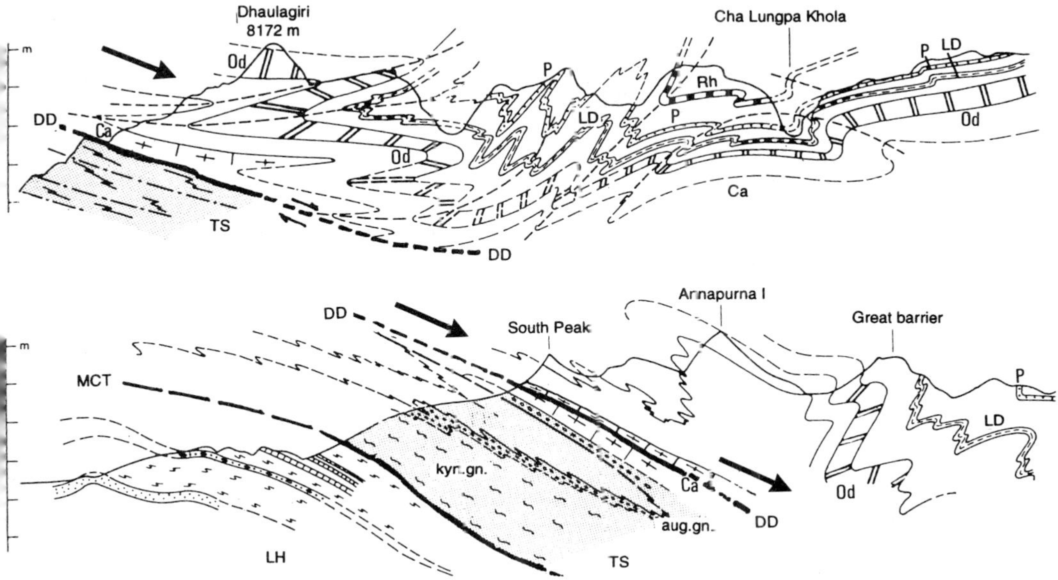

Fig. 163. Tectonic style of the Tibetan slab and its sedimentary cover in the Dhaulagiri and Annapurna blocks (Nepal Himalaya) (after Lefort and Pêcher (1989) *J. Metam. Geol.*, 7: 31–42).
Ca: Cambrian Larjung limestone; LD: Lower Devonian (blackschists); LH: different formations of the Lesser Himalaya; Od: Ordovician Nilgiri limestone; P: Permian (quartzites); Rh: Rhetian (pink and white sandstones); TS: Tibetan slab (kyn. gn: kyanite gneiss; aug. gn: augen gneiss).
Black arrows indicate the gravitational movement of sedimentary cover of the Greater Himalaya northwards, above the decollement surface (DD).

They comprise elements of the Indian margin which underwent metamorphism to low-temperature eclogite facies (17 kb, 550°) at Tso Morari and which were exhumed very early, before the Oligocene, isolating ophiolitic klippes (Amlag La nappe).

4. *The Indus-Tsang Po suture zone* is the ophiolitic region, marked by a depression along which flows the upper course of the Indus River in the west and the Tsang Po River in the east. One can observe here a suite of multideformed dense thrust slices, sometimes vertical, comprising flysch-facies units (derived from the base of the ancient Indian margin), radiolarite slabs, blueschists, peridotites, greywacke flysch units and continental conglomerates. With the exception of the last two members, all these facies are also found farther south, as klippes floating on the Greater Himalayan sequence in two regions (Ladakh and Amlag La).

This zone obviously consists of the remnants of an ancient oceanic region that disappeared by subduction, later deformed by collision of its margins.

5. *The Dras-Kohistan arc sequence* (Figs. 161, 165): West of the chain (Kashmir, Kohistan), the suture zone is divided into two parts by a thick suite of basic magmatic and volcanic rocks of tholeiitic nature and highly deformed. It represents an ancient intraoceanic arc trapped by Tertiary collision.

6. *The Transhimalayan range* consists of magmatic and volcanic formations of calco-alkaline tendency and Upper Cretaceous to Eocene age, emplaced on the margin of a continental region. The latter, southern Tibet, shows a metamorphic basement overlain by a sedimentary sequence generally deposited at shallow depth, which emerged at the beginning of the Upper Cretaceous.

Tectonic history (Fig. 164)

1. At first the oceanic region disappears by the coming together of the Indian block and the Euro-Asiatic margin because of the subduction of the Tethys ocean below the future Transhimalayan range. This part of evolution unfolds starting from the Middle Cretaceous up to the Palaeocene-Lower Eocene, the age of collision. The Transhimalayan magmatism, of Cretaceous-Eocene age, is typical of an active margin.

This subduction affects the thinnermost parts of the Indian margin which undergoes a low-temperature eclogite facies metamorphism (inner crystalline massif of Tso Morari).

2. At the end of this process a collision occurs between two continental margins, one active, in the north (Tibet), and the other passive, in the south (Indian plate). Thus the Indus suture zone is formed whose content is imbricated and ejected southwards as nappes and then backthrust northwards. The age of collision is Eocene (52 Ma), during which the rate of sea-floor spreading of the Indian ocean falls from 10 to 5 cm/yr. Starting from the Middle Eocene, major mammiferous fossils also become common to India and Mongolia: the sea gap might have disappeared or was sufficiently reduced so that it was no more an obstacle for the migration of these animals.

3. At this stage the Indo-Eurasia block is sutured but compression continues with the spreading of the Indian Ocean. Rethickening of the Indian margin occurs by shortening and slicing of the sedimentary cover and probably reversal of basement fractures; units of eclogite and blueschist facies are brought to outcrop. Then a cleavage is produced in the middle of the Indian continental crust along a major flat shear zone. Thus, the Tibetan slab is formed, overthrusting southwards. This shear zone is situated at the base of the granito-gneissic layer[1]. This feature is compatible with the thickness of the Greater Himalayan block and with the microstructures, as well as

[1]But the behaviour of deep crust remains conjectural (see Fig. 165).

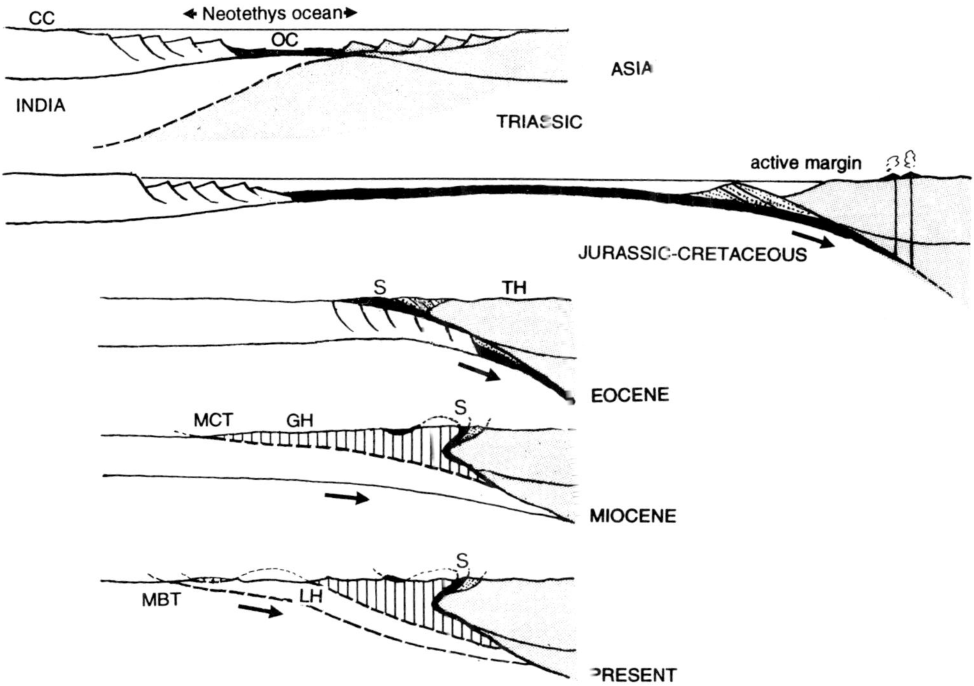

Fig. 164. Evolution of the Himalayan region since the Triassic.
In the Triassic a crustal fracture is opened up in Gondwana continent, resulting in the formation of two passive continental margins, viz., the Indian margin in the south and the Asiatic margin in the north. The Neotethys is thus created.
In the Jurassic the Neotethys ocean is opened and sinks in the Cretaceous by subduction below the Asiatic margin, which became an active margin, with accretionary prism and magmatic arc.
In the Eocene the Neotethys ocean is completely absorbed: the two margins enter into collision. A little oceanic material is preserved in the suture zone (S) or overflows on the Indian margin by an obduction phenomenon.
In the Miocene continuous pushing of the Indian plate under the effect of the opening of the Indian ocean leads to cleavage and thickening of the Indian crust. Thus originates the Greater Himalaya (GH) slab above the Main Central Thrust (MCT).
At Present, under the effect of the same stresses, a new cleavage zone operates farther away, because the friction forces do not allow further underthrusting of the Indian plate below the MCT. A new slab, viz., the Lesser Himalayan block (LH), has hence been formed above the Main Boundary Thrust (MBT), below which the Indian plate continues to sink.
CC: continental crust; OC: oceanic crust; S: suture; TH: Transhimalaya.

with the metamorphism observed at the base of this slab (temperature and pressure conditions indicated earlier).

Plastic deformation is very intensive. The MCT zone is especially represented by a suite of stretched schists, several hundred metres thick, having a major lineation parallel to the direction of displacement (almost N-S). Slaty cleavage affects the two parts over an average thickness of 10 km, on either side of the MCT, evidently with its maximum in the latter.

Displacement of the Tibetan slab was accompanied by a supply of heat responsible for metamorphism of its basement, as the base of this slab was already at high temperature. The basement metamorphism shows a reverse disposition of isograds, the hottest being on the top ('hot iron' effect). This has sometimes been interpreted as the result of immense recumbent folding with preserved reverse limb.

Moreover, the overthrusting of a hot block on a relatively colder one liberates a certain quantity of fluids which cross the MCT and induce partial anatexis of the Tibetan slab, resulting in the emplacement of small massifs of leucogranites (Manaslu, for example).

The *age* of this major thrust has been established from radiometric data, for want of stratigraphic markers. The minerals of metamorphism give ages of 10 to 25 Ma (Miocene). The emplacement of the Tibetan slab hence occurs between the end-Oligocene and the end-Miocene.

The extent of overthrusting is difficult to determine, mainly due to lack of stratigraphic markers. Based on the occurrence of klippes of the Tibetan slab and some inliers, it is estimated to be about 200 km.

4. The underthrust of the Lesser Himalaya below the Greater Himalaya is blocked in turn and a new cleavage plane is thus produced in an outer region, giving rise to the MBT and its annex, the MFT, in the Siwaliks. The thrust which has already attained nearly 100 km for the MBT is still active (Siwalik earthquakes).

The resultant crustal overthickening below the Greater Himalaya leads to a disequilibrium of the sedimentary cover which is detached from the metamorphic basement along a major normal fault, sometimes strike-slip fault, leaving the Tibetan slab bare and causing large folds with a northward inclination such as the major Annapurna fold (Fig. 163).

These successive cleavages are facilitated by the fact that the weight of the chain being built up induces a flexure of the lithosphere of the forelands, conforming to a process described earlier (p. 99), and thus gives rise to a molassic basin, viz., the Siwalik basin, 200 to 300 km in width and 7 km deep at places. As stated earlier (p. 8) the vast area over which this isostatic compensation is operative implies a certain rigidity of the lithosphere. This explains the relatively moderate root of the Himalayan chain (about 50 km).

The whole evolution ends in creation of a montane prism which, to a first approximation, can be compared to a Mohr-Coulomb prism. In this process the first stage is the rethickening of the thin crust of the margin (Indus and MCI nappes) which mainly occurs in marine environment. When this prism attains its equilibrium profile around 30 to 35 Ma, continuous shortening leads to an outward propagation (prograde) deformation.

The latter is localised in the region of the MCT. A new montane prism is formed; it is entirely aerial and subject to erosion and the higher the mountains, the more intense the erosion; the latter has the effect of continuing or even accelerating the phenomenon since it tends to reduce the height of the prism. Undoubtedly, it was not very efficient because a tectonic denudation occurs (north Himalayan normal faulting around 18 Ma). Deformation is localised farther south (MBT), building the present prism which is still active (Siwalik prism), facilitating very detailed analysis of relations between tectonics, erosion and sedimentation.

The Siwalik prism comprises recent sediments (Miocene to Present) deformed in front of the Himalaya and incorporated in the chain by boundary overthrusting. It is formed by a succession of slices and transported basins. Field study coupled with numerical modelling (Chalaron, 1994) shows that erosion plays a major role in the tectonic evolution of the system. Overthrusting leads to incorporation of new slices in front of the chain, resulting in modification of the topographic profile near this front. If erosion is sufficiently intensive to restore it, it facilitates the outward propagation (Gangetic plain) of the montane system which maintains a constant profile; this situation presently characterises the Nepal front south of Khatmandu. If erosion is very intensive at the level of the front, deformation is displaced southwards and new overthrusts and folds are developed; this mechanism is in operation west of Nepal. Finally, when erosion is intensive higher in the chain (Greater Chain), deformation is localised at the rear of the front; this is indicated by the recent activity in the region of MCT (2 Ma).

Secondary Effects of Collision on the Asian Continent

The Tibetan plateau extends north of the Himalaya with a mean elevation of the order of 5000. Beyond the plateau, varied structures of compression, extension and strike-slip movements are produced.

The Tibetan plateau (Fig. 165) is especially remarkable for the thickness of the continental crust that supports it, nearly 70 km, i.e., double the normal thickness. This has been interpreted as an indication that the Indian continental crust is thrust below that of Asia up to the northern boundary of the plateau. Such a vast extent of thrusting seems less probable in view of the energy required. However, the INDEPTH profiles (Fig. 166) show that the crust is effectively doubled below southern Tibet up to the level of Tsang Po suture.[2]

The Tibetan plateau is, in fact, made up of several microcontinents of Gondwana origin which entered into collision amongst themselves and with the Asian plate before and during the Middle Cretaceous. The contacts,

[2]The synthesised data of INDEPTH (*Science* (6.12.96) 274: 1684–1696) show that the depth of the crust of southern Tibet is affected by general anatexis.

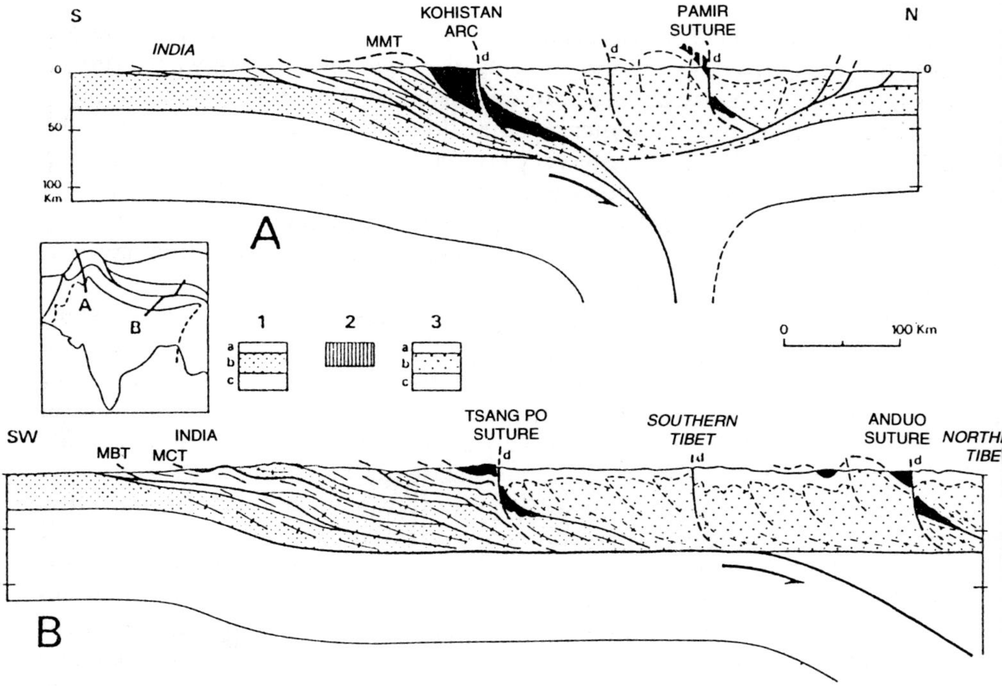

Fig. 165. Interpretative cross-sections of Western (A) and Eastern (B) Himalayas on a crustal scale (after Mattauer, 1983).
1. Indian lithosphere (a. sediments, b. ancient basement, c. mantle); 3. Asian lithosphere (a, b, c: same); 2. ophiolitic suture (oceanic crust and mantle); d. major strike-slip faults.

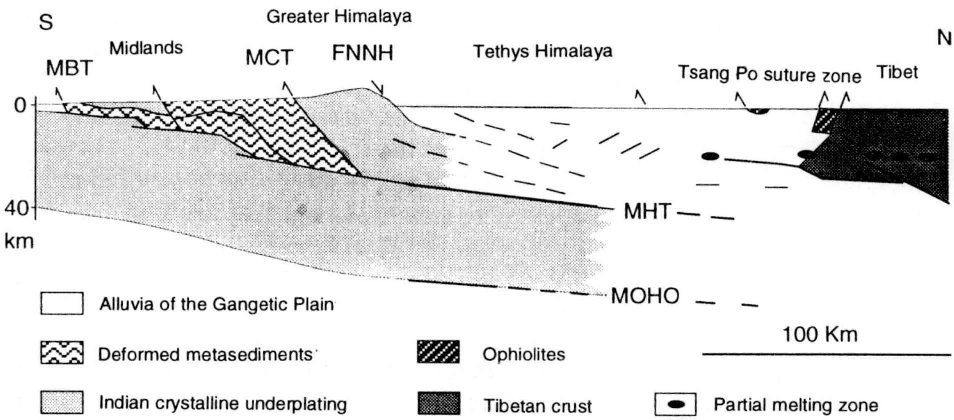

Fig. 166. Thickening of the crust below the Himalaya south of Tsang Po suture zone from the results of INDEPTH profiles (after Zhao et al., 1993; Makovsky et al., 1996, Nelson et al., 1996).

sometimes marked by ophiolites, are very straight and suggest a venetian-blind type layered structure, partly acquired during the Tertiary compressions. The overthickening thus obtained is maintained by the present Himalayan compression.

However, due to its extreme fracturing or an intensive heat flow, which certain authors associate with a thinning of the lithosphere due to detachment of the mantellic part, the Tibetan basement has no rigidity. The isostatic compensation is maintained only by its root, in an hydrostatic type equilibrium (Airy's model). This implies a deeper root than below the Himalayan range proper.

North of Tibet a number of varied structures represent the consequences of the Himalayan collision. An example is the *Tien Shan intracontinental chain* (Fig. 167A), an ancient basement chain of dual inclination, strongly rejuvenated during the Himalayan folding.

Continental-scale strike-slip faults also exist, roughly E-W, identifiable on aerial photos and satellite images. They represent a lateral spill of material east or westwards, a movement which in turn can induce local compressions of N-S orientation, such as the small intracontinental ranges thrusting westwards, south of Shan Si and in Se Shuan.

This dynamics of lateral movement leads to a general extension which is manifest in the Tibetan plateau by the presence of small N-S grabens and, on a larger scale, i.e., in the region of the Siberian craton, by larger structures such as the Baikal and Shan Si rifts. The origin of major basaltic plateaus that separate the preceding rifts (Fig. 167B) can also be related to this extension state.

Variscan Chain of Europe

European Variscan folding is a sinuous orogen which extends discontinuously from Spain to Czechoslovakia over a length of nearly 3000 km and a width of 700 km (Fig. 168). It then disappears below the Carpathians.

This chain emerged from the complex collisions between two major continental masses, viz., the North Atlantic continent in the north and the Gondwana continent in the south (on which are dependent a majority of the ancient cratons of France). While the North Atlantic margin can readily be reconstructed, the southern margin is more problematic to model because it must have had a complex form consisting of apophyses or microcontinents that complicate the palaeogeography considerably, and also because it has been deeply dislocated by alpine deformations. In particular, the Iberian and Corsica-Sardinian blocks should be relocated into their Permo-Carboniferous positions, which is relatively easy. On the contrary, zoning of the Variscan chain in the Alps still remains almost unknown.

Fig. 167A. Tectonics of the Asian plate north of the Himalaya (according to the interpretation by Tapponnier et al., 1975; 1986).

A. *Effects of the Himalayan collision on South-Eastern Asia*
(white arrows: movements of continental blocks; black arrows: extension resulting from strike-slip faulting; grey: extension zones; black: oceanic zones).
These effects have varied times of origin:

1. *From the Oligocene to the Middle Miocene* (–35 to –15 Ma), under the effect of collision of India, the Indochina block slid laterally (arrow 1) and moved over nearly 600 km relative to the China block at the rate of 3 to 5 cm/yr, at the same time as the South China Sea (SCS) opened up and, farther south, the Andaman basin (AB) was produced.
2. *After the Middle Miocene*, field observations and satellite-image data indicate that the expulsion of the Tibetan block towards the E or SE is still continuing at a rhythm of 2 to 3 cm/yr. The China block slides relative to Mongolia along the Altyn Tagh (AT) fault (arrow 2), thus opening the Shan Si and the Baikal rifts on the one hand, and on the other, relative to Indochina along the Red River (RR) fault, but in a direction opposite to phase 1, thus opening the Yunnan grabens.

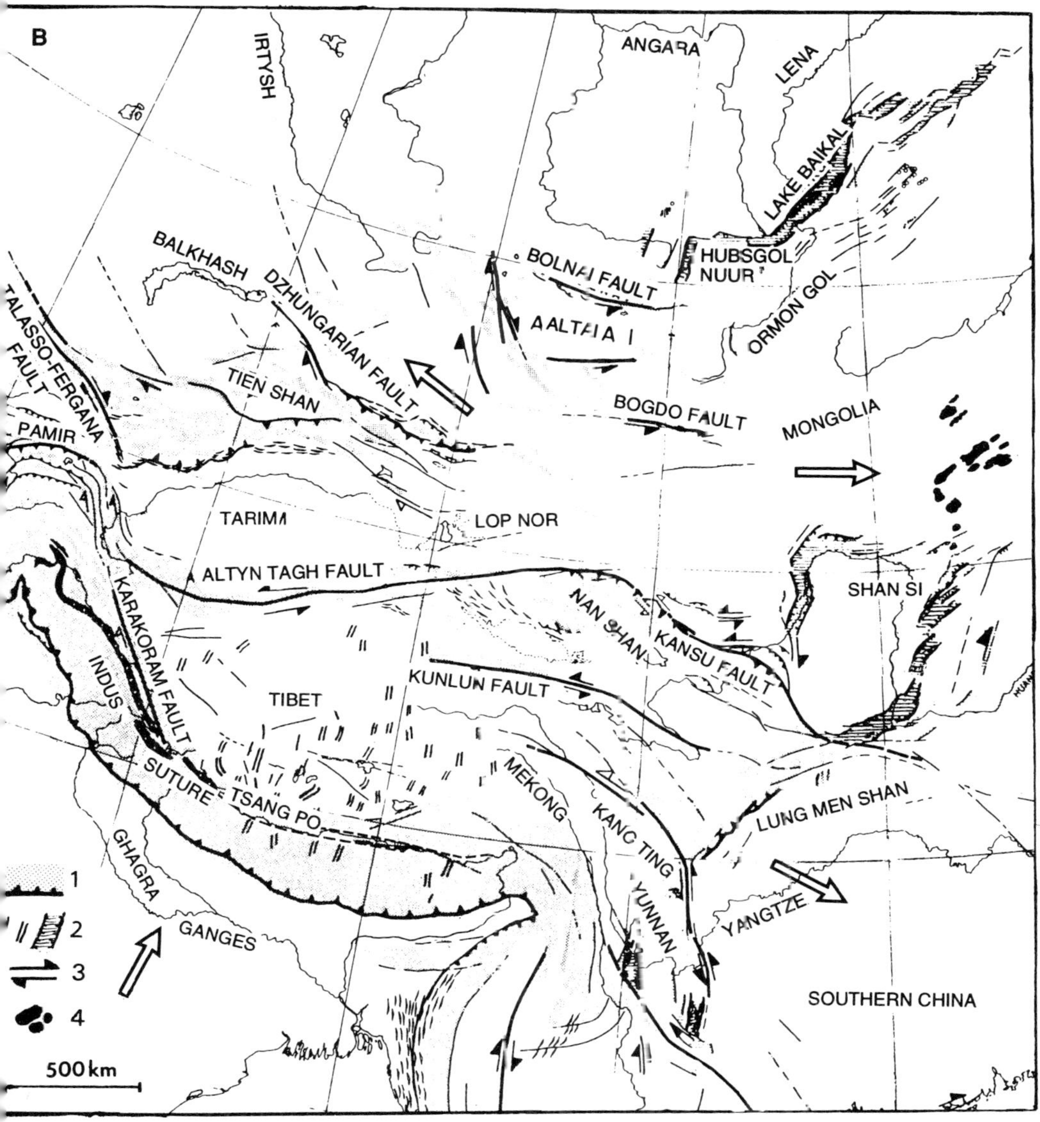

Fig. 167B. Details of structures north of the Himalayan chain.
1. Folded chains; 2. grabens; 3. strike-slip faults (the direction indicates the present movement); 4. basaltic plateaus. White arrows: continental displacements.

Variscan orogeny is characterised by a large central part (axial zone) in which occur the oldest (Precambrian to Lower Palaeozoic), the most metamorphosed and the earliest folded formations, as well as the greater part of granitoids. Westphalian and Stephanian formations are present here in restricted basins in the form of limnic deposits, sometimes very thick (up to 4000 m), lying unconformably on the folded and metamorphosed basement. The degree of metamorphism of the latter indicates that locally an almost 25-km thick continental crust disappeared before the Westphalian by erosion or following a significant crustal thinning (detachments).

On either side of the central zone occur the Devonian-Carboniferous basins that are barely or unmetamorphosed and devoid of granitoids, in which the transition to limnic coal-bearing facies takes place almost continuously through a flyschoid sequence; these represent foreland basins.

This relative symmetry is also indicated, on a tectonic plane, by a general fan-shaped, dual inclination disposition of structures and migration of the deformation outwards from the chain over the course of time.

The northern margin of the chain is distinctly marked, from Ireland to Germany, by the 'Variscan front', a major flat overthrust with northward vergence, clearly distinguishable from Wales to Belgium (Fig. 168) and Aix-la-Chapelle (Aachen). This thrust causes a folded and schistose Palaeozoic formation to lie on top of the paralic Carboniferous basin, with a non-deformed Devonian basement.

Beyond this zone, the Variscan front is intersected by the NW-SE dextral transverse faults forming the southern boundary of the Russian platform. Its continuation, thus offset, is not visible as it is masked by the Carpathians.

The southern boundary of the chain is clearly distinguishable only in the SE part of the Bohemian massif (Fig. 168), where one can observe a series of folds recumbent southwards, which thrust the folded Devonian-Carboniferous formations onto a non-deformed platform, and in northern Spain where the Cantabrian nappes rest on a basement that is equally non-deformed in the midst of 'Iberian-Armorican virgations', but the contact is rarely visible because masked by Tertiary basins.

This contact is nowhere visible throughout the alpine region. Moreover, it is also known that the Variscan basement of the Western Alps, highly metamorphosed, belongs to the core of the Variscan chain, whereas a major part of the Eastern Alps pertains to the southern Devonian-Carboniferous basin.

Major structural features

Two major structural features of collision chains are observed, viz., ophiolitic sutures and crustal cleavages of the margins.

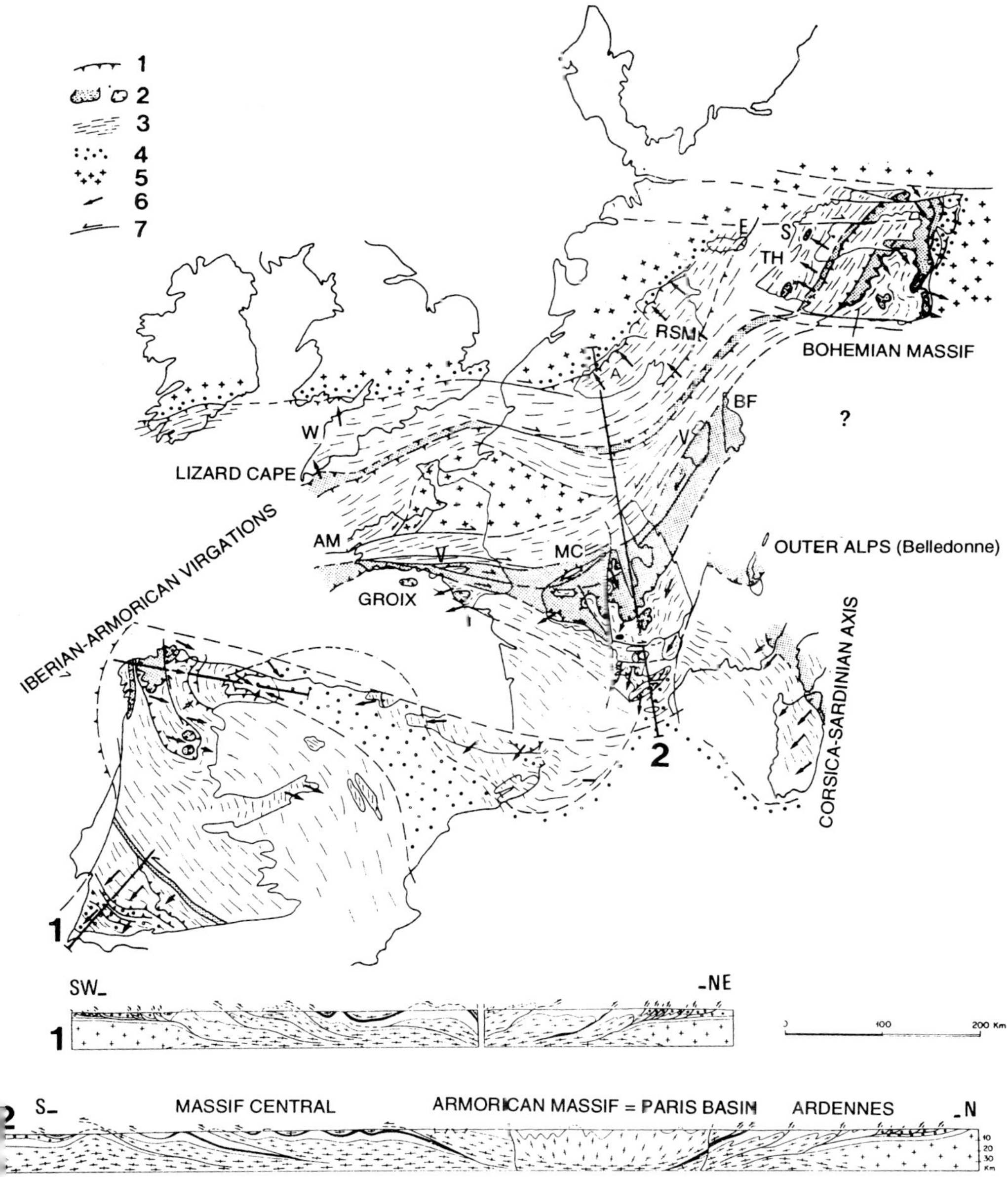

Fig. 168. Structural map of the Variscan chain in Western Europe (after Matte, 1986).
1. Major thrust zones; 2. inner crystalline nappes and ophiolitic sutures; 3. regions of slaty cleavage; 4. outer Devonian-Carboniferous basins; 5. slightly or non-deformed blocks; 6. vergences of structures; 7. strike-slip faults.
A: Ardennes; AM: Armorican Massif; BF: Black Forest; E: Erzgebirge; MC: Massif Central; RSM: Rhine schistose massif; S: Saxony; TH: Thuringe; V: Vendée; W: Wales.

Ophiolitic sutures

Long ignored because of intensive metamorphism and tectonics, the Variscan ophiolitic sequences, though highly fragmented, are presently known in the form of basic to ultrabasic rocks whose oceanic origin has been established from rare-earths geochemistry. All the data indicate ages of formation (oceanic opening?) of the order of 450 to 500 Ma (Ordovician). Their location suggests the existence of at least two regions with oceanic crust that are nearly parallel to each other.

— *A northern belt* which is exposed only in Cape Lizard (SW tip of England) and in the northern part of the Bohemian massif. The obduction of ophiolites, with northward vergence, appears to have been produced here in the Lower to the Middle Devonian since these rocks were metamorphosed to eclogite facies at 380 Ma and furthermore, chromite and detrital chromiferous spinels are found in the Upper Devonian flysch of the Saxo-Thuringian zone, situated below the roots of the ophiolitic nappe.

— *A southern belt with southward vergence.* The corresponding rocks, always intensely metamorphosed, are dispersed within leptyno-amphibolitic gneisses and form large nappes in the NW part of the Iberian peninsula, in the Massif Central and in the Alps (the Chamrousse massif, near Grenoble, and the Tauern basement)[3]. In all these exposures, tholeiites are found to be associated with basic and ultrabasic cumulative members. It should be noted, however, that, except for Belledonne, interpretation of these magmatic rocks as ophiolitic complexes remains controversial.

A major part of this oceanic material, both in the north as well as in the south, has been intensely metamorphosed to high-pressure facies (granulite, eclogite, glaucophane schists of which the most well known are those of Groix Island). A general retrometamorphism to amphibolite facies affected the suite around 380 Ma in the north and 430 to 380 Ma in the south. Obduction here is thus a little older.

In any case, and in spite of some uncertainties, it seems probable that the first stage of evolution of the Variscan chain was the disappearance by subduction of the existing oceanic regions during the Silurian or Lower Devonian.

Major intracrustal cleavages

Three main types of cleavages, i.e., major nappes that characterise the three successive zones of the chain can be distinguished (Figs. 168, 169).

— *Inner crystalline nappes* (essentially found in the southern half of the orogeny). These nappes characterise the axial (= inner) zone of the chain. They consist of highly metamorphosed formations, of polyphase (high pres-

---

[3]In view of the difficulty in structurally connecting the Variscan ophiolites of the Alps with the Southern belt, several authors prefer to consider a third belt there, independent of the second.

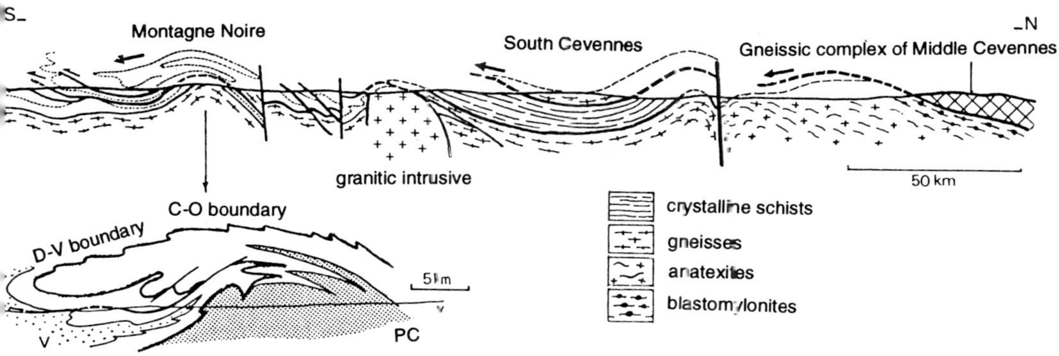

Fig. 169. Tectonic style of the southern Massif Central.
Inner crystalline nappes (on the right) and synmetamorphic recumbent folds (Montagne Noire, on the left) (general section, after Arthaud and Matte, 1975; section of Montagne Noire, after Mattauer, 1983, 1986).
C-O: Cambro-Ordovician; D: Devonian; PC: Precambrian; V: Visean; x-y: erosional surface of the section.

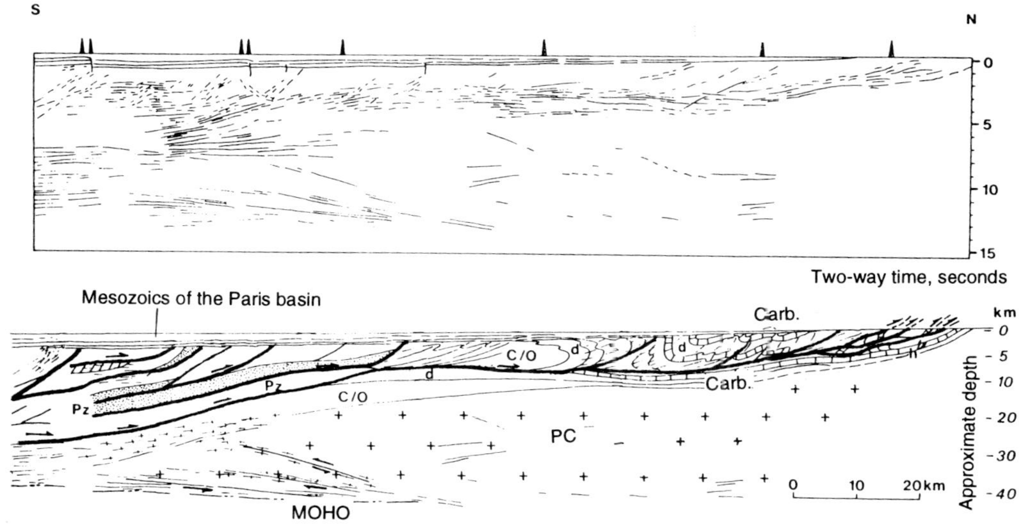

Fig. 170. Cross-section of the Variscan front in northern France according to the ECORS profile (after Raoult, 1986, simplified).
Carb: Carboniferous; C-O: Cambro-Ordovician; D: Devonian;
PC: Precambrian; Pz: Palaeozoic undifferentiated metamorphics.

sure to Barrovian) metamorphism, superposed on less-metamorphosed rocks. The amplitude of thrusting might reach 200 km in north-western Spain and the Massif Central, which is close to that of the Himalaya. Development of these nappes is accompanied by a synmetamorphic ductile deformation, epizonal to mesozonal, sometimes with reverse disposition of isograd contours, as in the Himalaya.

This metamorphism and the associated tectonics have been radiometrically dated in Vendée and in the Massif Central as 380–390 Ma (Lower Devonian) and stratigraphically by the Upper Devonian unconformity on its basement in the Massif Central (Brevenne), Black Forest and Portugal.

— *Synmetamorphic recumbent-folded nappes* situated in front of and below the preceding nappes whose formation is slightly later (300 to 350 Ma, i.e., Carboniferous). Two types of nappes correspond to this definition:

a. Large zones with horizontal schistosity, without major folded structures (nappes of Vendée, Albigeois, Cévennes), characterised by an intense shear deformation, with a strong stretching parallel to the direction of movement, and rare folds subparallel to the direction of stretching.

b. Major recumbent folds, clearly identified in the Palaeozoic sequences in which the stratigraphy is distinct. They have been described in NW Spain, with reversed limbs of 20 km, and in the southern part of the Massif Central (Fig.169). These folds show an axial-plane schistosity and their axes are nearly perpendicular to the direction of movement.

— *Decollement nappes of the outer basins*. The best example of such zones is the nappe of Ardennes, clearly demarcated by the ECORS profile (Fig. 170). In this region the Ordovician to Devonian sequence is detached at the Cambrian and Devonian horizon, along a horizontal contact situated at a depth of 5 km. The thrust sequence may also be strongly schistosed. The overthrust is produced on the Carboniferous horizon, thus indicating that it represents the most recent tangential structures.

It may thus be noted that cleavages commenced in the surroundings of the suture zone, as in the Himalaya, and developed outwards of the chain over the course of time.

Associated phenomena

1. *Variscan granitic magmatism*. Granitic magmatism is especially abundant in the core of the chain, between 370 and 280 Ma (Upper Devonian to Late Carboniferous), during which period it was manifestly associated with Barrovian metamorphism and thickening of the crust. Outer basins with normal crust are devoid of magmatism.

These granitoids can be grouped into three types:

— *Granodiorites*, often calco-alkaline, with basic enclaves. They are probably derived from the melting of the base of the crust, along with part of the mantle. These are the most widespread. The oldest (370–360 Ma) are found in the innermost parts of the chain.

— *Anatectic granites*, associated with syntectonic metamorphism, which often facilitates their dating.

— *Aluminous leucogranites*, associated with strike-slip faults and resulting from aqueous fusion of sediments or granites in the contact. They are rich in sedimentary inclusions and impoverished in basic enclaves. The classic examples are those of the Armorican Massif.

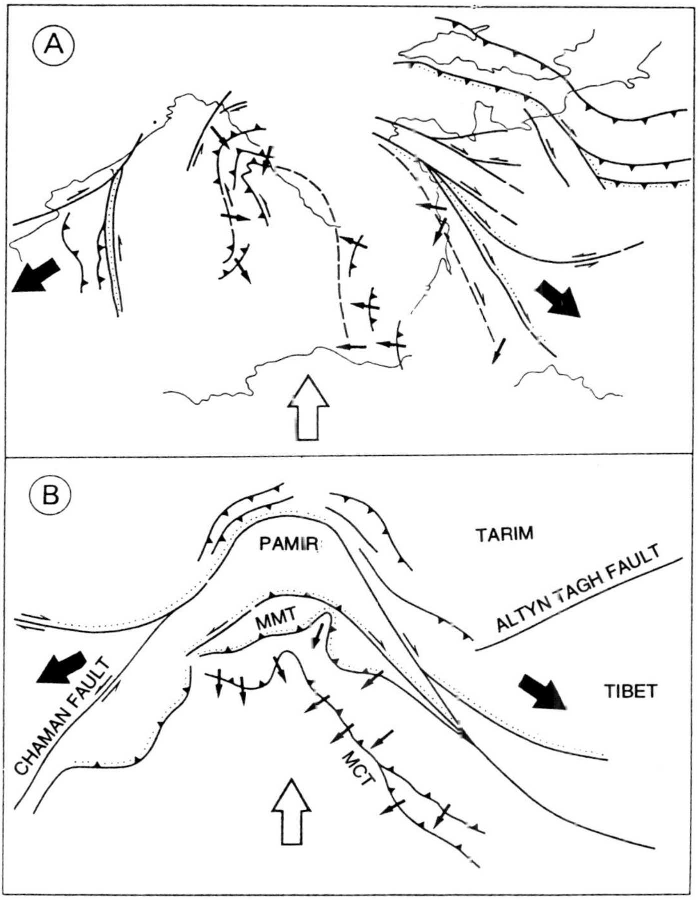

Fig. 171. Ductile strike-slip faults and overthrusts in the Ibero-Armorican (A) and West-Himalayan (B) virgations, to the same scale. Light dots indicate suture zones (after Matte, 1986).

2. *Ductile strike-slip faults.* These structures, generally parallel to the direction of the chain, are mainly observed on the two sides of the Iberia-Armorican virgations (Fig. 171). They are sinistral in the Iberian branch and dextral in the Armorican Massif.

These faults represent an intracontinental deformation subsequent to the formation of the nappes because the leucogranites that accompany their origin (as clearly indicated by the form of plutons) are dated 340 to 310 Ma. They haye often been compared with those of the Himalaya (p. 239). They suggest, like the latter, lateral escape of material on either side of the collision front of one of the masses present.

Structural evolution (Figs. 172 and 173)

The characteristics of the Variscan chain are similar to those of a collision chain after closure of two oceanic gaps, known as 'Rheic' in the north, and 'Mesogean' in the south. The ages of 450 to 500 Ma obtained on the ophiolitic material or the immediately preceding alkaline magmatism suggest a Cambro-Ordovician opening. These figures must be reduced because the limited palaeomagnetic data do not indicate their existence.

Subduction and obduction are produced during the Silurian and the Lower Devonian. This is the age of the 'ophiolitic melanges' of Galicia. It is also in the Saxo-Thuringian flysch of the Upper Devonian that chromite and detrital chromiferous spinels are observed, indicating the erosion of ophiolitic masses. Finally, high-pressure metamorphism (eclogites, blueschists), whose traces exist almost everywhere, give Silurian (430–380 Ma) ages.

The *collision* of continental margins (and of intraoceanic continental strips such as the 'Moldanubian zone' of some authors, Fig. 173) is produced in the Devonian, which conforms with a Barrovian metamorphism (that might extend up to anatexis) dated 380 Ma, and which may be superposed on the preceding one in zones where the latter exists. Upper Devonian unmetamorphosed formations are also observed in the Armorican Massif, lying unconformably on the structures developed during this phase.

Subsequently, cleavages increasingly affect outer zones of the chain, generally accompanied by metamorphism and granitisation dated 380 to 300 Ma, thus resulting in a dual-inclination structure.

The corresponding slices are gradually piled up outwards of the chain, thus inducing a flexure of the forelands and consequently the origin of foreland basins in which flyschs and molasse would be deposited. Crustal thickening and partial melting of the deepest slices also take place, along with production of granodiorites, as already mentioned above.

The compressional regime seems to have passed through some episodes which earlier authors described as phases, e.g. the *Breton phase*, in Devonian-Carboniferous transition, especially distinct in the Armorican Massif, and the *Sudete phase*, between the Dinantian and Namurian. The latter is complex because during the Carboniferous the corresponding deformation progressively encompassed outer zones more, with the result that this phase differs in age in each region.

Only the Moldanubian zone, situated between two oceanic sutures and represented in France by the northern part of the Armorican Massif, escaped this cleavage; it was subjected only to a moderate shortening because it was already a thickened zone that had endured an earlier orogeny, viz., the Cadomian orogeny. Its sedimentary cover was affected by a folding with vertical axial planes, of post-Dinantian and early Upper Westphalian, which can hence be qualified as Sudete *sensu lato*.

Progression of the effects of collision outwards of the chain has been imprinted in the *flyschs*, which will be described below.

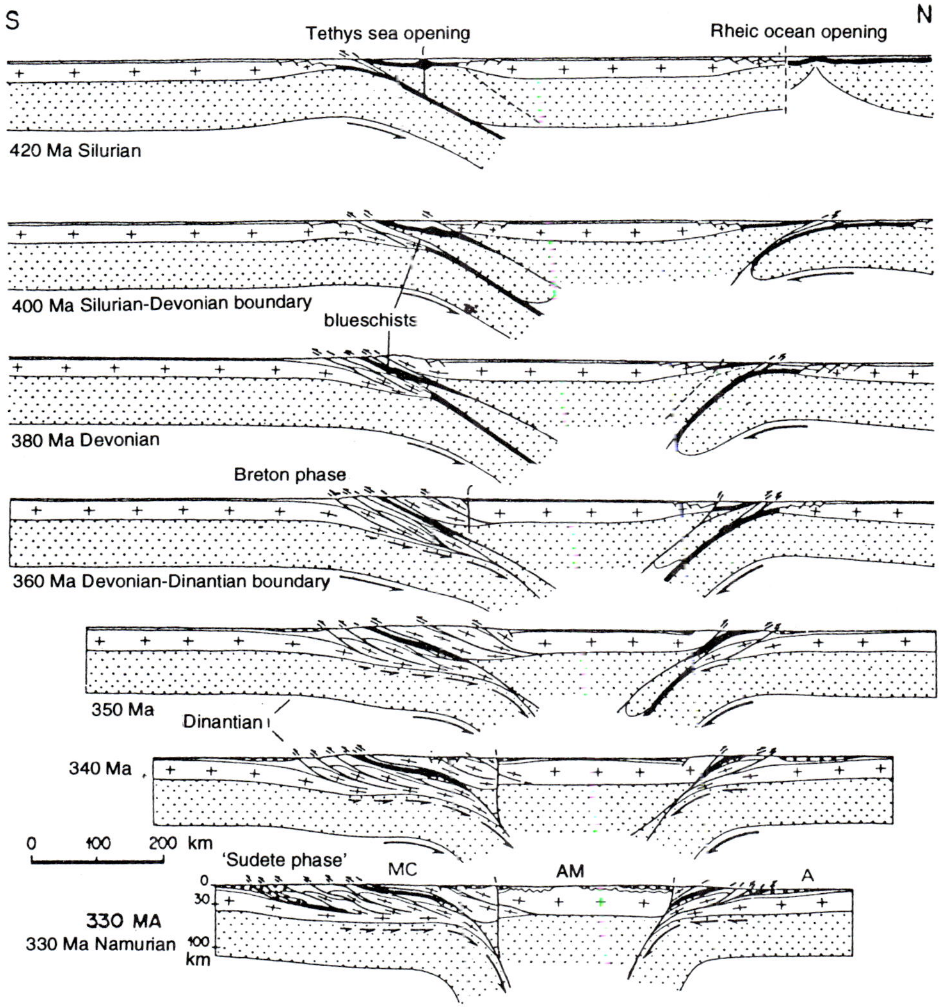

Fig. 172. Model of evolution of the Variscan chain on the Ardennes-Massif Central section: Mechanical behaviour of the continental crust and position of 'phases', simple stages in the progressive development of a tectonic accretionary prism (after Matte, 1986).
Light dots: lithospheric mantle; Crosses: continental crust;
Black: oceanic crust; White: Palaeozoic sediments; Bold dots: Devonian-Carboniferous sedimentary basins of forelands.
A: Ardennes; AM: Armorican Massif; MC: Massif Central.

These flyschs are distributed in two parallel bands (Fig. 174) reflecting the double polarity of the chain.

The Northern band, the best known, corresponds to the so-called Rheno-Variscan zones, which extend from Portugal to Harz, and the inner Saxo-Thuringian zones, bounded in the north by the Bohemian Massif and

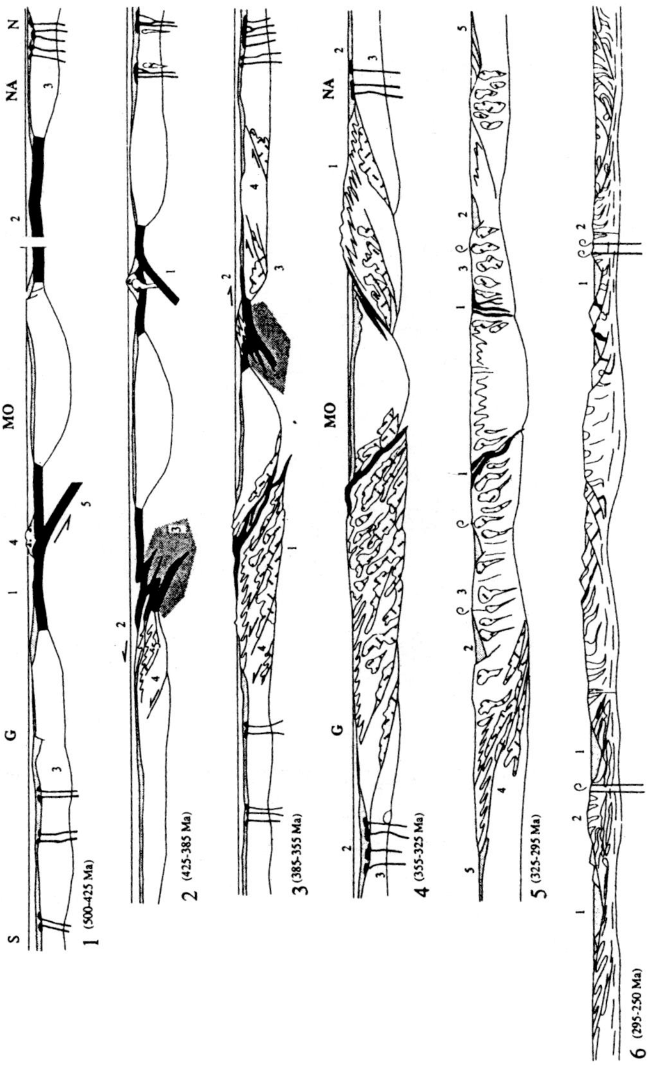

Fig. 173. Model of evolution of the Variscan chain on the Ardennes-Massif Central section (modified after Bard et al., 1980).

1. *Ordovician-Lower Silurian* (500–425 Ma). Two oceanic regions (1 and 2) separate three continental blocks (G: Gondwana; MO: Moldanubian zone; NA: North Atlantic continent), in which volcanic activity (3) is of the spilite-keratophyre type. In the south oceanic region (1), probably a volcanic arc (4) exists (assuming that the leptyno-amphibolitic complexes of the Massif Central started in the Ordovician) below which a subduction zone (5) occurs.

2. *Upper Silurian-Lower Devonian* (425–385 Ma). Subduction commences in the northern region (1) and ends in the southern region (2), with high-pressure metamorphism (3, grey) in the oceanic crust and the volcanodetrital sediments subducted (leptyno-amphibolitic complexes). Cleavage of the extreme Gondwana margin commences

separated from the preceding by the 'mid-Germanic crystalline ridge', which is probably the front of a basement slice. It might be this Saxo-Thuringian zone that later formed the basement of the Paris basin.

The Southern band is observed in southern France and NE Iberia, i.e., in the concave zone of the Iberian-Armorican virgation, as well as in some sectors of the Eastern Alps (Carnic and Karawanken Alps, Fig. 174).

Both in the North and the South, the directions of sedimentary transport are always turned outwards of the range. Moreover, these flyschs, once deposited, are rapidly affected by the deformation that progresses outwards, and are broken into slices more especially allochtonous as they are in a more inner position. Their age gradually decreases from the interior to the exterior of the chain, which conforms well with the notion of a syntectonic deposition occurring in front of a true prograde accretionary prism.

In the northern branch of the orogeny, the oldest known formations are of Middle to Upper Devonian age (but perhaps they are not the oldest). The most frequent are the Dinantian (*culm* facies of some authors) which change in the north to lacustrine molasse with coal beds of Namurian to Westphalian age.

In the southern part of the orogeny, the oldest rocks are of Visean to Namurian age and pass on to coal-bearing molasse, as in the north, but their deposition is interrupted by folding such that the coal basins here are post-tectonic.

---

(4), accompanied by intrusion of mesocrustal alkaline granites resulting from the aqueous fusion of the crust in the cleavage planes (5).
3. *Middle-Upper Devonian* (385–355 Ma). Accentuation of cleavage of the Gondwana margin and pile up of slices along with intrusion of granites of meso- to infracrustal origin (1). Blockage of the system in the south accelerates subduction in the north and obduction of the corresponding oceanic crust (2) with high-pressure metamorphism (3, grey). Cleavage of the North Atlantic margin commences, with generation of infracrustal granitoids.
4. *Visean* (355–325 Ma). Cleavage of the margins extends outwards (1) ('Sudete phase' of some authors), preceded by deposition of flyschs (2) in flexure basins where a spilitic to rhyodacitic volcanism is produced (3). At depth, the piling up and sliding of slices determines emplacement of infracrustal granites and, by uplift of isogeotherms, induces a Barrovian metamorphism and an important anatexis (see p. 237 note 2).
5. *Namurian-Westphalian-Lower Stephanian* (325–295 Ma). Areas earlier subjected to tectonics (in Lower and Middle Devonian) undergo a last compression ('Asturian phase' of some authors) that renders the structures vertical (1), are uplifted by isostasy (crustal thickening) and hence eroded. The products of erosion supply coal-bearing molasses of the limnic basins (2) set on the first normal faults accompanying the beginning of the late crustal thinning out. A post-orogenic, calco-alkaline to rhyolitic volcanism accompanies the uplift of infracrustal granites (3 and 4). In the outer parts of the range, the contraction affects the Visean and Westphalian basins (5).
6. Permian (295–250 Ma). Crust thins out by generalised gravitational stretching of the chain along detachment faults, formation of intramontane sedimentary basins (1), structuring of layered lower crust, fracturing and breaking up of the chain due to major strike-slip faults, accompanied by ultimate volcanic manifestations (2).

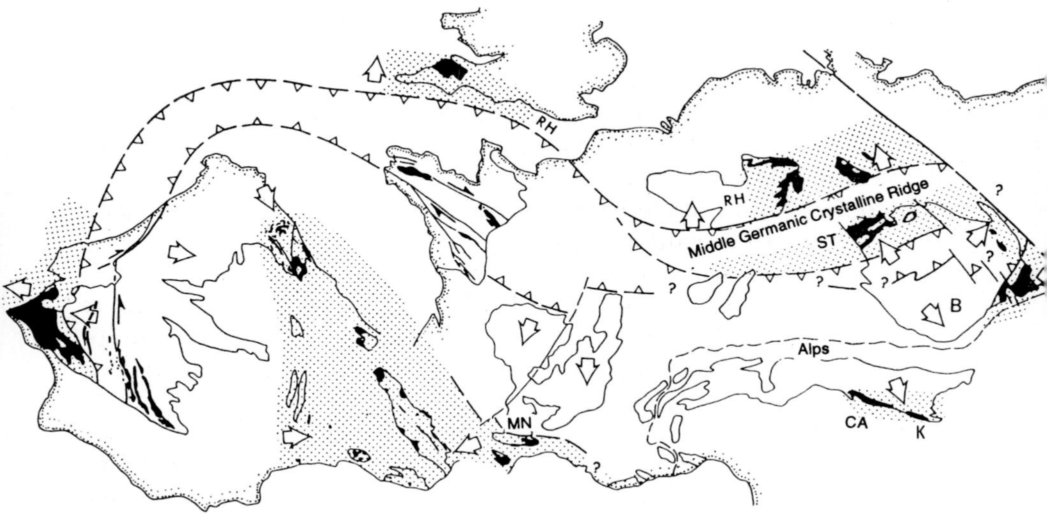

Fig. 174. Principal outcropping zones of Variscan flyschs in Western Europe (after Franke and Engle, 1986).
B: Bohemian massif; CA: Carnic Alps; K: Karawanken massif; MN: Montagne Noire; RH: Rheno-Variscan zone; ST: Saxo-Thuringian zone.

Study of these flyschs is equally important from a *sedimentological point of view*, because they indicate a source that is gradually modified proportionate to piling up and growth of nappes, erosion of which 'feeds' sediments. Grain-size distribution of these formations is also significant. For example, between Mouthoumet Massif and the southern Montagne Noire, the typical flysch gradually varies to a coarser and coarser facies, with olistoliths, and finally a true synsedimentary gravitational nappe appears, with sedimentary klippes of more than 10 km^2 (the 'Cabriéres slice'). Further, it is the old flyschs, allochthons, that are found redistributed in the most recent flyschs ('cannibalism' of flyschs).

Post-Variscan movements

After the classic Variscan foldings, always more or less synmetamorphic, a last period of compression occurs which, acting on an already severely hardened structure, induces an intensive fracturing of the West European block due to numerous transverse faults intersecting earlier structures (Fig. 175).

In fact, this compression regime and the connected strike-slip faults commenced in a ductile regime and with large throws as soon as the intrusion of synkinematic alkaline leucogranites, which ceased around 310 Ma (Westphalian-Stephanian boundary).

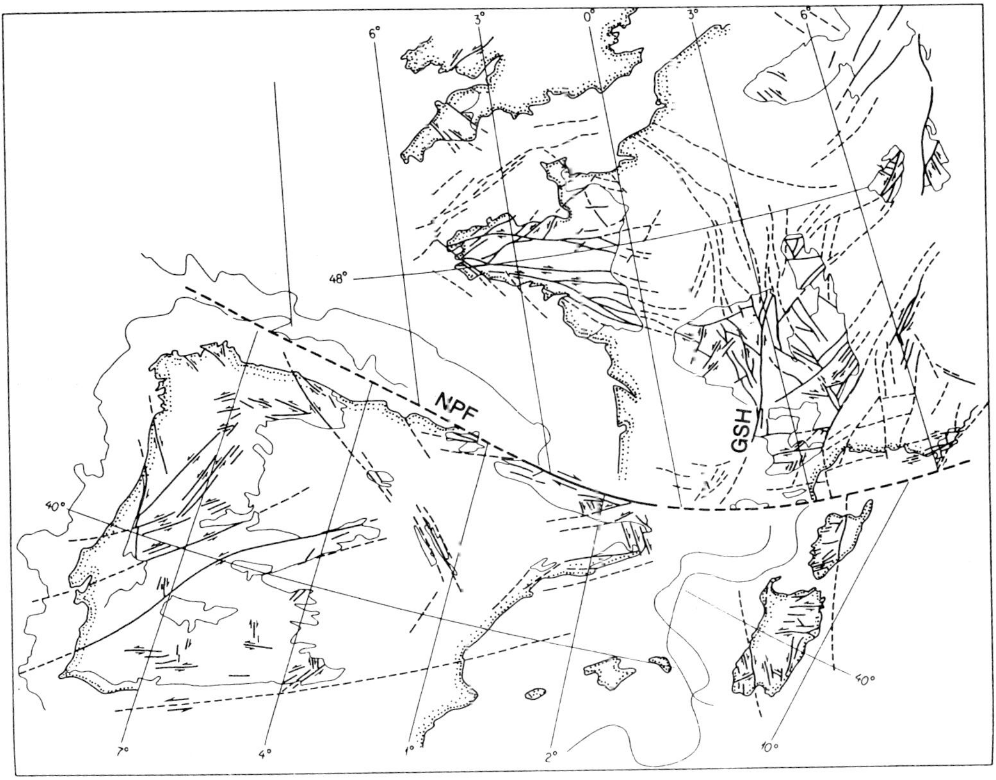

Fig. 175. Late Variscan transverse faults of SW Europe (after Arthaud and Matte, 1975). GSH: Grand Sillon Houiller; NPF: North Pyrenean fault.

In the Stephanian and Lower Permian strike-slip faulting continued in a fragile regime but with smaller throws. Some Stephanian coal basins are in effect distributed on these strike-slip faults and were subsequently affected either by fracturing or even by folding if their orientation was convenient for the same (Saint Etienne basin, for example), at the end of the Permian ('Saalian phase' of some authors). After this, the extension regime becomes common and is accompanied by a major acidic volcanism (Vosges, Esterel, Corsica).

These late transverse structures are distributed in two conjugate systems, NE-SW (sinistral) and NW-SE (dextral). Such a disposition is compatible with a NNW-SSE direction of compression, which also causes the folding of the Saint Etienne and Gard coal basins.

This fracturing is very significant because it strongly influences the subsequent geological history, especially from the Upper Permian, since the faults are reactivated as normal faults, bordering grabens. We have already mentioned (p. 85) this special feature which is the cause of the origin of the Paris or Aquitaine basins.

In the case of France, two major structures are the *North Pyrenean fault* and the *Grand Sillon Houiller* (Great Coal Belt) of the Massif Central, so named because it is marked by coal basins.

The *North Pyrenean fault* produces a 150 km long dextral offset across the Iberian-Armorican virgation (not to be confused with the sinistral offset of the Mesozoic, mentioned on p. 130).

The *Grand Sillon Houiller* cut in two parts the French Massif Central and continues below the basins that surround it in the north and south, as indicated by geophysical data. It thus extends over a length of 600 km and has a faulting of 70 to 100 km, as estimated from the Variscan structures that it offsets.

Variscan Syntectonic Extensions of the Massif Central

The European Variscan chain shows similarities with the Basin and Range chain of western America, notably presence of gneissic domes comparable to the metamorphic core complexes (MCC) associated with a detachment fault with stretched and tilted blocks as well as syntectonic carboniferous granites.

The Montagne Noire (Fig. 176) constitutes a good example with an axial dome of Precambrian to Lower Palaeozoic gneiss, highly stretched in a ductile regime and intruded by Upper Visean (330 Ma) granites. This axial zone is bounded on the north by a detachment fault (see the new map of France in 1,000,000), against which the small coal basin of Graissessac and that of Lodève of Permian age are supported. Farther north, a system of normal faults that probably developed on the older Variscan thrust surfaces form the boundaries of blocks within Lacaune mountain. This is followed by the faulted Permian basin of Saint Affrique, embedded on the top of Albigeois nappes, all followed by a disposition that mechanically accords well with the extension of Montagne Noire dome.

This extension may have commenced in the Upper Visean (age of the granite intruded into Montagne Noire dome), i.e., at the moment when the Variscan crust had attained its maximum thickness. Since the Upper Carboniferous, a detachment fault started functioning on its northern slope and continued its activity in the Permian along somewhat different directions as the framework of Variscan collision had changed and major strike-slip faults reoriented the stresses.

Viewed in the scale of the entire Massif Central, conditions are not so distinct in the absence of sedimentary basins associated with crustal-extension structures and due to difficulties in identifying detachment faults in the metamorphic formations. In general, only gneissic domes comparable to metamorphic core complexes, along with Carboniferous syntectonic granites, attest extension having affected the chain during its crustal thickening.

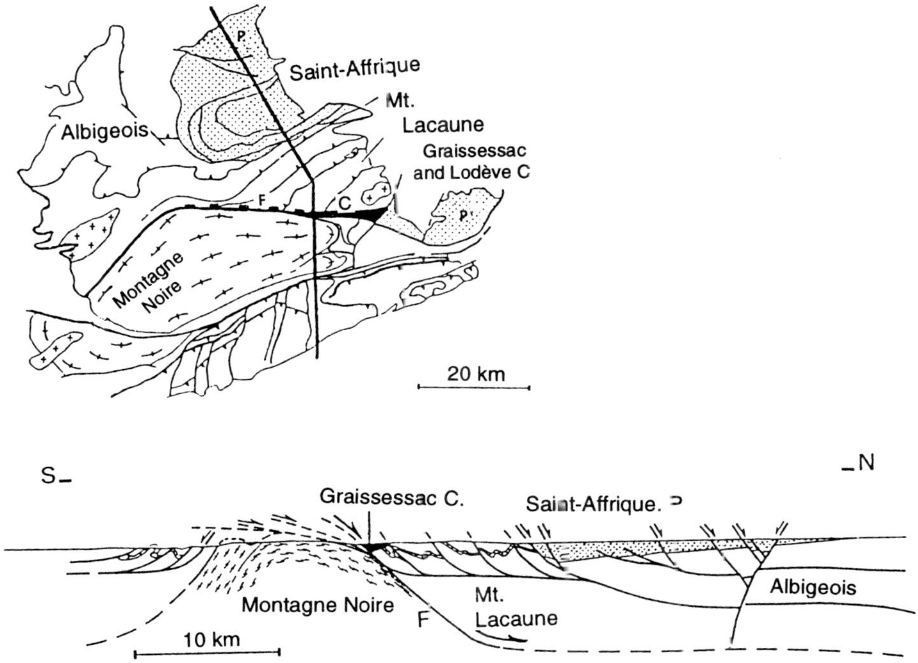

Fig. 176. Metamorphic core complex of Montagne Noire, with its detachment fault (F) and associated sedimentary basins. This structure attests the Permo-Carboniferous extension of the Variscan chain (after Burg et al., 1994, simplified).

Thus, we may observe the Rodez dome (flanked in the northern part of a Permian basin) and Haut Allier dome (also associated in the north with Brioude coal basin) in the northward continuation of the Montagne Noire section, even if the relations of these basins with the corresponding domes are masked by recent formations.

If these extension phenomena are removed from thought, more than one-third width of the present orogeny would be reduced (of the same order of magnitude as the Basin and Range chains), which gives an idea of the importance of this process long ignored or underestimated. It shows that the extensional stretching of the mountain chains that occurs when they have attained a limiting crustal thickness is a general phenomenon. Modalities of this stretching evidently vary depending on the conditions at the boundaries. In the Massif Central the magnitude of Variscan granitisations indicates a high thermal regime resembling that of Tibet (see p. 237).

The Alps

The Alps resulted from the collision between the Euro-Asian plate and the African plate or, to be more precise, the Adriatic (Insubric) promontory of the latter, sometimes considered an independent microplate that at present constitutes the basement of the Po and Adriatic plains.

At the initiation of the alpine cycle, i.e., in the Triassic, only a feebly submerged continental platform existed at the place of the future Alps. It is during the Jurassic that an extension phenomenon opens an oceanic region in this platform known as the Tethys sea (Fig. 177).

Whatever its form and structure, this oceanic region became closed during the Upper Cretaceous and, in the collision produced subsequently in the Eocene, the concerned continental margins were cleft into granito-gneissic slabs, all inclined in the same direction, viz., N or NW. This region thus

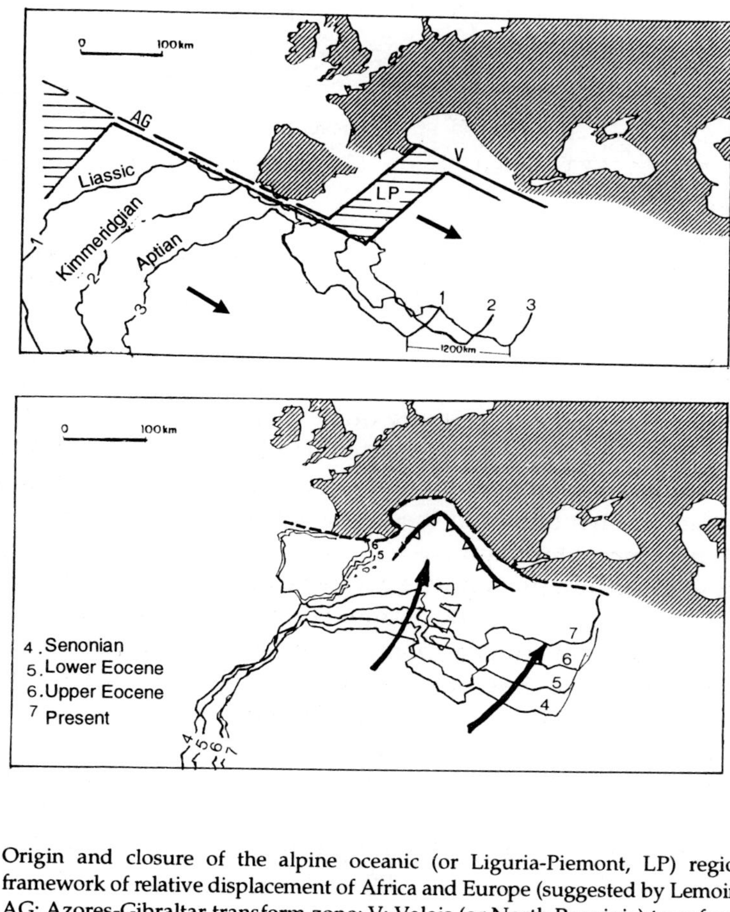

Fig. 177. Origin and closure of the alpine oceanic (or Liguria-Piemont, LP) region in the framework of relative displacement of Africa and Europe (suggested by Lemoine, 1984). AG: Azores-Gibraltar transform zone; V: Valais (or North Penninic) transform zone.

consists of a pile of nappes of the same vergence, the ophiolitic zone being just one of them.

Present structure (Fig. 178)
The structure of the Alps is described in the same manner as for the Himalaya, i.e., going from the forelands to the oceanic region.

1. ***European (or North) continental margin***: This consists of two parts which have always been designated as external and internal zones.

The external zone shows a continental crust of normal thickness (about 30 km). It is thus still on the European continent. Moving from the outer part of the chain, we first come across the French 'subalpine chains', i.e., the sedimentary cover (Mesozoic and Palaeogene) of a Variscan basement. This is a platform sequence more or less subsident at some points. It was folded at the end of evolution, end-Miocene, concomitant with the Jura mountains and in the same mechanical regime. In other words, contraction probably led to underthrusting of the outer basement below the inner alpine units along older listric fault planes. The overlying cover, detached in the Triassic evaporites, Liassic, Oxfordian or Cretaceous marls, is piled up as slices or stretch-thrusts (duplex of balanced sections, see Fig. 179), with NW, W or SW slope depending on their position in the alpine arc.

In the neighbourhood of internal zones the tectonic activity becomes more intensive and affects the basement, which thus appears in the *external crystalline massifs* (Argentera, Ecrins, Belledonne, Mont Blanc, Aar). The ages given by the Variscan granites of these massifs confirm a Neogene tectonic activity but the latter seems to have commenced much earlier than in the subalpine chains. Moreover, geophysical data show that they are very probably broken into major thrust slices in outer parts of the chain (Figs. 179, 180). Hence, the shortening was much more intensive and was especially spectacular in Switzerland because it induced the expulsion and folding of the sedimentary cover from the innermost elements of the basement (Mont Blanc, Gothard). The *Helvetic nappes*, inclined N or NW, have thus been emplaced over the external crystalline massifs and their autochthon or parautochthon cover (Fig. 180). These nappes were emplaced before the subalpine chains, probably in the Oligocene, immediately after formation of the internal zones, in which some units (known as prealpine nappes) have even overthrust them, thus constituting an overload of 2000 to 3000 m thickness.

The Neogene contraction subsequently completed the deformation and folded the suite into an anticline of nappes (Fig. 180).

Internal (or Penninic) zones, everywhere separated from the external zone by a major thrust plane (called the Frontal Penninic Thrust), form a complex of nappes with N or NW vergence, with a double vergence.

These are interpreted as the ancient margin proper of the European continent, i.e., a zone with thin crust, which is more and more intensively fragmented when approaching the Tethys sea due to the usual rifting of the

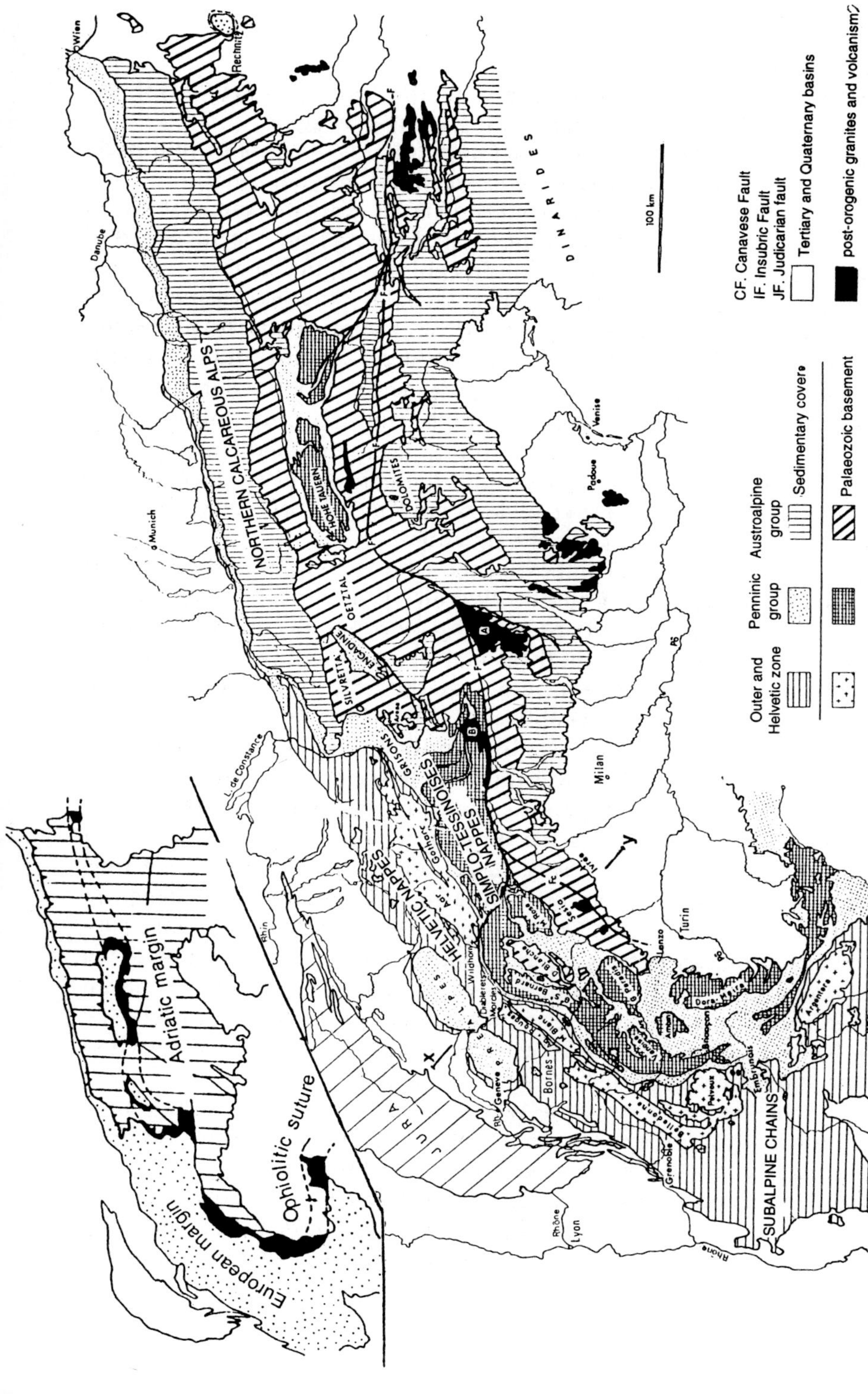

Fig. 178. Major geological features of the Alps and location of the geographic names cited in the text.
X-Y. Section of Fig. 180.

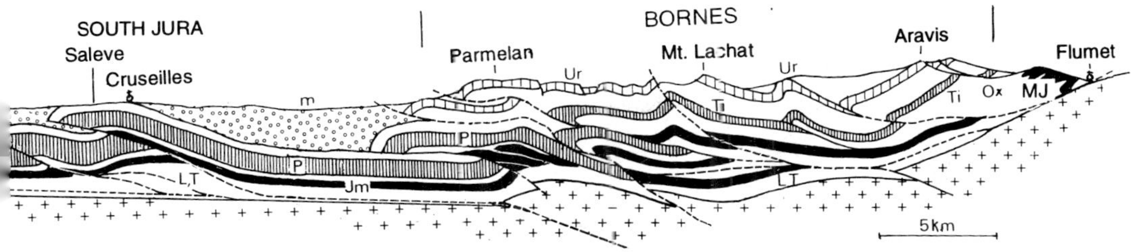

Fig. 179. Tectonic style of Bornes massif (subalpine chains of Savoie). Depth and style of the top of the basement are constrained by the ECORS-Alps profile (Fig. 187) (after Guellec et al., simplified, 1989, *C.R.Ac. Sc.*, 309: 172).
This section highlights the continuity of the sedimentary cover between the Jura and the subalpine chains, as also its break up into superposed slices: this structure is associated with the underthrusting of the Jura and Dauphinois basement below external crystalline massifs.
m: Miocene or Oligocene molasse; LC: Lower Cretaceous; Ur: Urgonian; P: Portlandian; Ti: Tithonian; Ox: Oxfordian; MJ: Middle Jurassic; L: Liassic; T: Triassic.

passive margins (see Fig. 103). Thus, in this zone there were more or less tilted and juxtaposed blocks, horsts and grabens, separated by normal faults which could have been rejuvenated into reverse faults during the Tertiary compression (the phenomenon of 'tectonic inversion').

Thus, a succession of domains can be distinguished that are both palaeogeographic and structural, among which the important are:

— *Briançonnais domain,* representing the major block, most uplifted during rifting (Liassic emergence);

— *Piemont domain,* representing, on the contrary, the block or blocks nearest to the Tethys sea, i.e., the most submerged. Sedimentation here is very monotonous (and will be the most metamorphosed at the time of alpine folding, giving rise to the classic Piemont 'Schistes lustrés').

The *Penninic basement* is always granito-gneissic and forms the *internal crystalline massifs* (Vanoise, Grand Saint Bernard, Gran Paradiso, Monte Rosa etc.), with a structure very different from that of their outer homologues since here the 'penninic fold' style is common, even for the gneisses (very flexible, long recumbent folds with preserved reverse flanks and axial-plane schistosity (Fig. 180). A-type lineations are the rule in this context of ductile shear movements. Their mean direction represents that of the major thrust movements and forms a fan-shaped structure across the arc of the Western Alps.

The sedimentary cover of these Penninic cores is almost always detached (Briançonnais and Piemont nappes). In any case, the contacts are stretched in such a manner that their original nature most often remains uncertain, all the more so as a greenschist facies affects the whole of these units, associated with one of two or three superposed schistosities. Various minerals

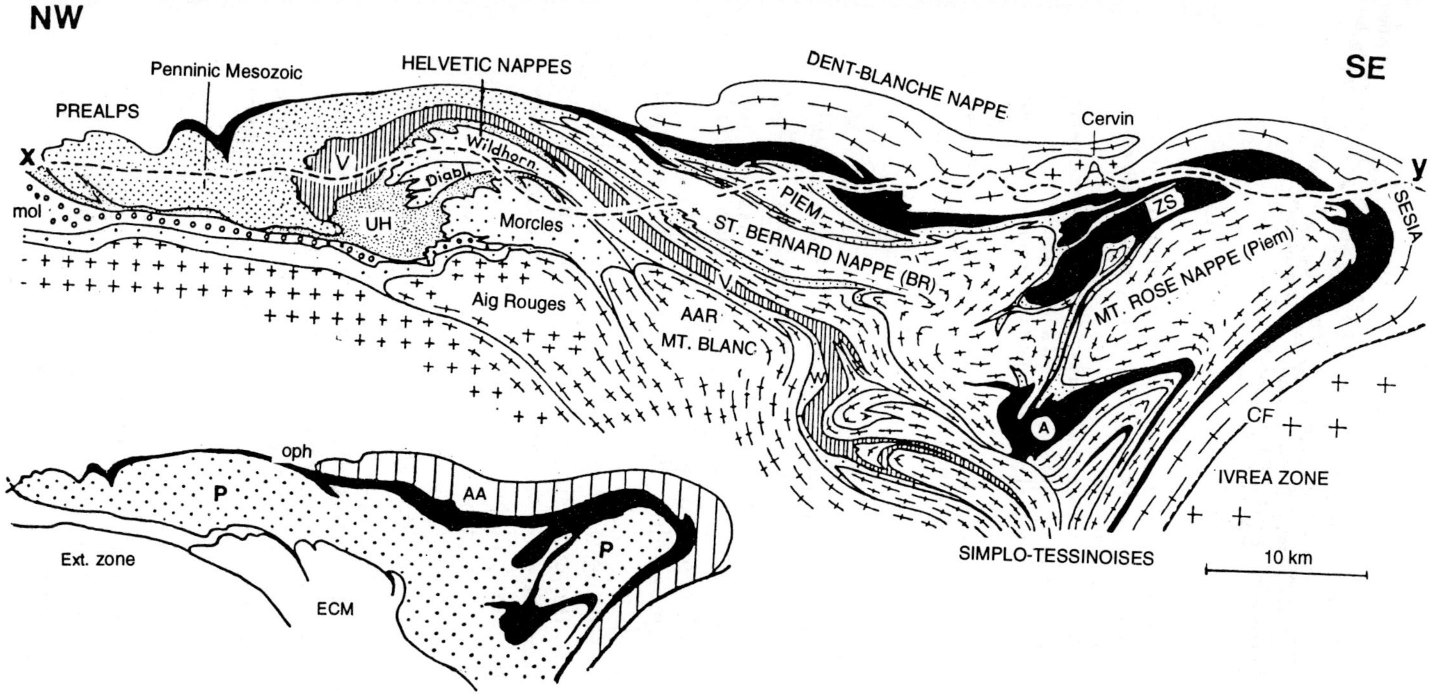

Fig. 180. Penninic nappes (after Escher et al., 1988).
The cross-section clearly indicates the specific tectonic style of these glove-finger-shaped nappes, with preserved reverse limbs and constant axial plane schistosity, all these characteristics indicating a highly deformable (ductile) environment.
Below the Austroalpine ranges, the following rocks are observed, from top to bottom:
— *Piemont units* (Piem): nappe of Mont Rose,
— *Briançonnais units* (BR): nappes of Grand Saint Bernard,
— *Valais units* (V): nappes of Simplo-Tessinoises basement and nappes associated with Schistes Lustrés (called 'Grisons'),
— *Ultra-Helvetic units* (UH).
The oceanic units (in black) are represented by ophiolitic nappes of Zermatt-Saas (ZS) and Antrona (A).
Other symbols: CF. Canavese Fault; X-Y: Present topographic surface.
Simplified scheme: AA: Austroalpine; mol: molasse; ECM: External Crystalline Massifs; oph: ophiolitic zone; P: Penninic.

produced from this metamorphism have given an age of 38–40 Ma, corresponding to Upper Eocene, for this so-called 'meso-alpine' paroxysmal phase. Later, the Neogene events also affected the Penninic structures.

Among the formations constituting this Penninic cover, the best known are the 'Piemont Schistes Lustrés', devoid of ophiolites, whose sequence starts immediately after a thick Triassic dolomitic layer and continues as monotonous facies up to the end of the Cretaceous (dated). While their nappe structure is evident, their internal configuration is much more difficult to determine, particularly so because these schists are intricate with their Liguria-Piemont ophiolitic homologues, i.e., the ancient sedimentary cover of the alpine oceanic region, which is described below.

2. ***Ophiolitic formations*** (Fig. 178): The ophiolite massifs and their sedimentary cover occur from the Mediterranean to Grisons, and then continue below the overthrust of the Adriatic margin (the *Austroalpine* nappes). This is the essential difference compared to the Himalaya. They are represented by the particular ophiolites in which peridotites and gabbros predominate. Pillow basalts are often absent as already mentioned (p. 5). They are also associated with the *Lanzo peridotitic massif*, sometimes considered to be the upper mantle of the south alpine (= Adriatic) region.

The sedimentary cover of these ophiolites is known as the *Liguria-Piemont* in France and Italy because it is also found in the Ligurian Apennines. In Switzerland it is called *Zermat-Saas zone*. It commences from the end of the Middle Jurassic and thus gives the age of the oceanic opening. The formations of this cover are radiolarites followed by pelagic limestones and black shales. The sequence continues up to the Upper Cretaceous after which it is folded and metamorphosed into high-pressure facies (radiometric ages of 110 to 80 Ma), and hence *before the Penninic region*. However, the whole group is involved in Eocene folding and concomitant greenschist facies metamorphism, thus making it difficult to distinguish, as already mentioned, the Piemont Schistes Lustrés of the Liguria-Piemont (except for their basements, viz., Triassic-Liassic for the former and ophiolitic for the latter).

Locally, part of this cover of the oceanic region (and its immediate boundaries) escaped metamorphism by detaching itself very early and sliding outwards of the Alps as gravitational nappes. This is the *Helminthoid Flysch* (Fig. 181) which is now found in the major klippes on the external zone (Prealps, Embrunais, Fig. 178).

This flysch is an interesting sediment because it was deposited in the alpine ocean in the Upper Cretaceous, i.e., at a moment when this ocean was reduced by subduction, such that it could locally record the consecutive deformations. Notably, olistoliths of the ophiolites which indicate the emplacement of slices of the oceanic crust are found in the Middle Cretaceous clastic horizons of 'black shales' facies. In fact, palinspatic reconstruction of the depositional basins of these flyschs suggests at least two of them, separated by a rise of ophiolitic rocks (ancient ridge ?), whose extension has been traced in the Ligurian Apennines by Italian geologists (Fig. 181).

3. ***Adriatic (= Insubric) margin***: It is customary to divide the Adriatic margin into two parts (Fig. 178):

— The northern part, overthrust on the preceding formations. It is represented by the *Austroalpine nappes*, with an ancient basement, Precambrian to Palaeozoic (the *Silvretta-Oetztal nappe*) and Mesozoic cover, which is detached and thrust independently northwards (the *North Calcareous Alps*, mainly Triassic).

The overthrust nature of this group is manifest by the windows of Tauern and Engadine in which appear the Penninic Schistes Lustrés with ophiolites or, inversely, by the klippe of the Dent Blanche and Cervin (whose roots, though uncertain, may have been the Sesia zone or the Ivrea zone according to some authors), and by the position of the North Calcareous Alps upon the peri-alpine Miocene formations (through an intermediate Penninic or Helvetic pack of slices).

— The southern part, known as the *'south alpine'* zone, whose overthrust nature cannot be demonstrated but must be assumed because it is separated from the Austroalpine region only by a major vertical fracture zone, called the *Insubric fault or line* (Figs. 178, 188). This fault, compressive and transverse (dextral strike-slip fault at least over 80 km), forms one of the sides of the Adriatic indenter, the other being constituted by the Judicarian fault. The most compressed zone of the eastern Alps and hence the most uplifted and the most eroded (Tauern window in which Penninic formation outcrops) is at the front of this indenter. Heating produced by this overthickening of the crust has induced partial melting and emplacement of granodioritic plutons of Oligo-Miocene age, the most important of them being the Adamello massif (Fig. 178). A calco-alkaline volcanism also exists in the Padova region. All this magmatic activity indicates melting of the deep crust in the core of this structure of nappes.

The south alpine zone consists of an ancient basement, Precambrian to Palaeozoic, and a sedimentary cover indicating quite a varied palaeogeography of this Insubric margin of the alpine ocean, due to the existence of tilted blocks or horsts as in the case of all passive margins.

Its tectonic structure, apparently simple on the surface, in fact reveals large cleavages leading to flat overthrust zones with southern vergence (Fig. 182). They are of Neogene age and thus give the completed chain, symmetrical with Dauphinois inclination, a double-vergence structure as in the case of classic intracontinental chains.

Structural Evolution

The structural evolution of the Alps consists of two parts:

— An *extension period (Triassic-Aptian)*, during which an oceanic domain developed between two continental margins that gradually moved away

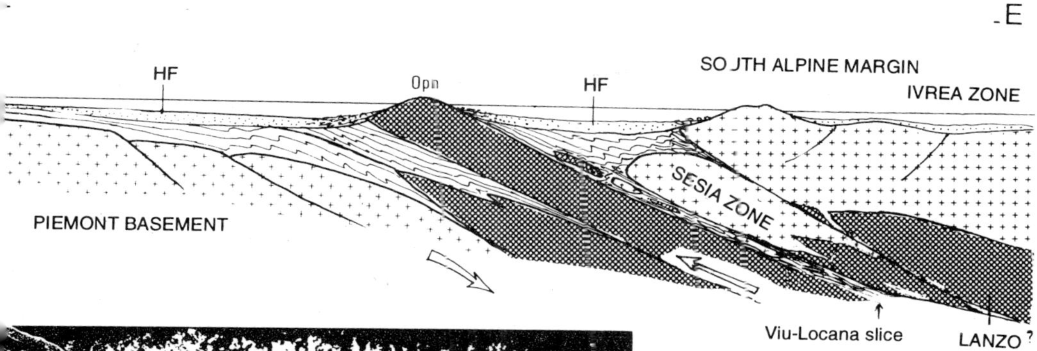

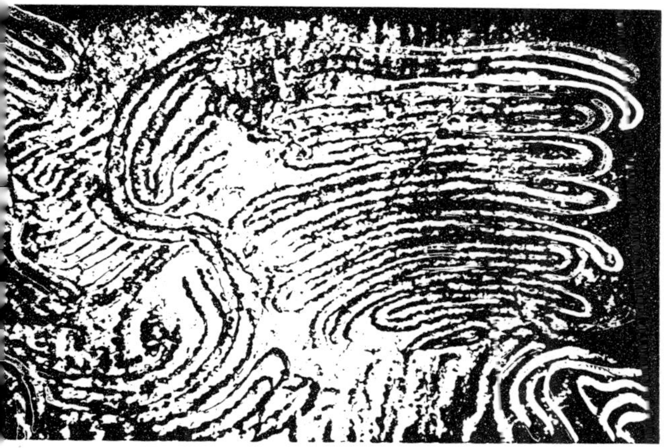

Fig. 181. Helminthoid flysch (HF) and its original structural framework.
As this flysch is entirely detached from its zone of origin, it can be reconstructed only very hypothetically. Various sedimentological evidence (olistoliths and ophiolitic or granito-gneissic pebbles) suggest the existence of two basins surrounding an ophiolitic ridge. The latter was perhaps an obducted slab or in the process of obduction.
In any case, this flysch is a syntectonic sediment accompanying the closure of the alpine ocean. It is deposited above the subduction plane and its accretionary prism, in which occur the Liguria-Piemont sediments (future Schistes Lustrés), gneissic slabs of the Adriatic margin (south alpine), like the Sesia massif, and the ophiolitic slices of the ancient oceanic crust (Lanzo massif).

from each other in the Upper Jurassic and disintegrated into partly tilted blocks and strips which correspond to major palaeogeographic zones.

Prior to the opening, i.e., in the Liassic and Middle Jurassic, a rifting process thus occurred across a continuous platform, resulting in the partial emergence of the rift borders (Briançonnais region, for example, on the European side) (see Fig. 103).

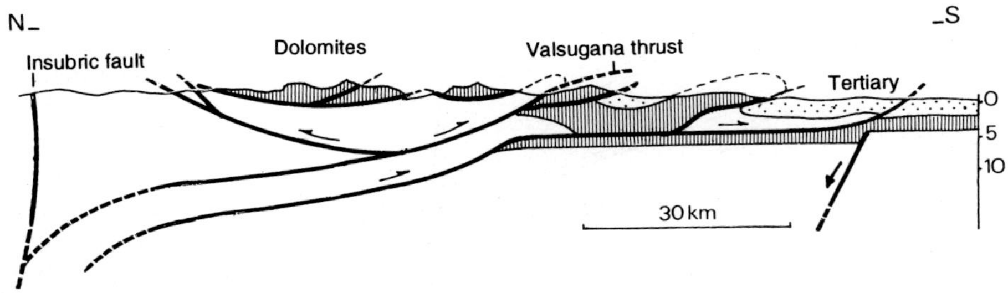

Fig. 182. Schematic cross-section of the south alpine zone in the region of Dolomites (after Doglioni, 1987).
Neogene crustal shortening gives rise to structures with a southward vergence, complicated by an antithetic thrust fault in northern Dolomites.
Palaeozoic basement in white, Mesozoic cover (including Upper Permian) in vertical hatching.

After the opening, these borders were submerged by thermal debulging, as indicated by the general occurrence of the pelagic facies in all the alpine zones at the end of the Jurassic and in the Lower Cretaceous.

— A *compression period (Aptian-Present)*, which is obviously the most complex. It consists of several stages during which deformation gradually extended to the forelands as in the Himalaya.

1. At first, the *principal oceanic (Liguria-Piemont) domain disappeared* by subduction below the south alpine margin from the Aptian to the end-Cretaceous. This subduction was accompanied by a high-pressure metamorphism dated by its minerals as 110 to 80 Ma (Fig. 190). The gradual disappearance of this oceanic lithosphere led to the detachment and piling into nappes of its sedimentary cover, associated with ophiolites, as a true tectonic accretionary prism in front of an active margin. These initial movements evidently caused the formation of the Austroalpine nappes on the disappearing oceanic domain. The absence of active-margin volcanism in the latter as well as in the south alpine range has not been satisfactorily explained to date.

2. After the disappearance of the entire alpine oceanic crust, the *convergence of the two margins, henceforth in contact, leads to their tectonic deformation.* The normal faults that separate the blocks and strips of the European margin are rejuvenated into reverse faults, but essentially, major flat cleavages are produced within the crust which rise to the surface by means of ramps consisting of these reverse faults (Fig. 184). The slabs thus delineated are overthrust (Fig. 183) and the bottommost are then metamorphosed to amphibolite to greenschist facies, radiometrically dated as 38 to 40 Ma (*Upper Eocene*). Deformation is of the Penninic type (long ductile recumbent folds), both in the case of gneissic material as well as the sedimentary cover.

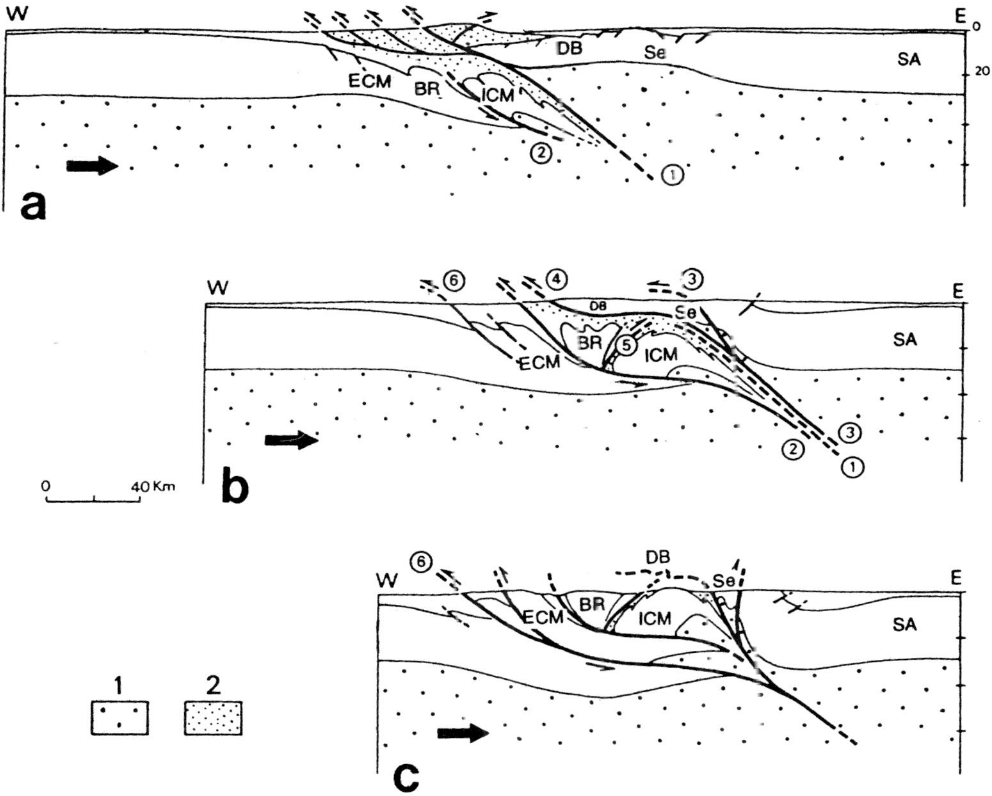

Fig. 183. Model of evolution of collision in the Western Alps (cross-sections of the Alps of Savoie and Gran Paradiso) (after Mattauer, *in* Malavieille et al. (1984) *BSFG*, modified).
1. Upper mantle, 2. ophiolites and their sedimentary cover.
Stages a and b: End of collision (Lower Oligocene ?):
— contact 1: trace of ancient subduction plane and obduction surface of residual oceanic material on the European margin.
— contact 2: first cleavages of the European margin.
The contacts 1 and 2 evolved into thrust planes in the midst of the formations of the inner zones.
— contacts 3 and 4: cleavages within the Austroalpine block (Adriatic margin), leading to formation of the Sesia-Dent Blanche unit (Se, DB).
— contact 5: backthrusting of Briançonnais (BR) block onto the internal crystalline massifs (ICM).
— contact 6: First imbrication of the external crystalline massifs (ECM).
Stage c: From the Neogene contraction to the Present:
Contact 6 has become the principal thrust plane.
Abbreviations: BR: Briançonnais; DB: Dent Blanche; ECM and ICM: external and Internal crystalline massifs; SA: south alpine; Se: Sesia.

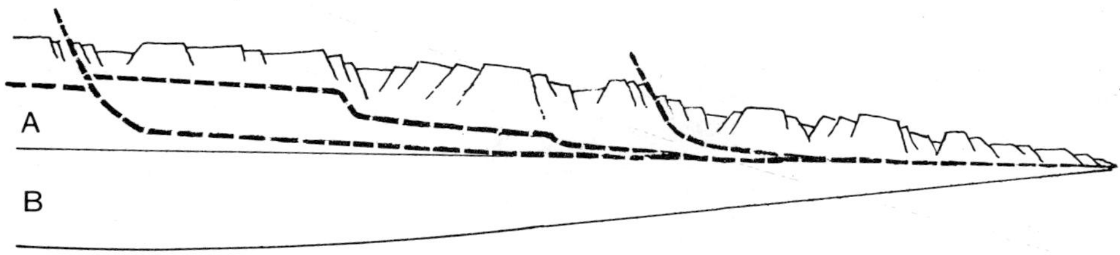

Fig. 184. Hypothetical model of the origin of flat cleavages in the alpine continental margin. These cleavages are probably produced in the low-velocity zones that separate the rigid crust (A) and the ductile crust (B). They may be continued by a listric fault acting as a ramp and thus isolate a compartment which, during contraction of the margin, will give a gneissic nappe.

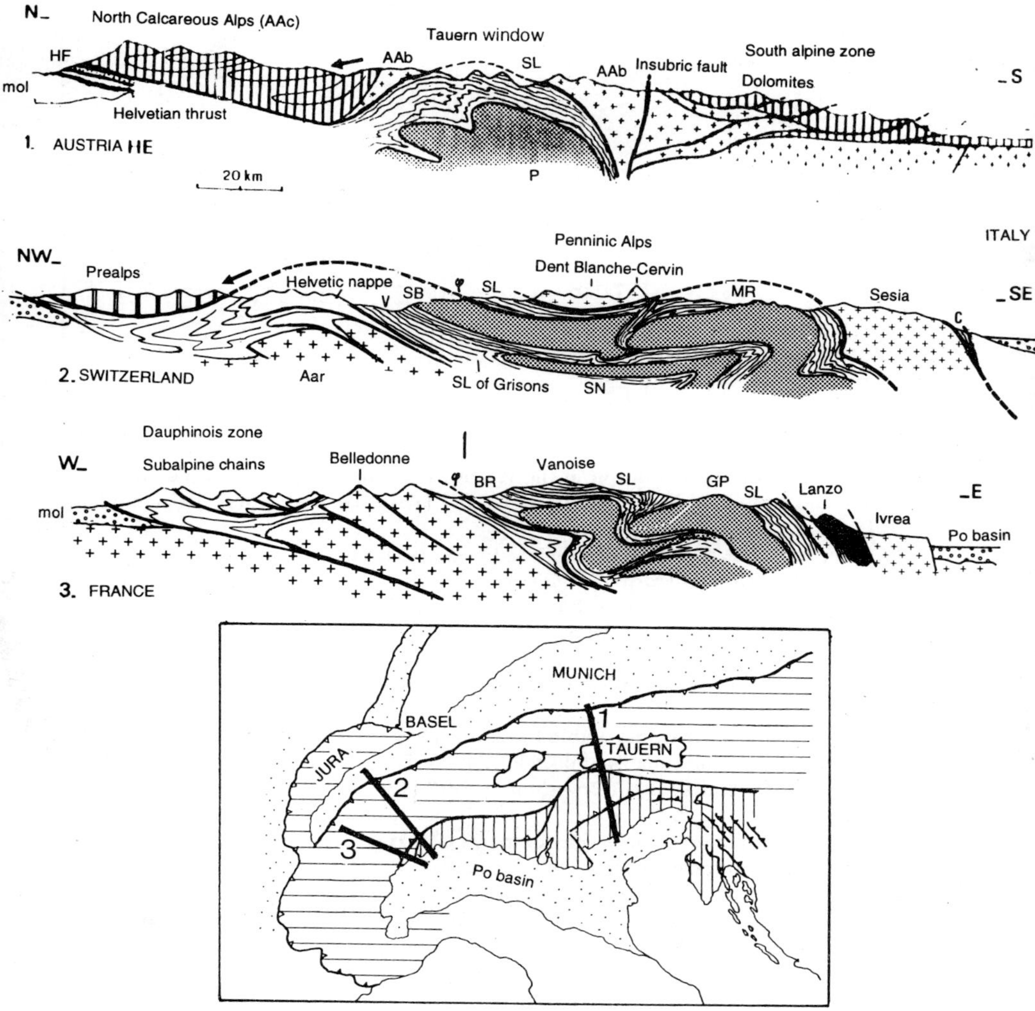

3. *In the Oligo-Miocene,* deformation extends beyond the European margin proper (in other words, the future internal zones) and reaches the boundary of the future external zone. The basement of the latter is compressed and shortened, but without Penninic-style folds. The corresponding sedimentary cover, detached from the basement, gives the Helvetic nappes with N or NW vergence.

In fact, analysis of this deformation is rendered complex by the arcuate form of the chain which must have been accentuated at that time. In the Austrian and Italo-Swiss Alps the chain is practically E-W and hence perpendicular to the direction of movement of the Insubric plate. Hence, deformation is maximum there. In the Franco-Italian Alps the chain appears as a lateral effect of the collision. Deformation there is less intensive, such that the external zone is less affected.

This Oligo-Miocene accentuated compressive state produces, on the surface, a double-vergence structure, whereas at depth, a high temperature is maintained and causes partial melting of the gneisses piled up there. The result is emplacement of the Adamello type granodioritic plutons along the Insubric line and, perhaps, contemporary calco-alkaline volcanism which is manifest farther south, in the boundary of the Po plain.

Fig. 185. Comparative schematic structures of the Franco Italian, Italo-Swiss and Italo-Austrian Alps.
AA: Austroalpine (b. basement, c. cover); C: Canavese; P: Penninic (BR: Briançonnais nappes, GP: Gran Paradiso nappe, MR: Monte Rosa nappe; SB: Grand Saint Bernard nappe; SL: Schistes Lustres; SN. Simplon nappes; V: Valais); HF: Helminthoid flysch; mol: Miocene molasse.
1. *In the case of the Austroalpine ranges,* section 1 is the most representative: The Austroalpine nappe appears as a major thrust slab whose sedimentary cover, being detached, gives rise to the North Calcareous Alps. Observe the Tauern window and the Insubric fault slightly sloping southwards. In section 2, the Austroalpine nappe is reduced to the Dent Blanche-Cervin klippe. In section 3, the Austroalpine nappes have disappeared. Only their root, i.e., the Sesia or the Ivrea zone remain (according to some authors).
2. *In the case of the Penninic range,* section 2 is the most representative. It clearly indicates the gneissic Penninic nappes, enveloped by the Schistes Lustrés units. A part of the cover has escaped this metamorphism by early decollement and gravitational sliding and forms the Prealps westwards. This section also shows the late 'backthrusting' of the Penninic roots.
The Penninic material reappears, in section 1, in the Tauern window (gneisses and Schistes Lustrés with ophiolites) and in the slices situated below the overthrusts of the North Calcareous Alps.
3. *In the case of the external zone,* section 3 is the most significant. The external crystalline massifs form large overthrust slabs whose cover is highly folded in Switzerland (Helvetic nappes) and gradually passes on to subalpine folds in France. In this section the Penninic material is already developed but the disposition of its basement is poorly known.

The Insubric fault then appears at the contact of that part of the Alps which is uplifted (by thermal intumescence and isostatic response to the crustal thickening) and the part which remains downward, with even a tendency of collapsing southwards (the Po basin, see p. 103). The slightly backthrust disposition of the Insubric fault indicates the persistence of the compression regime.

4. Starting from the *Upper Miocene*, contraction advances outwards as in the Himalaya with replacement of the MCT by the MBT and causes a major basement cleavage of the forelands and its underthrusting below that of the inner zones, with slicing of the external crystalline massifs and folding of the French subalpine chains.

This underthrusting enhances the crustal thickening, whereas the internal zones, hardened by metamorphism or by deep granitisation, are folded into a major basement fold. These two simultaneous actions lead to a major uplift of the chain.

The Neogene contraction also causes termination of the emplacement of Helvetic and Prealpine nappes in Switzerland, and that of the North Calcareous Alps in Austria, on the Miocene perialpine basin. Finally, it is this contraction which terminates the south alpine thrusting with southward vergence (Fig. 182).

Thus, we come to the present structure summarised in the section of Fig. 185.

The present deformation, as deduced from seismicity or focal mechanisms, has a main compressive component always perpendicular to the chain, i.e., northwards in the Austrian and Swiss Alps, to NW, W and finally SW in the French-Italian arcs, the latter directions not at all related to the Africa-European convergence. The causes of such a deviation are still conjectural.

Alpine metamorphism

Deformations leading to the structures just described are accompanied by metamorphism as in the case of all the collision chains. The Western Alps, extensively studied, can serve as a good example or model. Thus it can be stated that:

1. Within these metamorphisms, there is *a pressure gradient inwards of the chain*, i.e., from the external zone to the collision proper: rocks vary from non-metamorphosed in the subalpine chains to eclogites in the internal crystalline massifs and certain units of Schistes Lustrés, which shows that the innermost units were the deepest subducted.

2. *Metamorphic evolution occurred in several stages*, the first ones of high pressure and the latter ones of low pressure. Superposition of these stages is observed in the rocks of the innermost zones, the outer zones undergoing metamorphism only at the end of orogeny.

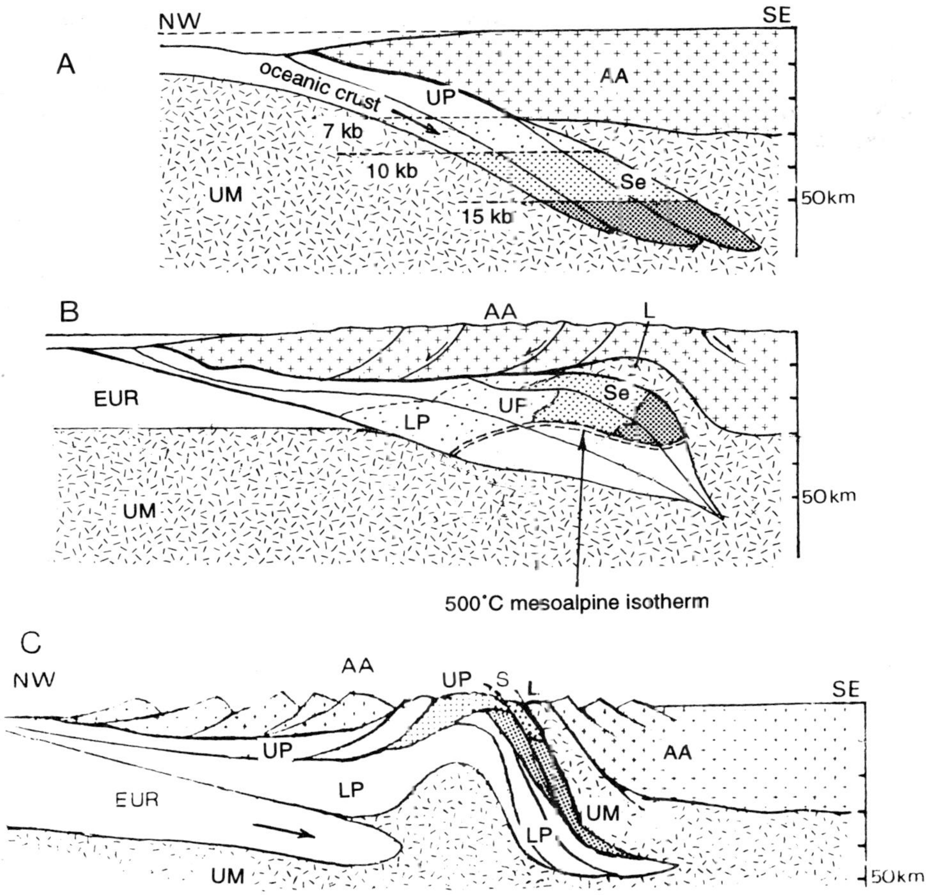

Fig. 186. Three stages of alpine metamorphism in the internal zones of the western Alps (after Platt, 1986, modified and completed by stage C).
A. *Upper Cretaceous*: subduction of the oceanic crust is in progress. The Sesia zone (Se, considered here as the extreme boundary of the Austroalpine margin AA) and the alpine oceanic material (Upper Penninic, UP) are dragged to depths of 50 to 70 km. There they are metamorphosed to high-pressure (grey shaded) and low-temperature (cold material) facies.
B. *Upper Eocene*: The European continental margin (EUR) is in turn involved not in the subduction process, which had ceased, but in the piling up of slices of the accretionary prism. It corresponds to Lower Penninic (LP, i.e., Briançonnais and Piemont belts). This arrival of new underthrust units led to the rapid uplift of the Upper Penninic formations and the Sesia zone already metamorphosed to high-pressure facies. This uplift may have induced or facilitated, on the surface, gravitational collapse of the Austroalpine nappes. In any case, this phenomenon led to the rise of isotherms and hence to facies metamorphism which became superposed on the earlier high-pressure paragenesis.
C. *Oligocene*: Ascent of the pile of nappes along ductile faults, whereas the Austroalpine structure, eroded, stretches into faulted blocks.
L: Lanzo peridotites; UM: Upper Mantle.

3. In the succeeding zone, *increase in the degree of metamorphism is not gradual*, each abnormal contact corresponding to a jump of metamorphism that may extend up to several kilobars. The contacts of presently visible nappes are no longer those of primitive piling up of units at the moment of metamorphic processes.

The highest degree of HP metamorphism is variable depending on the units. The inner Briançonnais units have reached blueschist to glaucophane facies, in which a greenschist facies has succeeded in the Upper Eocene (or around 38 Ma). In the innermost units as well (oceanic Schistes Lustrés and internal crystalline massifs), still higher pressures were reached, with eclogite facies and even (Italian massif of Dora-Maira and Zermatt zone) 'white-schist' facies with coesite (a variety of silica whose experimental formation requires a pressure of 30 kb) (Fig. 191). This high-pressure metamorphism of the innermost units is complex and dating of these different phases between the Lower to Upper Cretaceous-Palaeocene and the Eocene is presently the object of discussions.

However, in any case, the units metamorphosed in these high-pressure conditions later rise as a whole towards the surface (Fig. 186) as quite a general retrometamorphosed unit is observed in the greenschist facies (temperature preliminarily of the same order of magnitude and then decreasing, but pressure gradually increasing). Metamorphic lineations show that this retrometamorphism was contemporary with emplacement of the Eocene nappes which intersect randomly, so to speak, the boundaries of high-pressure facies.

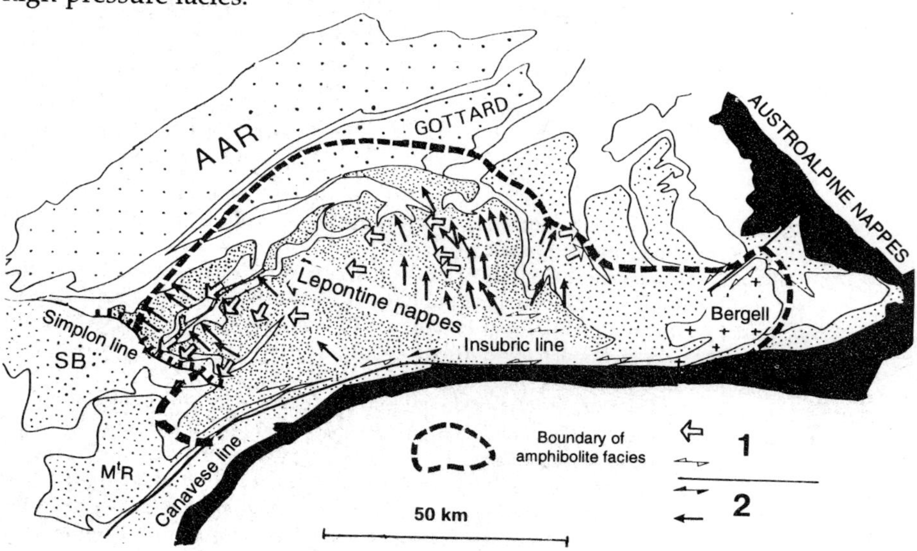

Fig. 187. Lepontine thermal dome (after Merle et al., 1986, in Jolivet, 1995, simplified). Arrows indicate lineations and displacements of Upper Eocene (1) and Neogene (2) age.

In the **Swiss central Alps** (Lepontine or Simplo-Tessinoises Alps), the model is more complex since the pile of Eocene nappes there undergoes, starting from the Oligocene and due to its E-W orientation, a much stronger compression than in the Western Alps because of the northward movement of the Adriatic margin which is produced then. This margin, cold and rigid, indents into the European margin, thick, hot and ductile, thus making the pile of nappes rise towards the surface. Evidently, there is at the same time transfer of heat, mainly along the vertical zone situated at the contact of the two plates, which triggers mobilisation of granitic magmas (Val Bregaglia).

Thus, in this part of the Alps, conventional metamorphic evolution (eclogites or glaucophane schists changing to greenschist facies of Upper Eocene-Lower Oligocene age) terminates by the appearance of a thermal dome (Fig. 187) inside which the temperature reaches 650°C, for a pressure not exceeding 6 kb (conditions of amphibolite facies, locally even those of granulites and anatexis).

This process also leads to an uplift in which the ultimate relation with a gravitational extensional stretching of the structure will be examined later.

Deep structure

The Alps indicate a thick 'root' extending to a depth of 50 to 60 km below the Penninic region.

Seismic data, particularly that obtained from the ECORS profile, shows that this root is probably made up of superposed slabs within which two slices of the upper mantle are believed to be identified, at least in the northern section of the Western Alps (Figs. 188, 189, 190). They are combined with the third slice which arrives on the surface, farther south, to give the Lanzo massif, and is accompanied by a strong positive gravity anomaly (known as the Ivrea anomaly).

The ECORS profile further confirms the existence of a cleavage surface inclined at 30° below the Belledonne massif, a structure generally assumed for all external crystalline massifs. This cleavage seems to be associated with the underthrust of the Dauphinois basement eastwards. The profile also seems to show recent faults in the region of the Penninic front which are probably of extension origin.

In Switzerland, profile NFP 20 obtained on the cross-section of the Aar massif, almost parallel to the ECORS profile, revealed a quite similar structure though more compressed (Fig. 189).

Syntectonic and Post-tectonic Extensions in the Alps

Like all the orogenies with sufficiently thick accretionary prism, the Alps exhibit surficial phenomena of extensional type detachment, which must have occurred since the uplift of the prism in question. We have already men-

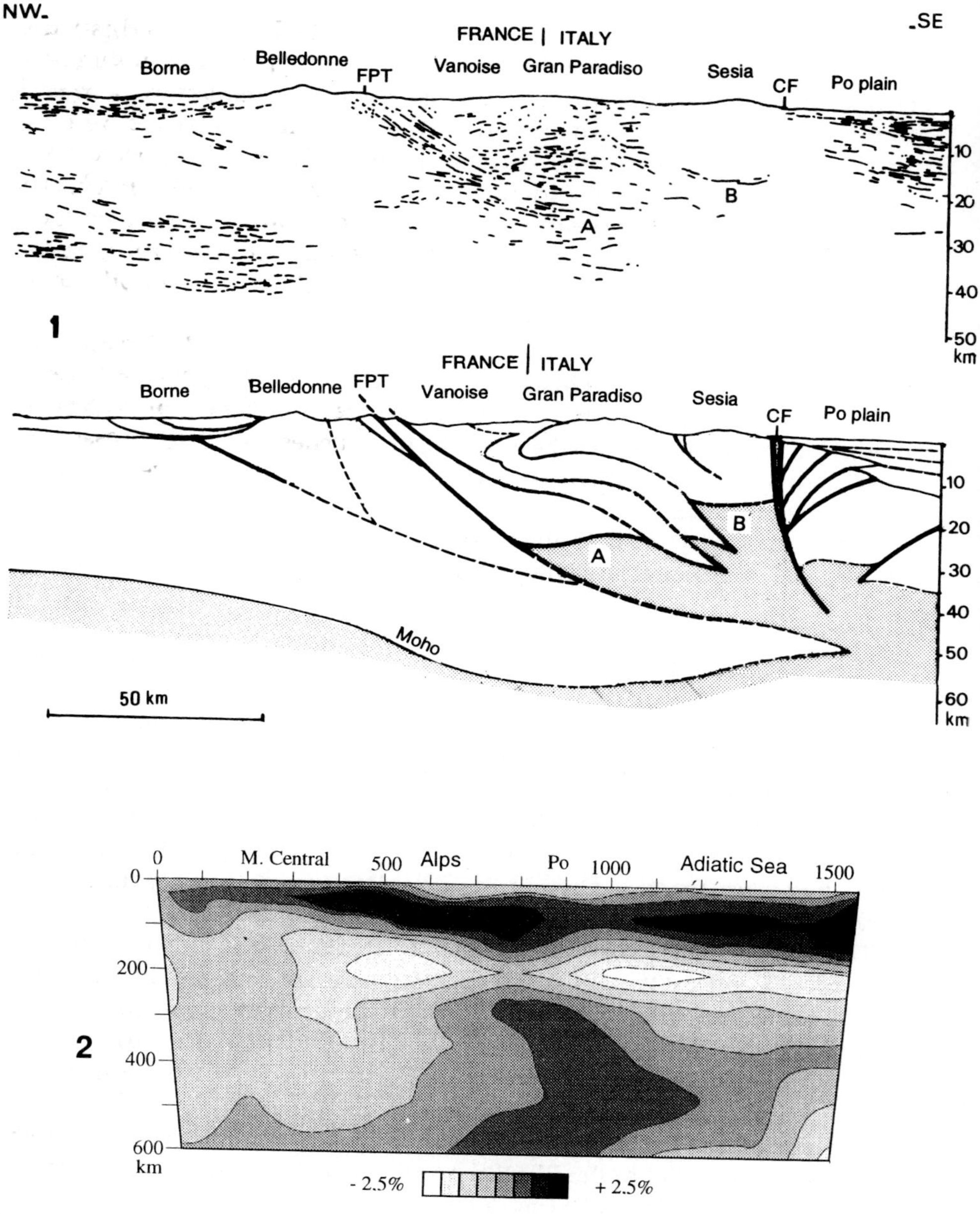

Fig. 188. 1. Deep structure of the Alps. Structure of the crust in the region of the Western Alps according to the ECORS profile (after Tardy et al., 1990, in: *Mem. SGF*, no. 156).
The interpretation presented here prefers the existence of large slices derived from the cleavage of the ancient European margin and resting on the European continental crust that is curved by flexure. Their piling up determines the deepening of the Moho up to a depth of about 60 km. These are, from bottom to top (from left to right):
— the *lower slice*, of the external crystalline massifs, which is overthrust onto the European

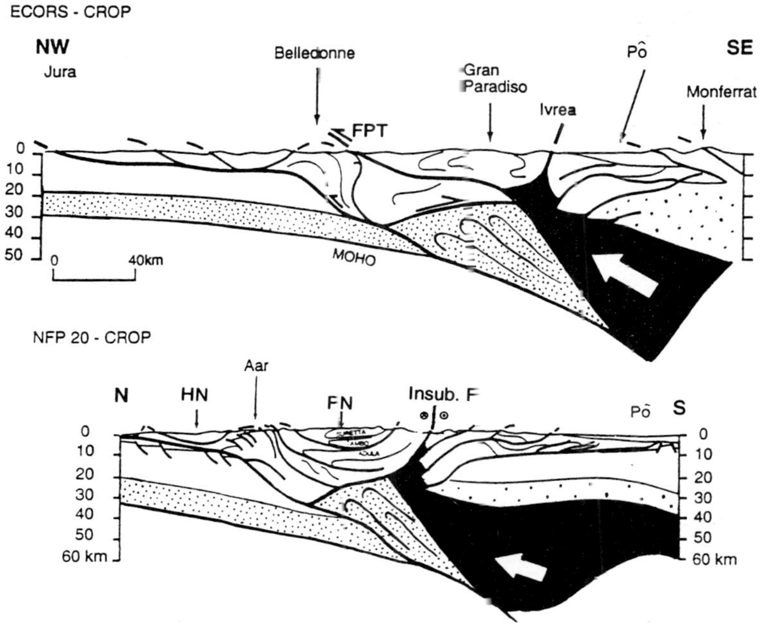

Fig. 189. Comparison between ECORS-CROP (France-Italy) and NFP20 projects (Swiss-Italy) (after Roure et al., 1966, simplified).
White: Upper crust; dotted: Lower crust; black: Insubric (= Apulian) mountain.
FPT: Frontal Penninic Thrust, HN: Helvetic nappes, PN: Penninic nappes.

---

crust over a length of about 25 km, thus compressing outwards the sedimentary cover of the latter (subalpine chains and the Jura).
— the *middle slice*, bounded by the Frontal Penninic Thrust (FPT), in which the upper mantle (A) participates.
— the *upper slice*, equally affecting the mantle (B).
— (CF) Canavese or Insubric fault.
2. The lithosphere beneath the Alpine region: seismic tomography along a WNW-ESE section (Rennes, Lyon, Milan, Dubrovnik), after Spakman, Geol. Ultraiect, 1990, simplified. By observing the variations in the velocity of P wave propagation, one can illustrate the thermal structure along a profile; slow velocities correspond to hot zones and accelerated velocities to deep and hence more rigid zones. A thickenning of the lithosphere below the Alps and its continuation eastwards can be noticed. The cold zone situated beyond 250 km depth below the Padan plain may represent a fragment of the detached lithosphere.

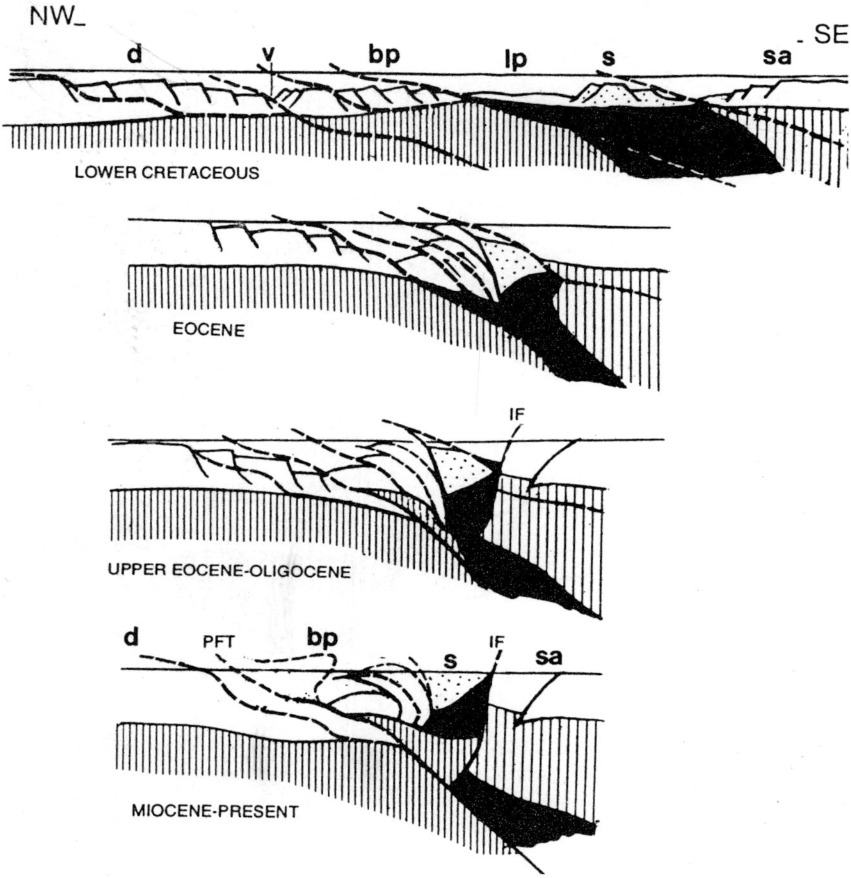

Fig. 190. A possible evolution of Western Alps along ECORS-CORP profile, indicating a lithospheric and crustal slicing. These slices are produced in the weakness zones in the crust which are the Valais (v) and Liguria-Piemont (lp) troughs. Existence of an east-Sesia trough is suggested.
bp: Briançonnais and Piemont regions; d: Dauphinois region; lp: Liguria-Piemont region (oceanic); s: Sesia zone; sa: south alpine zone; v: Valais region.
PFT: Penninic Frontal Thrust, IF: Insubric Fault.
Vertical dense hatching: European mantle; Vertical wide hatching: Insubric (or south alpine); black: suboceanic mantle.
(after Nicolas et al. (1996) *BSGF*, 170: 100, modified).

tioned them in connection with metamorphism because it is these phenomena which, through the denudation they produce, exhume the most metamorphosed parts of the pile of nappes and that, too, so rapidly that the high-pressure minerals do not at all undergo retrograde thermal metamorphism. However, spreading may continue for some more time as long as the conditions at the boundaries permit the same.

The **Lepontine** or **Simplo-Tessinois** massif (see Fig. 187) is a particularly striking example according to which a veritable dome of nappes, accompanied by a high thermal flux, forms in the Upper Eocene and lasts up to the end of the Oligocene, an epoch in which even granitoids derived from magmas generated in the upper mantle form in proximity. This dome is bounded in the SW by a true transverse detachment fault, the Simplon fault, which led to isolation of the dome beyond its cover of Penninic nappes. It may be considered as belonging to a metamorphic core complex (MCC) type evolution, even though this alpine episode is situated in a compressive context associated with the nearing of African and European plates to each other. It cannot be ruled out that a rapid tectonic denudation phenomenon occurs there operating on normal fault type shear surfaces.

An almost similar case is that of the inner crystalline massif **Dora-Maira**, in the Franco-Italian Western Alps where a true dome of nappes is also uplifted starting from the Upper Eocene through its cover, resulting in rapid outcropping of coesite facies occurring over a depth of about 100 kilometres. There also, the dome is bounded on its only visible western side by ductile shear surfaces of normal fault type (Fig. 191) separating gneisses and eclogitised ophiolites of Viso from their envelope of Schistes Lustrés. They clearly represent rapid uplift of the massif accompanied by detachment of its cover of nappes. Differences of 2 to 3 kb and 100 to 150° between eclogitised ophiolites of Viso and Schistes Lustrés with blueschist facies imply a vertical displacement of at least 8 km along these faults, evidently later to metamorphism.

In fact, the subsequent behaviour of these detached units remains the most difficult to analyse. One case is, however, quite spectacular, viz., the **Eastern Alps**, east of Tauern (Fig. 192). This part of the Alps effectively underwent, in the Oligo-Miocene, a vigorous compression which affected the pile of nappes dating Upper Cretaceous (see p. 266). The compression was especially more vigorous at this point as we are at the head of the 'Insubric' or 'Adriatic' indenter i.e. the basement of Southern Alps east of the Judicarian fault (Fig. 192). The sinistral activity of this fault and the associated en echelon folds clearly represent this northward indentation. This process led to a crustal overthickening (50 km) of the Tauern region, which contrasts with the thinning of the Pannonic basin east of the Eastern Alps, where the crust does not exceed 25 km in thickness. It would thus promote gravitational spreading of the Eastern Alps towards this basin.

The Tauern dome, where highly metamorphosed (more than 10 kb) Penninic gneisses with ductile behaviour outcrop as inliers, is interpreted as a veritable (MCC) bounded in the north and south by very straight ductile faults and in the west by a detachment fault above which the central part of the Eastern Alps slides eastwards, with an altitude effectively decreasing up to the Pannonic basin where it sinks down.

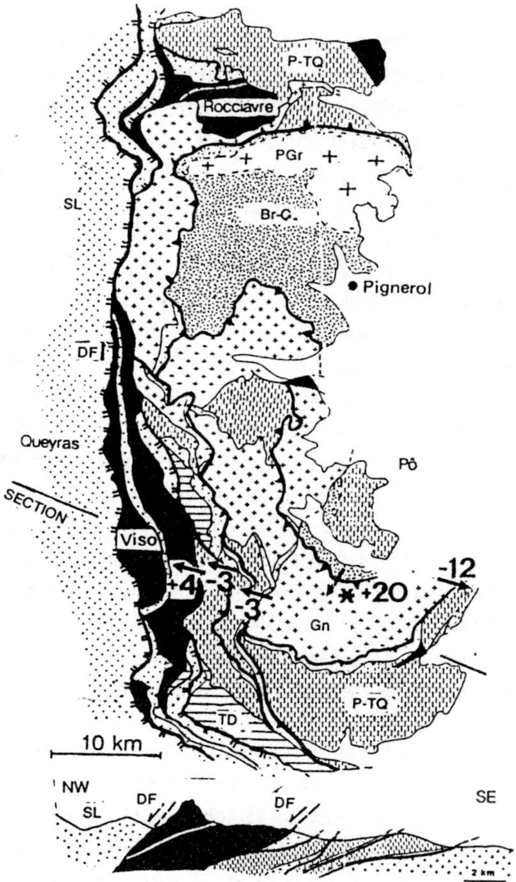

Fig. 191. Internal crystalline massif of Dora-Maira (SW of Turin) (after Philippot, 1990, modified and simplified).

This massif shows a pile of ductile nappes, generally gneissic, of Briançonnais and Piemont origin later refolded into a dome whose western flank at least presents detachment faults (DF) along which there has been outcropping of gneissic and ophiolitic nappes relative to Schistes Lustrés of Queyras (SL).

Bold numbers depict the sudden increase in metamorphic pressure (in kb) which corresponds to the passage of a thrust surface in direction indicated by the arrow. The irregularity in succession of units shows that the present pile was formed subsequent to this HP metamorphism: piling up must have taken place concomitantly by underthrusting of units at the base of the Upper Eocene-Lower Oligocene accretionary prism (for example, underthrusting of the Briançonnais unit of Pignerol below the coesite schists) and by dispersion of units of an older, Upper Cretaceous prism.

The star indicates the presence of rocks with coesite.

Br-C: Briançonnais carboniferous; Gn: Precarboniferous gneisses; P: Permian; PGr: Permian granite; TD: Triassic Dolomites; TQ: Triassic Quartzites; Black: ophiolites.

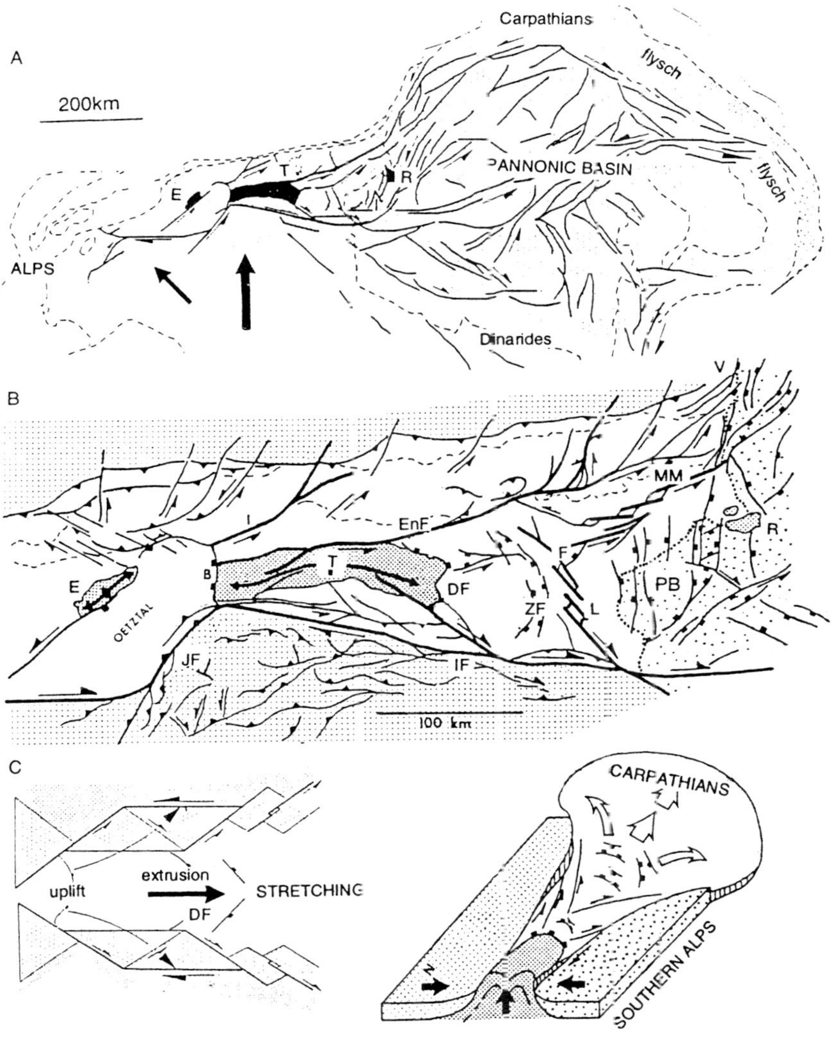

Fig. 192. Post-tectonic extensions and gravitational collapse of the Eastern Alps (after Ratschbacher et al., 1991, simplified and modified).

A—Alpinocarpathic suite and the Pannonic basin (PB). Black: Penninic inliers of eastern Alps (E: Engadine; T: Tauern; R: Rechnitz). Black arrows indicate the direction of shortening associated with the nearing of African-European plates.

B—Neogene fracturing of Eastern Alps. In grey, Penninic inliers (E: Engadine; T: Tauern; R: Rechnitz). DF: Detachment fault on the eastern flank of Tauern (Brenner fault, B, is its symmetric counterpart on the western flank). EnF: Ennstal Fault; IF: Insubric fault; JF: Judicarian fault; FMM: Fault and pull-apart basins of Mur-Mürz; F, L, ZF: Fohnsdorf, Lavanthal and Zollfeld basins; V. Vienna.

C—Model of uplift, extrusion and gravitational collapse of the Carpathians in the Neogene.

The central part is bounded in the north and the south by two major strike-slip faults (Fig. 192): in the south a dextral insubric fault, in the north a sinistral fault of a slightly divergent disposition. A network of secondary faults is developed between the two, giving rise to small grabens (Zollfeld, Fohnsdorf) or pull-apart basins (Mur-Mürz. Lavanthal). This fan-shaped swarm of secondary faults is more open eastwards than that generated by the major faults, which clearly represents extensional stretching of this group.

In the **Western Alps** such phenomena are less common or not known. Certainly this part of the chain exhibits a radial disposition of thrusts which may have induced a gravitational spreading, but this disposition cannot be coherently connected to an extensional denudation of external or internal crystalline massifs. Contrarily, a quite dense network of post-schistose normal faults is observed here, indicating a regional shear extension later to ductile extension associated with outcropping of HP-LT metamorphic assemblages. This extension becomes manifest although the general regime of the chain, compressive, is always in operation.

EXTENSION PHENOMENA AND TERMINAL COLLAPSE OF OROGENS: BALANCE AND SYNTHESIS

In the preceding pages we have mentioned several times the extension phenomena which frequently occurs in the major mountain chains as soon as the nappes pile up,i.e., when the accretionary prism that they represent reaches a certain volume and thus commences to be uplifted. These movements subsequently affect the top of the chain completed when its altitude is no longer balanced by the stress field exerted on its boundaries. These extension phenomena evidently have nothing in common with those which elsewhere gave rise to grabens or rifts starting from a thinned crust. Here, the crust is thickened by convergence and collision.

At the end of this thickening, surficial slabs are found to be detached from the pile and sliding over its flanks along detachment faults, a process that promotes rapid denudation of its deeper parts, which erosion alone would have taken much longer to achieve.

Extension phenomena are complicated by the fact that they commenced very early in the history of the chain, as soon as an accretionary prism of sufficient volume had built up. It is known that in collision and subduction chains this prism is most often built by underplating, i.e., by incessant arrival of material to the base of the prism, which has two consequences:

1. With regard to volume and mechanical consideration, there is an increase in volume of the prism which, by isostasy, would be uplifted relative to the host rock along shear surfaces functioning as normal fault. With a difference that the process takes place here in a hot and ductile environment, with metamorphism and stretch lineations. Complications may also

occur if the accretionary prism in the course of uplift is compressed by a later tectonic phase, accentuating its extrusive dome nature.

2. From the thermal point of view, the accretionary prism formed by underplating of cold units is itself relatively colder. When convergence ceases, the crust would be gradually heated by conduction. The shear crust-ductile crust transition would be uplifted, proportionately reducing the resistance of the former, stretched and dislocated, which hence can no longer support the weight of the orogenic structure. The latter would subside, more exactly spread, at least to the extent that the pressure field in which it is engaged allows it.

Modes of heat transfer associated with this process are complex and still not understood. They may be concentrated in certain points and hence induce, besides high-temperature metamorphic facies, veritable partial melting of the crust and mantle, leading to emplacement of intrusive granites (peri-Adriatic intrusions such as those of Val Bregaglia or Adamello in the Alps).

The concept of rapid uplift of the accretionary prism along ductile normal faults has emerged from observations of abnormal contacts in major chains. In fact, one can often see here nappes metamorphosed to high-pressure facies (hence at a depth of 30–35 km or more) directly overthrust by one or several little or non-metamorphosed units. When it can be measured from the ages of the minerals formed, the time gap between metamorphism of the lower units and thrusting of the overlying units is moreover very short. Such a disposition, implying rapid disappearance of intermediate units sometimes 10 to 20 km thick, obviously cannot be attributed to erosion, a much too long process. Hence the concept of ductile stretching, more or less shear, in the interior of the prism, which disturbs the disposition of the latter and transforms it into an extrusive dome.

In any case, at the end, the process of viscous raising of the prism leads to surface phenomena which may be of the MCC type if produced by detachment faults that bring domes strongly affected by thermal metamorphism into contact with surficial shear slabs, little or non-metamorphosed, which may hence slide and tilt. These movements are the irresistible consequence of post-metamorphic uplifts taking place at mean velocities of 5 to 6 mm/yr, figures obtained from cooling curves of minerals of metamorphism associated with uplift.

But what mainly determines the modalities of this gravitational spreading of the surface are the 'boundary conditions', viz., the regime in which the orogen in question occurs when its spreading commences.

First case: Regime is always compressive. Spreading is then limited and the orogeny acquires the form of a plateau with a high mean altitude, generally more than 3000 m (Fig. 193).

Such is the case of the **Tibetan plateau**. Following the Indo-Eurasia collision in the Early Eocene (52 Ma), the thickness of the crust has doubled below the plateau and this occurs over a width of more than 1000 km (Fig. 167). This doubling of the crust took place at the end of the Miocene: the mean altitude (3000 m) could not have been more for mechanical reasons and deformation extended farther north into Tien Shan and Altai, as well as farther south in the well-dated Himalayan thrusts.

Further, for 5 Ma, palynological studies of recent sediments have shown variations in flora indicating that Tibet may have been uplifted by 2 km, thus reaching its present altitude of 5 km, whereas an acidic volcanism manifested and N-S grabens (E-W extension) were produced. In other words, thrusts exerted mainly on the periphery of the plateau at altitudes less than 3 km, while extension of the plateau manifested above 5000 m. This suggests that the altitude of 5 km could not have been more and that the plateau tended to stretch.

The phenomenon is the consequence of an equilibrium between the stress resulting from convergence of plates and gravitational effects produced from isostatic uplift of a thickened crust. Another factor of equilibrium is also the viscosity of the lithosphere subjected to orogeny, viscosity itself dependent on the heat flow of mantle origin which seems significant in the particular case, associated with a specific behaviour of the asthenosphere of disputed origin.

Nevertheless, the tendency is to generalise the image of Tibet in other chains assuming that the latter start to spread when they exceed 3 km in height.

Another classical example of these elevated orogenic 'plateaus' is that of Peruvian and Bolivian **Altiplano** (4000 m mean altitude). Wedged between the western and eastern Cordilleras (Figs. 139, 193), it was also affected in a recent epoch by a major uplift corresponding to a compression episode, whereas a system of horsts and grabens developed at its surface.

Second case: The regime in the borders is not compressive or not compressive everywhere or ceases to be compressive. In other words, an 'escape' is possible in some direction. We have seen an excellent example of this in the case of the **Eastern Alps** whose central part slid towards the Pannonic basin (Fig. 192). This interpretation would pair, as we saw earlier, with the ductile rising of Tauern in the Oligo-Miocene into a thermal dome accompanied by emplacement of granitoids. A disposition similar to that of the MCC of the Basin and Range of western America, now considered the result of Laramian orogeny, is once again observed here (see p. 77 and Fig. 193).

These metamorphic core complexes may represent the normal consequence of tectonic denudation phenomena. We shall give two other examples of the same which also illustrate the final stage of an orogenic stretching, viz., that in which the chain subsides to the point of making place for marine basins.

The **Aegean sea** (Fig. 193) resulted from the subsidence of the chain connecting Greece and Turkey. In the Naxos-Paros island also one can observe beneath a detachment fault dipping north metamorphic formations

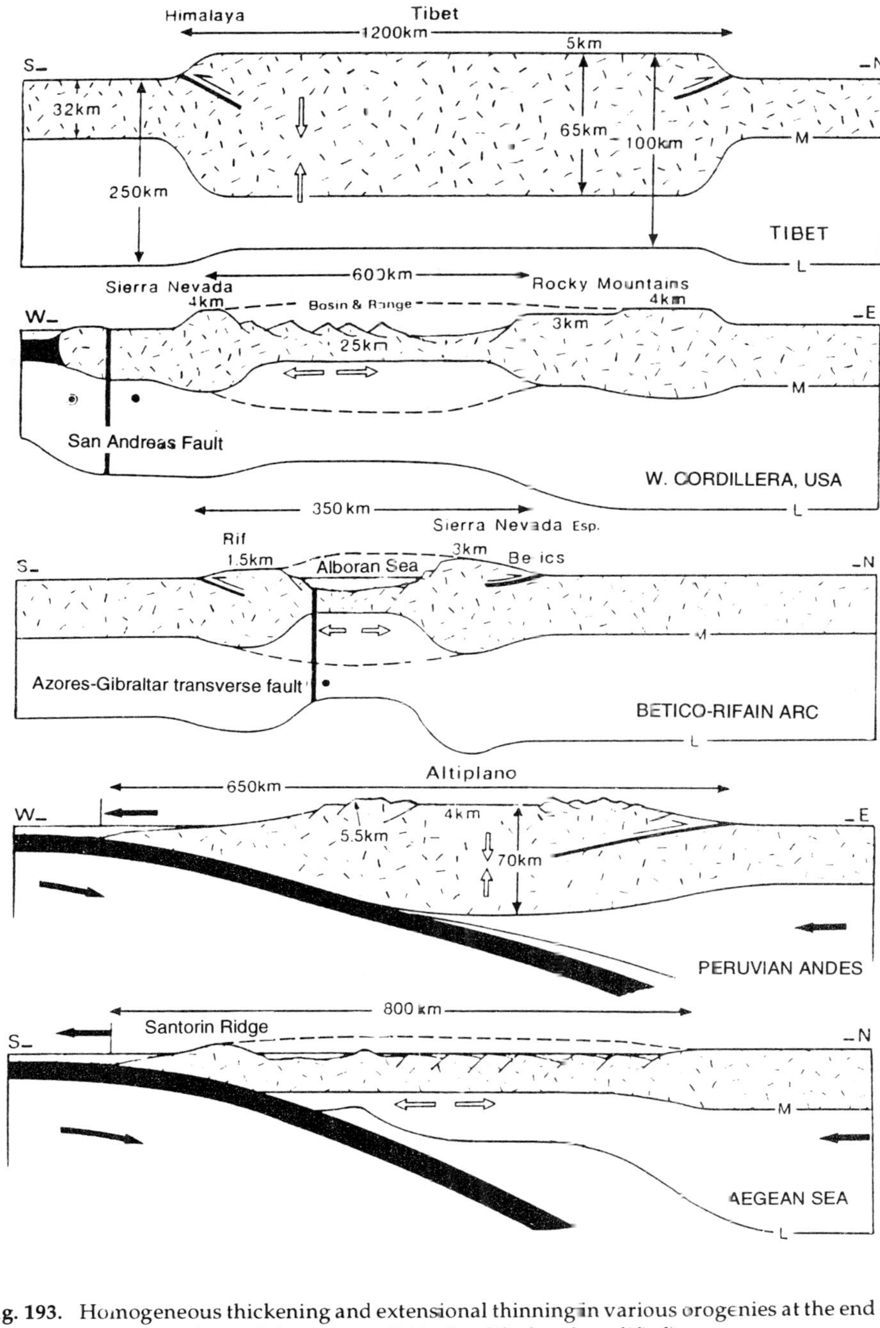

Fig. 193. Homogeneous thickening and extensional thinning in various orogenies at the end of tectonic evolution (after Dewey 1988, simplified and modified).

(greenschist to amphibolite facies) aged 25 Ma, intruded by a granite dated 11 Ma. The overlying unit, non-metamorphosed and sliding northwards, is itself broken into small blocks by normal faults. Whatever the mode, all of the Aegean region is segmented into tilted blocks (Fig. 193). Here it represents a still undeveloped stage which ought to continue by the creation of a back-arc basin associated with east-Mediterranean subduction.

The **North Tyrrhenian basin** resulted from the subsidence of the Corsica-Apennine chain which commenced in the Oligo-Miocene. From this basin emerged the metamorphic core complex of Elbe island (western part) with its syntectonic granite of the end-Miocene, on which the nappes of the eastern part of the island slid by detachment fault. Farther eastwards, the metamorphic massif of the *Apuane Alps* (famous for Carrara marble) may also be considered an MCC emerging from tilted blocks that slid from the Ligurian side.

In summary, the evolution of a collision chain passes through three successive stages (Fig. 194).

Stage 1. During collision, the thrusts and the underthrusts build an accretionary prism whose base is relatively cooled by the underplating effect (see Fig. 146). An HP metamorphism prevails there (blueschists or even eclogites) without magmatism. Isostatic uplift occurs and the chain is raised to a height beyond which the forces can no longer maintain the structure in equilibrium.

Stage 2. If the compressive state continues, new phenomena would develop (Fig. 194).

— *Tendency towards stretching of the crustal root,* since its internal pressure is stronger than in the neighbouring regions (see demonstration in Fig. 195). This stretching occurs in a ductile regime and may use older thrust surfaces.

— *Uplift of the shear/ductile boundary* which causes decoupling of these two media, the first commencing detachment and sliding relative to the second along shear faults functioning as normal faults (detachment faults). Thus detachment of more or less large slabs is promoted, which slide gravitationally along any directions. They may even become radial (Eastern Alps stage).

This tectonic denudation may also be locally associated with formation of MCCs (Lepontine Alps, for example).

This stage is accompanied by release and transfer of heat resulting in intrusion of granites.

Stage 3. If boundaries conditions permit, stretching results in thinning of the crust up to complete disappearance of the ancient crustal root (Fig. 194C). The processes of ductile extension previously initiated, lead to the establishment of typical MCCs (Basin and Range stage).

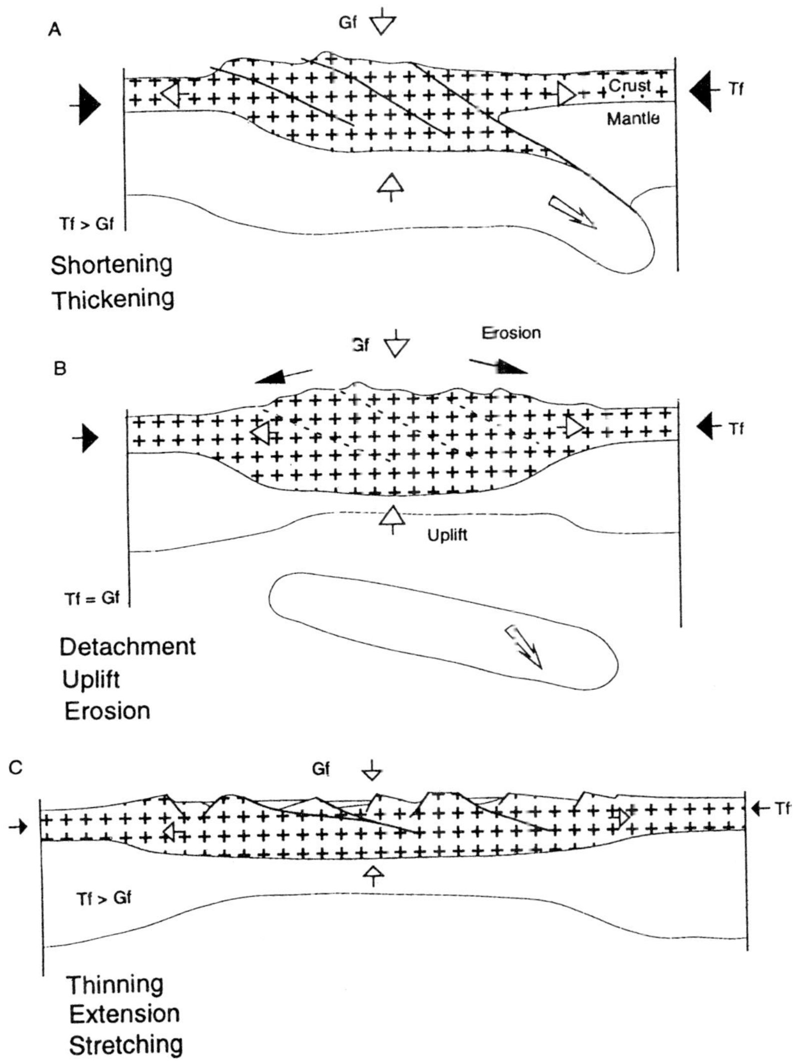

Fig. 194. Gravitational subsidence of a collision chain (after Malavieille et al. (1990) modified). Gf. Gravity forces; Tf. Tectonic forces.

Role of the mantle. In the process described above the role of the upper mantle is not mentioned because it remains conjectural. It may be considered that in proportion to the extent of gravitational stretching, the crustal root decreases, which indicates that the mantle rises. It may even rise very fast inside the MCC structures.

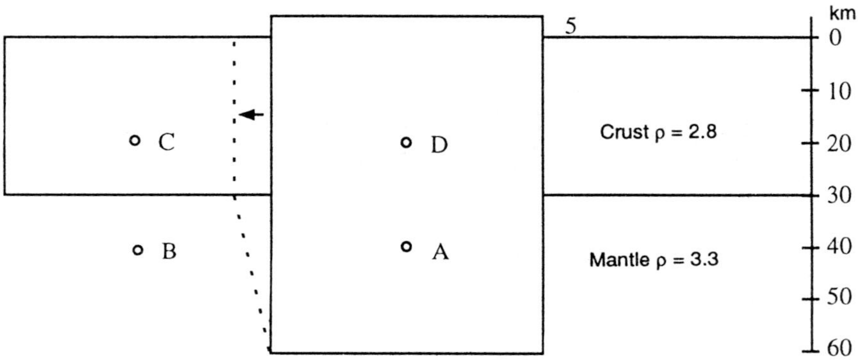

Fig. 195. Internal pressures at the root of a collision chain: Pressures at the points A, B, C, D can be calculated by multiplying the thickness of overlying formations by their density. Internal pressure of the root is greater and makes it flow laterally (arrow) (after Boillot, 1996).

Whatever the mode, uplift of the mantle induces an ascent of isotherms which, with the associated release, would promote partial melting, giving rise to products likely to contaminate crustal anatectic magmas, contamination readily observed.

It is possible, however, that more complex mantle phenomena occur. Various authors have developed hypotheses according to which the crustal root is extended by a mantle root whose basal part may be detached from the top part (Fig. 194B). This phenomenon may have evidently triggered rapid uplift of all the overlying structures thus made lighter. In fact, deep earthquakes exist beneath certain high-altitude plateaus (Colorado, Iran, Hindu Kush) which may induce a similar mechanism; this interpretation remains however debatable.

CHAPTER 5

COLLAGES

The concept of collage dates back to the 1970s and refers to structures in Alaska where a number of juxtaposed compartments were identified along strike-slip faults on the North American coast. It seemed impossible to logically interrelate these blocks, so different in nature and age, and hence they were thought to be of a distant origin, transported to this region by the movement of an oceanic lithosphere that is presently subducted.

Such blocks were designated as 'suspect terranes'[1], a term generally translated into French as 'exotic blocks'.

The identification of 'suspect terranes' is based on three groups of criteria:

—different remanent magnetic inclinations and declinations on either side of the collage zone;

—juxtaposition of fossil fauna belonging to different provinces, offset in latitude;

—absence of ophiolitic suture or overthrust zone because the juxtaposition results from strike-slip faulting (this criterion is more difficult to identify compared to the preceding ones since such a juxtaposition zone might be subsequently transformed into an overthrust zone).

The concept of collage has changed somewhat in the last few years. The term is in fact used by some authors to denote any structure resulting from strike-slip faulting that juxtaposes different parts, whatever be the structural context (see, for example, Frizon de Lamotte, *BSGF*, 1987).

The term is also employed to describe the collision of an island arc with the continent, which may induce overthrusts as well as ophiolitic sutures (see, for example, Mégard for the Colombian Andes, 1987). Finally, some authors go still further by treating it as a synonym of tectonic juxtaposition of different structural suites, irrespective of their composition and style of contact zone (see, for example, D. G. Howell's *Tectonics of Suspect Terranes*, Chapman and Hall (1989) 232 pp.).

---

[1]The term 'terrane', isolated, is a synonym of the 'structural zone' of European geologists, which means a palaeogeographic and tectonic group clearly distinctive in the midst of an orogen.

For the purpose of simplicity and clarity, we have here maintained the original meaning of 'collage' by reserving it for specific structures associated with a certain mode of displacement of the oceanic lithosphere relative to the neighbouring continent.

EXAMPLES

Wrangellia Block (Wrangell Mts.) (Fig. 196A)

The Wrangell mountains in Alaska and the Queen Charlotte and Vancouver islands in Canada, notwithstanding the long distance (2000 km) between them, show a highly remarkable similarity of original geological characteristics.

All three structures, in effect, exhibit a piling up of several thousand metres of Triassic tholeiitic lavas covered by Carnian and Norian platform carbonates. This group rests on an andesitic complex and Permian sediments. It thus probably represents the disintegrated remnants of an ancient island arc. But what is especially important is the fact that this sequence totally differs from the Triassic and Permian formations of regions in which these blocks are juxtaposed. On the contrary, the overlying Mesozoic formations, when they exist, show much less contrast.

One might think that this group represents, as in the case of many other structures of crustal shortening, the juxtaposition of different palaeogeographic regions, wherein the transition zones have been obliterated by overthrusting. However, the palaeomagnetic data of the Triassic tholeiites indicates that the latter were emplaced at a much lower latitude than their present position (15° N or S). Moreover, the Triassic palaeo-poles of these blocks are different from those of the neighbouring continent.

It emerges from the palaeomagnetic data that there may have been three segments of an ancient island arc, known as Wrangellia, its primitive position being farther south. The period of arrival of the fragments to their present position is not yet definitively established, more so because it is variable in different regions. On the basis of uncertain evidence, ages have been proposed ranging from the Upper Jurassic for the southernmost blocks to the Middle Cretaceous for the northernmost.

The emplacement mechanism induced major strike-slip faults. The latter are highly hypothetical because prior to the origin of the San Andreas fault. They are probably associated with the northward migration and subsequent disappearance of the Kula fossil plate (Fig. 197).

Yakutat Block (Fig. 196B)

The Yakutat block forms the boundary of the Gulf of Alaska at the foot of Saint Elias Mountain (5490 m), one of the peaks of the North American

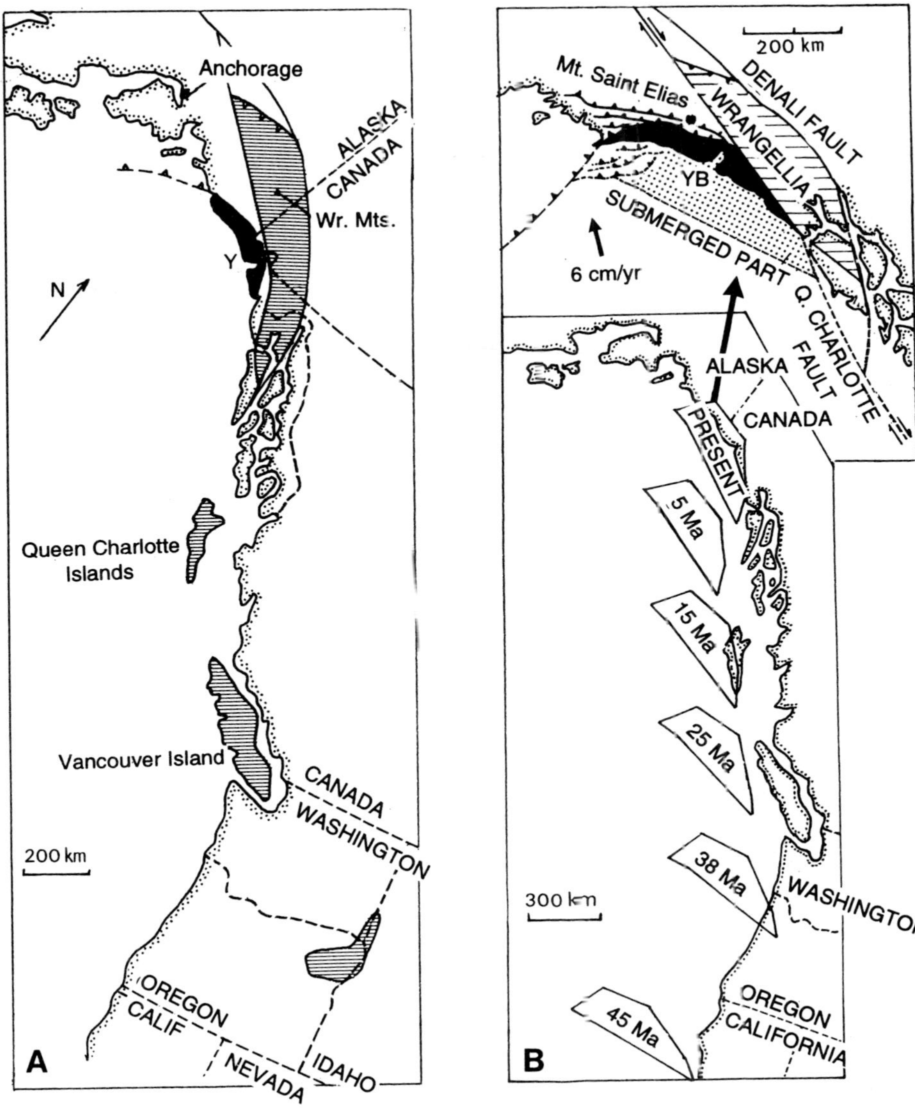

Fig. 196. Exotic blocks (suspect terranes) on the SW coast of Alaska.
A—Different outcropping zones of Wrangellia type sequences (Wr. mts.: Wrangell mts.) (after Jones et al., 1977); Y: Position of Yakutat (black).
B—Structural map of the Yakutat block and its successive positions since the Eocene (after Bruns, 1983). YB: Yakutat Bay.

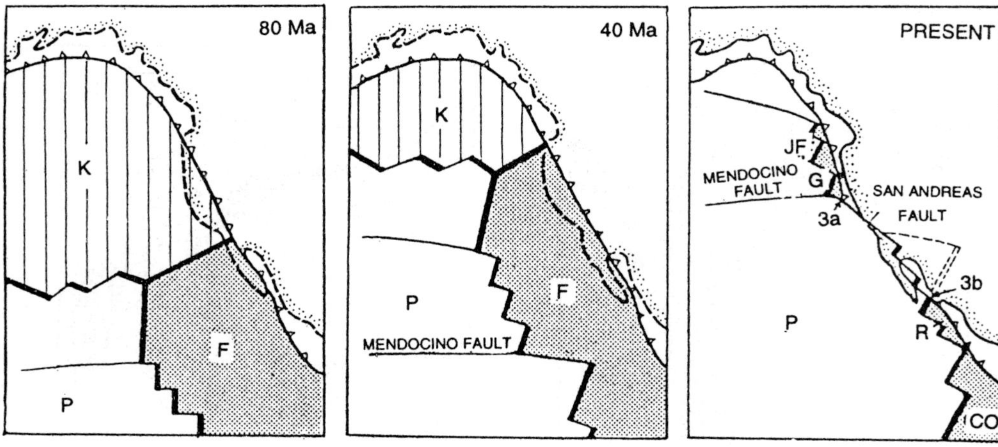

Fig. 197. Disappearance of the Kula (K) and Farallon (F) plates at the north-eastern margin of the Pacific plate (P).
The subduction phenomenon leading to the disappearance of the Farallon plate has not yet reached its end stage. Some residual fragments have remained which are the small plates of, from north to south, Juan de Fuca (JF), Gorda (G), Rivera (R) and Cocos (CO). Gorda and Rivera plates have, moreover, been offset by the San Andreas faulting since 20 Ma ago.
3a. Mendocino triple junction point (North American subduction plane + Mendocino and San Andreas transform faults).
3b. Rivera triple junction point (Central American subduction plane + East Pacific ridge + one of the transform faults of the Gulf of California).

Pacific Cordillera. As in the preceding case, this block is bounded on all sides by strike-slip faults except at the foot of Saint Elias Mountain where it is subducted below the North American continent, thus producing a compression system, part of which is the Saint Elias Mountain. This subduction zone is the same as that of the Aleutian arc.

The block comprises, in the east, a continental-margin sedimentary sequence and, in the west, an oceanic crust covered by 3 to 5-km thick Palaeogene sediments. The massifs that have provided the detrital material of these sediments are found only in the states of California, Oregon and Washington (i.e., in latitudes lower than 15 to 20°) and are totally absent in the mountains that presently border the block in question. Fossils confirm this data.

It is hence thought that the block was displaced northwards during the Tertiary. As in the case of Wrangellia, details of the mechanism are highly hypothetical, combining the sea-floor spreading of the East Pacific, the disappearance of the Kula plate by subduction in the north and the strike-slip faulting of the San Andreas fault and its northern extensions (Queen

Charlotte and annex faults). These are the faults which must have added to the moving oceanic crust a fragment of the North American continental margin, thus forming the composite block of Yakutat. The phenomenon is still in progress since the block is presently penetrating the Aleutian subduction zone, below the Saint Elias Mountain, and is subjected to some amount of folding.

Definitely, it now appears to be well established that a special type of collision exists on the North American coast, which can be denoted by the term 'collage', i.e., juxtaposition by strike-slip faulting. This phenomenon might be associated with the very special relations of the North American continent and the structures of the East Pacific ridges that are more or less oblique to the continent, the disappearing Kula and Farallon microplates). Further, this process might still be active in the sense that blocks continue to arrive from the south. The Baja California peninsula gives us a picture of detachment and commencement of migration of such blocks. In fact, its displacement along the San Andreas fault takes place at the rate of 4 cm/yr, with the result that if the movement continues, the Baja California block will come into contact with Alaska in 50 Ma.

Other Examples

Do we find such dispositions elsewhere?

Such structures have been indicated in *Colombia* (p. 215) and *California* (Salinian block, p. 226). Moreover, some authors have used them to explain the structure of *Japan*, which merits a short comment.

It is in fact known that if the edifice associated with the Neogene volcanic island arc is excluded from consideration, we observe that all the older structures in Japan are made up of slabs separated by subvertical tectonic contacts formed by recent (post-Upper Cretaceous) sinistral strike-slip faults. These compartments are made up of formations of different nature and age which might have hence been previously very far away from one another, with no transition zone existing between them.

In South Honshu (Fig. 198), where these strips outcrop, their age is found to decrease from NW to SE since the successive units observed are the *Hida-Sangun zone*, of Palaeozoic sediments, the *Tamba-Ryoke zone*, which is a Nevadan (Upper Jurassic) chain, and the *Shimanto zone*, a Tertiary chain, probably resulting from an intercontinental collision. but in which the southern continent, displaced northwards, has disappeared. A remnant of this continent might perhaps be found in one of the older massifs which border the North Honshu coast (Abukuma plateau and Kitakami massif) and which have no relation with their context.

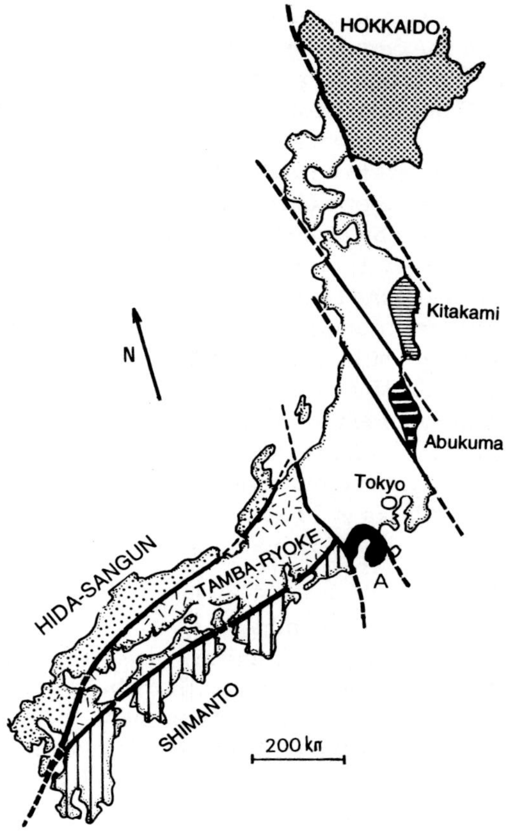

Fig. 198. Ultimate structure of the collage of Japan.
A: Northern edge of the Philippine plate, presently in collision with the Japanese plate.

However, one can definitely appreciate by this example the danger of the concept of 'collage', viz., utilising it as a solution of convenience: any compartment whose relations with the neighbouring zones are not clearly understood may be immediately declared exotic (suspect). Palaeomagnetic and palaeoentological verification is therefore absolutely necessary.

REFERENCES

(as of December 1996)

The titles given below were selected because they correspond to a recent, readily accessible compilation, or because illustrations have been borrowed from them, or because they contain sufficiently up-to-date and complete bibliographies pertinent to the corresponding chapter.

GENERAL WORKS

APBG. 1987. Bulletin spécial consacré à la géologie structurale de la France, 126 pp.

Allègre, Cl. 1983. *L'écume de la Terre.* Fayard, 366 pp.

Anonyme. 1979. La dérive des continents, la tectonique de plaques. Bibl. *Pour la Science,* 216.

Anonyme. 1995. L'Ecorce terrestre (Dossier). Rev. *Pour la Science,* 129 pp.

Boillot, G. 1996. *La dynamique de la lithosphère.* Masson, Paris, 129 pp.

BRGM. 1996. Carte géologique de France au 1/1,000,000, 6th ed

Caron, J.-M et al. 1989. *La Planète Terre.* Ed. Ophrys, 271 pp.

Choukroune, P. 1995. *Déformations et Déplacements dans la Croûte Terrestre.* Masson, Paris, 226 pp.

Condie, K.-C. 1989. *Plate Tectonics and Crustal Evolution.* Pergamon Press, 3rd ed., 476 pp.

Jolivet, L. 1995. *La Déformation des Continents.* Ens. Scient, Hermann, Paris, 410 pp.

Lliboutry, L. 1982. *Tectonophysique et Géodynamique.* Masson, Paris, 339 pp.

Mattauer, M. 1975. *Déformation des Matériaux de l'Ecorce Terrestre.* Hermann, Paris, 493 pp.

Moores, E. and R. Twiss. 1996. *Tectonics.* Freeman, 672 pp.

Soc. Geol. France (1984). Des océans aux continents. BSGF (7), XXVI (3): 325–568.

Souquet, P. 1986–1987. *Structure et Dynamique de la Lithosphère,* vol. 1: *Tectonique de Plaques* (83 pp.); vol. 2: *Tectonique Globale* (200 pp.). NCDP, CRDP, Toulouse.

Wilson, M. 1988. *Igneous Petrogenesis, a Global Tectonic Approach.* Ass. Hyman, London, 465 pp.

Windley, B.-F. 1977. *The Evolving Continents.* J. Wiley & Sons, New York, 386 pp.

OCEANIC STRUCTURES

General

Nicolas, A. 1990. *Les Montagnes sous la Mer.* BRGM, 186 pp.

Soc. Géol. France (1979). *BSGF* 5: 515–673.

UNESCO. 1980. Atlas géologique du monde:
Océan atlantique (f. 22, au 34,000,000)

Océan indien (f. 21, au 29,000,000)
Océan pacifique (f. 20, au 36,000,000)

Oceanic Crust

Allègre, Cl. 1979. Les ophiolites ou la recherche des océans perdus. *Pour la Science* 17: 109–121.

Boillot, G. 1987. Les fonds sous-marins du 3e type. *La Recherche* 193: 1294–1303.

Juteau, Th. 1980. La naissance de la croûte océanique. *La Recherche* 117: 1374–1384.

Juteau, Th. 1984. De la croûte océanique aux ophiolites. *BSGF*, pp. 471–488.

Tectonophysics. 1988. Special issue on the ophiolites of Oman, no. 151, 400 pp.

Ridges

MacDonald, K. et al. 1988. A new view of the mid-ocean ridge from the behaviour of ridge axis discontinuity. *Nature* 335: 217–225.

Smoot, C. 1989. N-Atlantic fracture zone distribution and patterns shown by multibeam sonar. *Geology* 17: 1119–1122.

Sykes, L.-R. 1967. Mechanism of earthquakes and nature of the faulting on the mid-oceanic ridges. *J. Geoph. Res.* 72: 2131–2153.

Zonenshain, L.-P. et al. 1989. Tectonics of the Mid-Atlantic rift valley between the TAAG and Mark areas (26–24°N): evidence for vertical tectonism. *Tectonophysics* 159: 1–23.

Transform Faults

Calais, E. and B. Mercier de Lépinay. 1990. Un modèle naturel de tectonique transpressive: les structures en échelons de la fosse de l'Oriente (N. Caraibes). *Bull. IFP* 45 (2): 147–160.

Intraoceanic Island Arcs

General

Hawkins, J. et al. 1984. Evolution of intra-oceanic arc-trench systems. *Tectonophysics* 102: 175–205.

Karig, D. 1972. Remnant arcs. *Bull. Soc. Geol. Amer.* 83: 1057–1068.

Packham, G. and D. Falvey. 1971. An hypothesis for the formation of marginal seas in the Western Pacific. *Tectonophysics* 11: 79–109.

Regional descriptions

Anonymous. 1981. *Studies in East Asian Tectonics and Resources*. Un. Nat. ESCAP, CCOP, Tech. publ. 7a, 2nd ed. 250 pp.

Auzende, J.-M. et al. 1988. Seafloor spreading in the North Fiji basin (S.-W. Pacific). *Tectonophysics* 146: 317–352. Also see: *Geology* (1988) 16: 925–929.

Blanchet, R. et al. 1979. Le système des Mariannes (Pacifique W.). *BSGF*, pp. 529–532.

Eissen, J.-Ph. et al. 1990. Pétrologie et géochimie de la zone d'accrétion du centre du bassin N.-Fidjien (S.-W. Pacifique). *Cpte. Rend. Acad. Sci., Paris*, 310: 771–778.

Hamilton, W. 1981. *Tectonics of the Indonesian Region*. USA Geol. Survey, Prof. Paper 1078, 2nd ed., 345 pp.

Houza, E. et al. 1989. Subduction erosion and accretion in the Solomon Sea region. *Tectonophysics* 160: 49–62.

Hussong, D.M. and S. Uyeda. 1981. Tectonics of the Mariane arc. CR 26th CGI Paris, 1980. *Mém. BRGM (C3) et Oceanologica Acta,* pp. 203–216.

Pelletier, B. and R. Louat. 1989. Seismotectonics and present-day relative plate motions in the Tonga-Kermadec region. *Tectonophysics* 165: 237–250.

Rangin, Cl. 1989. The Sulu arc, a back-arc basin setting with a Neogene collision zone. *Tectonophysics* 161: 119–141.

Ridgway, J. 1987. Neogene displacements in the Solomon island arc. *Tectonophysics* 133: 81–93.

Scott, R.-B. and L. Kroenke. 1981. Periodicity of remnant arcs and back-arc basins of the South Philippine Sea. CR 26th CGI Paris, 1980. *Mem. BRGM (C3) et Oceanologica Acta,* pp. 193–202.

Antilles

Beck, C. et al. 1988. Anatomie et physiologie d'un prisme d'accrétion: premiers résultats des forages du complexe de la ride de la Barbade. *Bull. SGF,* pp. 129–140.

Bouysse, P. and D. Westercamp. 1990. Subduction of Atlantic aseismic ridges and Late Cenozoic evolution of the Lesser Antilles island arc. *Tectonophysics* 175: 349–380.

Biju-Duval, B. et al. 1982. Multibeam bathymetric survey on the Barbados ridge complex. A key to the knowledge and interpretation of an accretionary wedge. *Tectonophysics* 86: 275–304.

Casey-Moore, J. et al. 1988. Tectonics and hydrogeology of the N. Barbados ridge. *Geol. Soc. Amer. Bull.* 100: 1578–1593.

Soc. Geol. France (1983). *Bull.,* 6: 797–910.

Westercamp, D. 1988. Magma generation in the Lesser Antilles: geological constraints. *Tectonophysics* 149: 145–163.

Marginal Seas

Jolivet, L. et al. 1989. Tectonic setting of Western Pacific marginal basins. *Tectonophysics* 160: 23–47.

Tectonophysics (1980). 160, 1–4. Particular case of the Sea of Japan.

Plumes and Hot Spots

Clague, D. and G. Dalrymple. 1987. The Hawaiian Emperor Volcanic Chain. Part L. Geologic Evolution. U.S. Geol. Surv. Prof. paper 1350.

Crough, Th. 1984. Seamounts as recorders of hot-spot epeirogeny. *Geol. Soc. Amer. Bull.* 95: 3–8.

Goslin, J. and M. Maia. 1993. Les apports des travaux recents de géophysique à la compréhension du volcanisme intraplaque du domaine océanique: évolution de l'hypothèse des points chauds. *Mem. Soc. Geol. Fr.,* 163: 113–128.

Moore, Th. et al. 1983. The Walvis ridge. *Geol. Soc. Amer. Bull.* 94: 907–925.

Sleep, N.-H. 1990. Hot spots and mantle plume: some phenomenology. *J. Geophys. Res.* 95 (B5): 6715–6736.

Vink, J., J. Morgan and P. Vogt. 1985. Panaches et points chauds. *Pour la Science,* pp. 80–85.

SEDIMENTARY BASINS

General

Anonymous. 1981. *Extensional Tectonics Associated with Convergent Plate Boundaries.* Geol. Soc., London, 442 pp.

Anonyme. 1989. *Dynamique et Méthodes d'Étude des Bassins Sédimentaires.* Ed. Technip, 446 pp.
Bull. Elf Aquitaine (1977), pp. 111–320, in particular:
Arthaud, F. et al. Cadre tectonique de quelques bassins sédimentaires, pp. 147–188.
Perrodon, A. Concepts, modèle et logique des bassins sédimentaires, pp. 111–130
Ingersoll, R. 1988. Tectonics of sedimentary basins. *Geol. Soc. Amer. Bull.* 100: 1704–1719.
Kingston, D.-R. et al. 1983. Global basins classification system. *AAPG Bull.* 67: 2175–2193.
Perrodon, A. 1983. Géodynamique des bassins sédimentaires et systèmes pétroliers. *Bull. Elf. Aquitaine,* pp. 645–676.
Perrodon, A. 1988. Bassins sédimentaires, provinces pétrolières et tectonique globale. *Bull. Elf Aquitaine,* pp. 493–512.

Grabens

General

Ann. Soc. Geol. Nord (1983). *Tectonique Cassante en Distension et Coulissement.* CIII, 386 pp.
Buck, W.-R. 1986. Small-scale convection induced by passive rifting: the cause for uplift of rift shoulders. *Earth Plan. Sci. Let.* 77: 362–372.
Bull. Elf Aquitaine (1983). Rifts et fossés anciens, 7: 127–448.
Illies, J.-H. 1981. Mechanism of graben formation. *Tectonophysics* 73: 249–266.
Quennell, A. (ed.). 1985. *Continental Rifts.* Nostrand Publ., New York, 345 pp.
Tectonophysics (1983). Processes of continental rifting (94), nos. 1–4.
White, R. and D. McKenzie. 1989. Le volcanisme de rift. *Pour la Science,* pp. 80–89.

Rhine rift

Brun, J.P., H.A. Gutscher and DEKORP-ECORS. 1992. Deep crustal structure of the Rhine graben from DEKORP-ECORS seismic reflection data: a summary. *Tectonophysics* 208: 139 and 147.
Larroque, J.M. and Ph. Laurent. 1988. Evolution of the stress-field pattern in the south of the Rhine graben from the Eocene to the present. *Tectonophysics,* pp. 41–58.
Sittler, Cl. 1974. Le fossé rhénan. In: *Géologie de la France*. Doin, Paris, pp. 78–104.
Villemin, Th. and Fr. Bergerat. 1987. L'évolution structurale du fossé rhénan au cours du Cénozoique. *Bull. SGF,* pp. 245–255.
Villemin, Th. et al. 1986. The Rhine graben: extension, subsidence and shoulder uplift. *Tectonophysics* 128: 47–59.

Gulf of Suez

Bull. Soc. Geol. Fr. (1990), 3: 371–446.
Chenet, P.Y. and Y. Letouzey. 1983. Tectonique du rift de Suez. *Bull. Elf Aquitaine* 7: 201–215.
Chorowicz, J. et al. 1987. Tectoniques superposées synsédimentaires des golfes de Suez et d'Aqaba. *Bull. SGF,* pp. 223–234.
Notes et Mem. Total-CFP (1987), no. 21, in particular:
Chorowicz, J. and N. Lyberis. Evolution tectonique de la péninsule du Sinai: la formation des plis de l'arc syrien et des golfes de Suez et d'Aqaba, pp. 199–209.
Ott d'Estevou, Ph. et al. Principaux aspects de l'évolution structurale du rift de Suez et de la mer Rouge N.-W. (Egypte), pp. 167–197.
Tectonophysics (1988). The Gulf of Suez and Red Sea rifting (153) nos. 1–4, in particular:
Bayer, H. et al. Sedimentary and structural evolution of the N.-W. Arabian Red Sea margin, pp. 137–153.
Colleta, B. et al. Longitudinal evolution of the Suez rift structure (Egypt), pp. 221–233.
Lyberis, N. Tectonic evolution of the Gulf of Suez and the Gulf of Aqaba, pp. 209–220.

Montenat, C. et al. Tectonic and sedimentary evolution of the Gulf of Suez and the N.-W. Red Sea, pp. 161–178.

Rift of East Africa Great Lakes

Bosworth, W. 1987. Off-axis volcanism in the Gregory rift, E.-Africa: implications for models of continental rifting. *Geology* 15: 397–400.

Chorowicz, J. 1983. Le rift E.-africain: début d'ouverture d'un océan. *Bull. Elf Aquitaine* 7: 155–162.

Ebinger, C. 1989. Tectonic development of the western branch of E.-Africa rift system. *Geol. Soc. Amer. Bull.* 101: 952–967.

J. Afr. Earth Sci. (1989). Africa Rifting, 629 pp.

Janssen, M.E. et al. 1995. Temporal and spatial correlations between changes in plate notions and the evolution of rifted basins in Africa. *Geol. Soc. Amer. Bull.* 107: 1317–1332.

Mondeguer, A. et al. 1989. Sedimentary basins in an extension and strike-slip background: the South Tanganyika trough complex (E.-Africa rift). *Bull. SGF*, pp. 501–522.

Morley, C.K. 1989. Extension, detachment and sedimentation in continental rifts, with particular reference to E. Africa. *Tectonics* 8: 1175–1192.

Sander, St. and Br. Rosendahl. 1989. The geometry of rifting in Lake Tanganyika (E.-Africa). *J. Afric. Earth Sci.* 8: 323–354.

Specht, Th. and Br. Rosendahl. 1989. Architecture of Lake Malawi rift (E.-Africa). *J. Afric. Earth Sci.* 8: 355–382.

Tiercelin, J. et al. 1988. East-African rift system: age and tectonic significance of the Tanganyika-Rukwa-Malawi intracontinental transcurrent fault zone. *Tectonophysics* 148: 241–252.

Lake Baikal

Hutchinson, D.R. et al. 1992. Depositional and tectonic framework of the rift basins of Lake Baikal from multichannel seismic data. *Geology* 20: 589–592.

Kiselev, A. 1987. Volcanism of the Baikal rift zone. *Tectonophysics* 143: 235–244.

Logatchev, N. and Y. Zorin. 1987. Evidence and causes of the two-stages development of the Baikal rift. *Tectonophysics* 143: 225–234.

Zorin, Y. 1981. Baikal rift. *Tectonophysics* 73: 91–104.

Zorin, Y. et al. 1989. Thickness of the lithosphere beneath the Baikal rift zone and adjacent regions. *Tectonophysics* 168: 327–337.

Pull-apart rifts

General

Aydin, A. and A. Nur. 1982. Evolution of pull-apart basins and their scale independence. *Tectonics* pp. 91–105.

Mann et al. 1983. Development of pull-apart basins. *J. Geology* 91: 529–544.

Dead Sea

Garfunkal, Z. 1981. Internal structure of the Dead Sea leaky transform (rift) in relation to plate kinematics. *Tectonophysics* 80: 81–88.

Garfunkel, Z. et al. 1981. Active faulting in the Dead Sea rift. *Tectonophysics* 80: 1–26.

Tectonophysics (1987). The Dead Sea and its environs, (141) nos. 1–3, in particular:

Kashai, E. and P.F. Crockes. Structural geometry and evolution of the Dead Sea-Jordan rift system as deduced from new subsurface data, pp 33–60.

Ze'ev Reckes. Mechanical aspects of pull-apart basins and push-up swells, with application to the Dead Sea transform, pp. 75–88.

Ten Brink, K. and Z. Ben Avraham. 1989. The anatomy of a pull-apart basin: seismic reflection observation of the Dead Sea basin. *Tectonics* 8: 333–350.

Salton Sea

Bilham, R. and G. King. 1989. The morphology of strike-slip faults: examples from the San Andreas fault, California. *J. Geophys. Res.* 94: 10204–10216.
Crowell, J.C. 1981. Juncture of S. Andreas transform system and Gulf of California rift. CR 26th CGI, Paris, 1980. *Oceanologica Acta*, pp. 137–141.
Wallace, R. 1990. The San Andreas Fault System (Calif.). U.S. Geol. Survey, Prof. Paper 1515.

Variscan pull-apart rifts, France

Bonijoly, D. and C. Castaing. 1994. Fracturation et genèse des bassins stéphaniens du Massif central français en régime compressif. *Ann. Soc. Géol. Nord,* CIII: 187–199.
Rolet, J. 1984. Graben losangiques (pull-apart) en régime de décrochement. Le rôle des coulissements hercyniens dans l'individualisation des bassins carbonifères du Massif armoricain. *Ann. Soc. Géol. Nord,* CIII: 209–220.

Basin and Range

Allmendiger, R.W. et al. 1983. Cenozoic and Mesozoic structure of the eastern Basin and Range Province (Utah) from COCORP seismic-reflection data. *Geology* 11: 532–536.
Allmendiger, R.W. et al. 1987. Overview of the COCORP 40° N transect, W. USA. The fabric of an orogenic belt. *Geol. Soc. Amer. Bull.* 98: 308–319.
Coney, J. 1980. Cordilleran metamorphic core complexes: an overview. *Geol. Soc. Amer. Mem.*, no. 153.
Lister, G. and G. Davis. 1989. The origin of metamorphic core complexes and detachment faults formed during Tertiary continental extension in the N.-Colorado River region, USA. *J. Str. Geol.,* II: 65–94.
Soc. Geol. France (1986). *Bull.* 6: 871–879.
Stewart, J. 1978. Basin-Range structures in Western N.-America, a review. *Geol. Soc. Amer. Mem.* 152: 1–30.
Wernicke, B. 1981. Low-angle normal fault in the Basin and Range Province: nappe tectonics in an extending orogen. *Nature* 291: 645–648.
Wernicke, B. et al. 1988. Basin and Range extensional tectonics at the latitude of Las Vegas, Nevada. *Geol. Soc. Amer. Bull.* 100: 1738–1757.

Basins Proper

Paris Basin

Brunet, M. Fr. 1986. La subsidence dans les trois bassins sédimentaires français (Paris, Aquitaine, Sud-Est). *Bull. Elf Aquitaine* 10: 23–32.
Curnelle, R. and P. Dubois. 1986. Evolution mésozoique des grands bassins, sédimentaires français. *Bull. SGF,* pp. 529–546.
Pomerol, Ch. 1980. Le Bassin de Paris. In: *Géologie des Pays Européens* (vol. France). Dunod, Paris, pp. 437–474.
Pomerol, Ch. et al. 1980. Les formations sédimentaires du bassin de Paris, reflet des principales étapes de son évolution. CR 26th CGI, Paris, 1980. *Mem. BRGM* 107: 257–273.
Soc. Geol. France (1980). *Bull.* 4: 629–695.

Other basins

Bertrand-Sarfati, J. et al. 1977. Géodynamique des aires sédimentaires cratoniques: quelques exemples sahariens. *Bull. Elf Aquitaine,* pp. 217–231.

Jackson, M. and S. Seni. 1983. Geometry and evolution of salt structures in a marginal rift basin of the Gulf of Mexico (E. Texas). *Geology* 11: 131–135.
Massa, D. and T. Delort. 1984. Evolution du bassin de Syrte (Lybie) du Cambrien au Crétacé basal. *Bull. SGF*, pp. 1087–1096.
Roure, F. et al. 1992. Geometry and kinematics of extensional structures in the Alpine foreland basin of SE France. *J. Str. Geol.* 14: 503–519.

Aulacogens

Aquitaine basin

Curnelle, R. et al. 1980. Le bassin d'Aquitaine. In: *Géologie des Pays Européens* (vol. France). Dunod, Paris, pp. 231–242.
Montadert, L. 1984. Le bassin d'Aquitaine. In: *Les Marges Continentales Actuelles et Fossiles autour de la France.* Masson, Paris, pp. 122–131.

Benue

Benkhelil, J. and B. Robineau. 1983. Le fossé de la Bénoué est-il un rift? *Bull. Elf Aquitaine* 7: 315–321.
Benkhelil, J. 1988. Structure et évolution géodynamique du bassin intracratonique de la Bénoué (Nigeria). *Bull. Elf Aquitaine* 12: 29–128.
Benkhelil, J. 1989. The origin and evolution of the Cretaceous Benue trough (Nigeria). *J. Afr. Earth Sci.* 8: 251–282.
Bull. Elf Aquitaine (1987). Special report on the Benue trough, 11: 153–180.
Guiraud, R. and J.-C. Maurin. 1991. Le rifting en Afrique au Crétacé inférieur. *BSGF*, pp. 811–823.
Popoff, M. 1988. Du Gondwana à l'Atlantique Sud: les connexions du fossé de la Bénoué avec les bassins du N.-E. brésilien jusqu à l'ouverture du Golfe de Guinée au Crétacé inférieur. *J. Afr. Earth Sci.* 7 (2): 409–431.
Popoff, M. et al. 1983. Approche géodynamique du fossé de la Bénoué (N.-E. Nigeria). *Bull. Elf Aquitaine* 7: 323–337.

Athapuscow basin

Hoffmann, P.F. et al. 1977. Geology of the Athapuscow Aulacogen (eastern arm of Great Slave Lake). Geol. Survey Canada, paper 77, 1A.
Zolnai, G. 1986. Les aulacogènes du continent N.-américain. *Bull. SGF*, pp. 809–818.

Basins of Strike-slip Fault Zones

Spain

Montenat, C. et al. 1987. Originalité géodynamique des bassins néogènes du domaine bétique oriental (Espagne). *Notes et Mém. Cie TOTAL (CFP)* 21: 11–49.
See also *Bull. Elf Aquitaine* 11: 1–22.

California

Harding, T.P. 1970. Tectonic significance and hydrocarbure trapping consequences of sequential folding synchronous with San Andreas faulting, San Joachim valley, California. *AAPG Bull.* 60: 356–378.
Van Mount, S. and J. Suppe. 1987. State of stress near the San Andreas fault: implications for wrench tectonics. *Geology* 15: 1143–1146.
Wilcox, R. et al. 1973. Basic wrench tectonics. *AAPG Bull.* 57: 74–96.

Molassic Basins

Switzerland

Bachmann, G. et al. 1982. Exploration in a classic thrust belt and its foreland: Bavarian Alps, Germany. *AAPG Bull.* 55: 2529–2542.

Homewood, P. et al. 1986. Dynamics of the molasse basin of Western Switzerland. Spec. Publ. Intern. Assoc. Sediment 8: 199–217.

Ménard, G. 1988. Structure et cinématique d'une chaîne de collision: les Alpes occidentales et centrales. Thèse. Grenoble, 268 pp. (pour le bassin molassique, pp. 123–129).

Mugnier, J.L. and G. Ménard. 1986. Le développement du bassin molassique suisse et l'évolution des Alpes externes, un modèle cinématique. *Bull. Elf Aquitaine* 10: 167–180.

Müller, M. et al. 1988. Tectonic style and pressure distribution at the northern margin of the Alps between Lake Constance and the river Inn. *Geol. Rdsch.* 77: 787–786.

Pfiffner, O. 1986. Evolution of the North Alpine foreland basin in the Central Alps. Spec. Publ. Intern. Assoc. Sediment 8: 219–228.

Po basin

Cassono, E. et al. 1990. Geophysical data along the North Italian sector of the European geotraverse. *Tectonophysics* 176–182.

Castellarin, A. and G.-B. Vai. 1986. South Alpine versus Po plain Apenninic arcs. In: *The Origin of Arcs* (Wezel, ed.). Development in Geotectonics. Elsevier, 21: 253–280.

Pieri, M. 1984. Storia delle ricerche nel sottosuolo padane fino alle ricostruzioni attuali. *Vol. giub. Centen. Soc. Geol. Italia,* pp. 155–176.

Pieri, M. and M. Groppi. 1981. Subsurface geological structures of the Po basin. *Geodinamica,* CNR, publ. 414, 23 pp.

Ricci-Lucchi, F. 1986. The foreland basin system of the N. Apennines and relative clastic wedges. *Giorn. Geol.* 48: 165–185.

Pannonic basin

AAPG (1988). The Pannonian Basin. *Mém.* no. 45.

Bergerat, Fr. et al. 1984. Etude de la fracturation dans le bassin pannonien. Mécanismes et étapes de sa création. *Ann. Soc. Geol. Nord,* CIII: 265–272.

Royden, L. et al. 1982. Transform faulting, extension and subduction in the Carpathian Pannonian region. *Geol. Soc. Amer. Bull.* 93: 717–725.

Sclater, J.-G. et al. 1980. The formation of the intra-Carpathian basin as determined from subsidence data. *Earth Plan. Sci. Let.,* pp. 139–162.

Tyrrhenian basin

Bertrand, H. et al. 1989. Evolution géochimique d'un bassin arrière-arc: la mer tyrrhénienne au Pliocène. *C.R. Acad. Sci., Paris* 308: 1423–1429.

Kastens, K. et al. 1988. ODP Leg 107 in the Tyrrhenian Sea: insights into passive margin and back-arc evolution. *Geol. Soc. Amer. Bull.* 100: 1140–1156.

CRUSTAL FRACTURES

Red Sea

Bayer, H. et al. 1989. Large tectonic and lithospheric structures of the Red Sea region. *J. Afr. Earth Sci.* 8: 565–587.

Bosence, D. and B. Burser (eds.). 1996. *Sedimentary and Tectonic Evolution of Rift Basins* (Red Sea, Gulf of Aden). Chapmann & Hall, London, 512 pp.

Jestin, Fl. and Ph. Huchon. 1992. Cinématique de la jonction triple mer Rouge-Golfe d' Aden-rift éthiopien depuis l'Oligocène. *BSGF*, pp. 125–133.

Juteau, Th. 1983. Structure et pétrologie du rift axial de la mer Rouge vers 18° N. *Bull. Elf Aquitaine* 7: 217–231.

Pautot, G. et al. 1986. La dépression axiale du segment Nord de la mer Rouge. *BSGF*, pp. 381–399.

Tectonophysics (1988), vol. 153, nos. 1–4, in particular:

Coleman, R. and A. McGuire. Magma system related to the Red Sea opening, pp. 77–100.

Garfunkel, Z. Relation between continental rifting and uplifting: evidence from the Suez rift and N. Red Sea, pp. 33–50.

Le Pichon, X. and J.M. Gautier. The rotation of Arabia and the Levant fault system, pp. 271–294.

Voggenreiter, W. et al. Low-angle detachment origin for the Red Sea rift system?, pp. 51–77.

Voggenreiter, W. and H. Hötzl. 1989. Kinematic evolution of the S.-W. Arabian continental margin: implications for the origin of the Red Sea. *J. Afr. Earth Sci.* 8: 541–564.

Afars

Boucarut, M. et al. 1980. Etude stratigraphique de la région de Djibouti, 50 pp.

Bull. Soc. Géol. Fr. (1980). 6: 797–1015.

Gaulier, J.M. and Ph. Huchon. 1991. Tectonic evolution of the Afar triple junction. *Bull. S. Geol. Fr.* 162: 451–464.

Mohr, P. 1989. Nature of the crust under Afar: new igneous not thinned continental. *Tectonophysics* 67: 1–11.

Gulf of California

Colletta, B. and J. Angelier. 1983. Tectonique cassante du N.-W. mexicain et ouverture du golfe de Californie. *Bull. Elf Aquitaine* 7: 433–44[illegible].

Ortlieb, L. et al. 1989. Geodetic and tectonic analysis along an active plate boundary: the central Gulf of California. *Tectonics* 8: 429–441.

Wallace, R.E. 1990. The San Andreas Fault System. USA Geol. Survey, Prof. Paper 1515.

Gulf of Genoa

Arthaud, F. and P. Matte. 1976. Détermination de la position initiale de la Corse et de la Sardaigne à la fin de l'orogenèse hercynienne grâce aux marqueurs géologiques antémésozoiques. *BSGF* 7: 833–840.

Bellon, H. et al. 1977. Le déplacement de la Sardaigne. Synthèse des données géochronologiques, magmatiques et paléomagnétiques. *BSGF* 7: 825–832.

Chabrier, G. and G. Mascle. 1975. Comparaison des évolutions géologiques de la Provence et de la Sardaigne. *Rev. Géog. Phys. Géol. Dyn.*, XVII: 121–136.

Gauthier, A. and J.-P. Rehault. 1986. Dérive de la plaque corso-sarde. Doc. de Géol. Régionale. CRDP-Corse, Ajaccio.

Westphal, M. et al. 1976. Le microcontinent corso-sarde, sa position initiale: données paléomagnétiques et raccords géologiques. *Tectonophysics* 30: 141–157.

Gulf of Gascogne

Boillot, G. 1984. Le golfe de Gascogne et les Pyrénées. In: *Les Marges Continentales Actuelles et Fossiles autour de la France.* Masson, Paris, pp. [illegible]–73.

CONTINENTAL MARGINS

General Works

Boillot, G. 1990. *Géologie des Marges Continentales.* Masson, Paris, 3rd ed., 134 pp.
Bull. BRGM et Oceanologica Acta (1981). C.R. 26th CGI Paris, 1980 (C3).

Passive Margins

General

Boillot, G. 1984. Les marges continentales stables et leur destin. BSGF, pp. 517–532.
Boillot, G. et al. 1984. *Les Marges Continentales Actuelles et Fossiles autour de la France.* Masson, Paris, 342 pp.
Boillot, G. et al. 1989. Undercrusting by serpentinite beneath rifted margins. *Nature* 341: 523–525.

Specific cases

Boillot, G. et al. 1989. Rifting of the W-Galicia continental margin: a review. *BSGF,* pp. 393–400.
Bull. Soc. Geol. France (1988). The Tethyan margin in the Alps, nos. 4–5: 597–787.
Lepvrier, Cl. and J. Geyssant. 1985. L'évolution structurale de la marge W du Spitzberg: coulissement et rifting tertiaire. *BSGF* 8: 115–125; *Ann. Soc. Geol. Nord,* CIII: 333–344.
Reyre, D. 1984. Remarques sur l'origine et l'évolution des bassins sédimentaires africains de la côte atlantique. *BSGF,* pp. 1041–1059.

Dissymmetric rifting

Coward, M. 1986. Heterogeneous stretching, simple shear and basin development. *Earth Plan. Sci. Let.* 80: 325–336.
Lister, G. et al. 1986. Detachment faulting and the evolution of passive continental margins. *Geology* 14: 246–250.
Malod, J. 1987. Dissymétrie des marges de l'Atlantique N: relation avec les modèles de distension. *BSGF,* pp. 465–472.
Stampfli, G. and M. Marthaler. 1990. Divergent and convergent margins in the NW Alps. *Geodinamica Acta, Paris* 4 (3): 159–184.
Wernicke, B. (*see* Basin and Range)

Active Margins

General

Blanchet, R. 1984. Les marges actives des océans. *BSGF,* pp. 353–550.
Hamilton, W. 1988. Plate tectonics and island arcs. *Geol. Soc. Amer. Bull.* 100: 1503–1527.
Tectonophysics (1989), vol. 160, nos. 1–4, in particular:
Aubouin, J. Some aspects of the tectonics of the subduction zone, pp. 1–21.
Zvi Ben-Avraham (ed.). 1990. *The Evolution of the Pacific Ocean Margins.* Oxford Monographs on Geol. & Geoph., no. 8, 246 pp.

Volcanism

Maury, R. 1984. Les conséquences volcaniques de la subduction. *BSGF,* pp. 489–500.
Ringwood, A.-E. 1974. The petrological evolution of island arc systems. *J. Geol. Soc.* 130: 183–204.

Trench and subduction

Bourgois, J. et al. 1982. La marge pacifique du Guatemala: un modèle de marge extensive en domaine convergent. *C. R. Acad. Sci. Paris* 295: 607–613.
Cadet, J.-P. et al. 1987. Erosion gravitaire et subduction de monts sous-marins: résultat des plongées par submersible dans la fossé du Japon (campagne Kaiko). *C.R. Acad. Sci. Paris* 305: 1327–1335.
Earth Plan. Sci. Lett. (January 1987). The Kaiko operation.
Lallemant, S. et al. 1986. Kaiko: l'exploration des fossés du Japon. *La Recherche* 182: 1345–1356.
Tectonophysics (1987). The Kaiko operation.
Thornburg, T. and L. Kulm. 1987. Sedimentation in the Chile trough. *Geol. Soc. Amer. Bull.* 98: 33–52.

Accretionary prisms

Casey-Moore, J. et al. 1982. Geology and tectonic evolution of a juvenile accretionary terrane along a truncated convergent margin, S. Mexico. *Geol. Soc. Amer. Bull.* 93: 847–861.
Mascle, A. 1989. Les prismes d'accrétion. *La Recherche* 215: 1308–1315.
McCall, G. and R. Kidd. 1982. The Makran, S.-E. Iran: the anatomy of a convergent plate margin active from Cretaceous to Present. In: *Trench-Fore-arc Geology* (J.K. Leggett, ed.). Geol. Soc., London, pp. 387–397.
McCarthy, J. and D. Scholl. 1985. Mechanism of subduction-accretion along the Central Aleutian trench. *Geol. Soc. Amer. Bull.* 96: 691–701.
Platt, J.P. et al. 1985. Large-scale sediment underplating in the Makran accretionary, S.-W. Pakistan. *Geology* 13: 507–511.
Sample, J.-C. and D. Fisher. 1986. Duplex accretion and underplating in an ancient accretionary complex, Kodiak islands, Alaska. *Geology* 14: 160–163.
Scholl, D. et al. 1977. Sediment subduction and offscraping at Pacific margins. In: *Island Arcs, Deep Sea Trenches and Back-arc Basins* (M. Talwani and W.C. Pitman, eds.). M. Ewing, series 1, Amer. Geophys. Union, pp. 199–210.
Von Huene, R. 1986. To accrete or not accrete: that is the question. *Geol. Rdsch.* 75 (1): 1–15.

Fore-arc basins

Dickinson, W. and D. Seely. 1979. Structure and stratigraphy of fore-arc regions. *Amer. Assoc. Petr. Geol. Bull.* 63: 2–31.
Ryan, H. and D. Scholl. 1989. The evolution of fore-arc structures along an oblique convergent margin, Central Aleutian arc. *Tectonics* 8: 497–516.

Back-arc basins and marginal basins

(*see* also Intraoceanic Arcs)

Boggs, S. 1984. Quaternary sedimentation in the Japan arc-trench system. *Geol. Soc. Amer. Bull.* 95: 669–685.
Tamaki, K. 1985. Two models of back-arc spreading. *Geology* 13: 475–478.
Tatsumi, Y. et al. 1989. Opening of the Sea of Japan back-arc basin by asthenospheric injection. *Tectonophysics* 166: 317–329.

MOUNTAIN CHAINS

General

Aubouin, J. 1984. Des marges continentales aux chaînes plissées. *BSGF*, pp. 501–516.

Mattauer, M. 1981. La formation des chaînes de montagnes. *Pour la Science* 46: 40–55.
Mercier, J.-L. 1984. La déformation des continents au voisinage des marges convergentes. *BSGF*, pp. 551–566.

Intracontinental Chains

Anti-Atlas

Choubert, G. and A. Faure-Muret. 1971. L'Anti-Atlas et la chaîne de l'Ougarta. In: *Tectonique de l'Afrique.* UNESCO, Paris, pp. 359–371.

Rocky Mountains, USA

Brewer, J.-A. et al. 1982. COCORP profiling across the Rocky Mountain front in southern Wyoming (part I: Laramide structures). *Geol. Soc. Amer. Bull.* 93: 1242–1252.

Lebanon, Anti-Lebanon-Palmyras

Bride, J. et al. Seismic reflection structure of intracratonic Palmyrides (Syria). *Amer. Assoc. Petr. Geol. Bull.* 74: 238–259.
Giannerini, G. et al. 1988. Déformations intraplaques et volcanisme associé: exemple de la bordure N.-W. de la plaque arabique au Cénozoique. *BSGF*, pp. 937–947.
Hgai, Ron. 1987. Deformation along the Yammuneh, the restraining bend of the Dead Sea transform. *Tectonics* 6: 653–666.
Le Pichon, M. and J.-M. Gaulier. 1988. The rotation of Arabia and the Levant fault system. *Tectonophysics* 153: 271–294.
Walley, O. 1988. A braided strike-slip model for the northern continuation of the Dead Sea fault and its implications for Levantine tectonics. *Tectonophysics* 145: 63–72.

Transverse Ranges (California)

Namson, J. and T. Davis. 1988. Structural transect of the Western Transverse Ranges. *Geology* 18: 23–25.
Weldon, R. and E. Humphreys. 1986. A kinematic model of Southern California. *Tectonics* 5: 33–48.
Yeats, R. 1987. South Island (New Zealand) and Transverse Ranges of southern California: a seismotectonic comparison. *Tectonics* 6: 363–376.

High Atlas (Morocco)

Laville, E. and C. Harmand. 1982. Evolution magmatique et tectonique du bassin intracontinental mésozoique du Haut-Atlas (Maroc): un modèle de mise en place synsedimentaire de massifs anarogeniques liés à dés decrochements. *BSGF*, pp. 213–227.
Mattauer, M. et al. 1977. Sur les mécanismes de formation des chaînes intracontinentales. L'exemple des chaînes atlasiques du Maroc. *BSGF*, pp. 521–526.
Michard, A. 1976. *Elements de géologie marocaine.* Notes et Mém. S. Geol. Maroc, no. 252, 408 pp.

Iberides

Guimera, J. and M. Alvaro. 1990. Structure et évolution de la compression alpine dans la chaîne ibérique et la chaîne côtière catalane (Espagne). *BSGF*, pp. 339–348.
Viallard, P. 1989. Décollement de couverture et décollement medio-crustal dans une chaîne intraplaque: variation verticale du style tectonique des Ibérides (Espagne). *BSGF*, pp. 913–918.

Pyrenees

Arthaud, Fr. and M. Séguret. 1981. Les structures pyrénéennes du Languedoc et du golfe du Lion (S. de la France). *BSGF*, pp. 51–63.

Boillot, G. 1980. Les Pyrénées, victime de la collision entre deux continents. *Pour la Science* 27: 57–75.

Boillot, G. 1984. Les modèles explicatifs des Pyrénées. In: *Les Marges Continentales Actuelles et Fossiles autour de la France.* Masson, Paris, pp. 57–72.

Bull. Soc. Geol. Fr. (1978), no. 5, pp. 578–764.

Bull. Soc. Geol. Fr. (1990). The ECORS programme in the Pyrenees, no. 2, pp. 209–348, in particular:
Mattauer, M. Une autre interprétation du profil ECORS-Pyrénées, pp. 307–312.

Choukroune, P. and A. Garrido. 1989. Les Pyrénées vues par ECORS: une image inattendue. *La Recherche* 206: 132–135.

Choukroune, P. et al. 1969. The ECORS Pyrenean deep seismic profile reflection data. *Tectonics* 8: 23–39.

Mattauer, M. 1985. Présentation d'un modèle lithosphérique des Pyrénées *C. R. Acad. Sci. Paris* 300: 71–73.

Mirouse, P. 1980. Les Pyrénées. In: *Geologie des Pays Européens* (vol. France). Dunod, Paris, pp. 212–223.

Roure, P. et al. 1989. ECORS deep seismic data and balanced cross sections of the Pyrenees. *Tectonics* 8: 41–50.

Souquet, P. and E. Debroas. 1980. Tectogenèse et évolution des bassins de sédimentation dans le cycle alpin des Pyrénées. CR 26th CGI Paris. *Mem. BRGM* 107–213–233.

Jura

Chauve, P. 1980. Le Jura. In: *Géologie des Pays Européens* (vol. France). Dunod, Paris, pp. 405–421.

Chauve, P. et al. 1988. Le chevauchement du Jura sur la Bresse. Données nouvelles et réinterprétation des sondages. *BSGF*, pp. 861–870.

Laubscher, H.P. 1986. The Eastern Jura: relation between thin-skinned and basement tectonics, local and regional. *Geol. Rdsch.* 75 (3): 535–553.

Mugnier, J.L. and P. Vialon. 1983. The mechanism of overlapping of the Bresse graben by the Jura formation in the Vignoble area (France). *Tectonophysics* 106: 155–163. Also *see: J. Str. Geol.* (1986) 8: 373–387.

Rocky Mountains, Canada

Bally, A. 1984. Tectogenèse et sismique reflexion. *BSGF*, pp. 297–285.

Geol. Survey of Canada (1978). Plates of the map. 1457 A, Calgary, 1/25,000,000.

Price, R.A. 1981. The Cordillera foreland thrust and fold belt in the Southern Canadian Rocky Mountains. Spec. publ. Geol. Soc., London, no. 9, pp. 427–488.

Subduction Chains

Chemenda, A. 1993. *Subduction: Insights from Physical Modeling.* Kluwer Publ., 215 pp.

Chemenda, A. et al. 1995. A mechanism for syncollisional deep-rock exhumation and associated normal faulting: results from physical modeling. *EPSL* 132: 225–232.

Andes, Peru and Bolivia

Bourgois, J. and D. Janjou. 1981. Subduction océanique, subduction continentale et surrection andine: l'exemple du Pérou nord. *C. R. Acad. Sci. Paris* 293: 859–864.

Dalmayrac, B. and M. Mattauer. 1980. Subduction et phases de compression dans la chaîne des Andes. *C. R. Acad. Sci. Paris* 290: 1345–1348.
Isacks, Br. 1988. Uplift of the Central Andean Plateau and bending of the Bolivian orocline. *J. Geoph. Res.* 93 (B4): 3211–3231.
Jaillard, E. 1987. Sedimentary evolution of an active margin during Middle-Upper Cretaceous times. *Geol. Rdsch.* 76: 677–697.
Jordan, T. et al. 1983. Andean tectonics related to geometry of subducted Nazca plate. *Geol. Soc. Amer. Bull.* 94: 341–361.
Mégard, F. 1984. The Andean orogenic period and its major structures in Central and Northern Peru. *J. Geol. Soc., London* 141: 893–900.
Mégard, F. 1987. Structure and evolution of the Peruvian Andes. In: *The Anatomy of Mountain Ranges* (J.P. Schaer and J. Rogers, eds.). Princeton Univ. Press, pp. 179–210.
Mégard, F. and H. Philip. 1976. Plio-Quaternary tectonomagmatic zonation and plate tectonics in the Central Andes. *Earth Plan. Sci. Let.* 33: 231–238.
Roeder, D. 1988. Andean-age structure of Eastern Cordillera (Prov. La Paz, Bolivia). *Tectonics* 7: 23–39.
Sébrier, M. et al. 1988. Tectonics and uplift in Central Andes (Peru, Bolivia and N. Chile) from Eocene to Present. *Géodynamique* 3: 85–106.

Obduction Chains

General

Coleman, R.G. 1971. Plate tectonic emplacement of upper mantle peridotites along continental edges. *J. Geoph. Res.* 76: 1212–1222.
Deway, J.F. 1976. Ophiolite obduction. *Tectonophysics* 31: 93–120.
Michard; A. 1987. L'obduction. *La Recherche* 186: 312–322.

Oman

Bernoulli, D. and H. Weissert. 1877 [*sic*]. The upper Hawasina nappes in the Central Oman mountains. *Geodinamica Acta 1:* 47–58.
Bull. Soc. Geol. Fr. (1989), 2, pp. 163–294, in particular:
Bechennec, Fr. et al. Géologie d'une chaîne issue de la Tethys, les montagnes d'-Oman, pp. 167–188.
Goffé, B. et al. 1988. A case of obduction-related high-pressure, low-temperature metamorphism in upper crustal nappes, Oman. *Tectonophysics* 151: 363–386.
Michard, A. 1987. L'obduction. *La Recherche* 186: 312–322.

New Caledonia

Cluzel et al. 1995. Dénudation tectonique du complexe à noyau métamorphique de haute pression d'age tertiaire de la Nouvelle Calédonie. *C. R. Acad. Sci. Paris* 321: 57–64.
Collot, J. et al. 1987. Overthrust emplacement of New Caledonia ophiolite: geophysical evidence. *Tectonics* 6: 215–232.
Paris, J.-P. 1981. *Géologie de la Nouvelle Calédonie.* Mem. BRGM, no. 119, 278 pp.
Rigolot, P. 1989. Evolution morphologique et structurale de la marge W de la ride de Nouvelle Calédonie (S.-W. Pacifique). *Bull. Elf Aquitaine* 13: 319–344.

New Guinea

Cullen, A. and J. Pigott. 1989. Post-Jurassic tectonic evolution of Papua New Guinea. *Tectonophysics* 162: 291–302.

Liminary Collision Chains

Andes, Colombia and Ecuador

Jaillard, E. et al. 1997. Les déformations paléogènes de la zone d'avant-arc sud-équatorienne en relation avec l'évolution géodynamique. *Bull. S. Geol. Fr* 168, pp. 403–412.

Mégard, p. 1987. Cordilleran Andes and marginal Andes: a review of Andean geology north of the Arica Elbow. In: *Circum-Pacific Belts and Evolution of the Pacific Ocean Basin* (J. Monger and J. Francheteau, eds.). Amer. Geophys. Union, Geodynamics series, no. 18, pp. 71–95.

Zvi Ben Avraham (ed.). 1989. *The Evolution of the Pacific Ocean Margin.* Oxford Monographs on Geol. & Geoph., no. 4, pp. 208–230.

Taiwan

Pelletier, B. 1985. De la fosse de Manille à la chaîne de Taiwan. Thèse, Brest, 267 pp.

Tectonophysics (1986), vol. 125, no. 1, in particular:

Pelletier, B. and J. Stephan. Middle Miocene obduction and Late Miocene beginning of collision registered in the Hengchun peninsula: geodynamic implications for the evolution of Taiwan, pp. 133–160.

Stephan, J. et al. Geodynamic evolution of the Taiwan-Luzon-Mindoro belt since the late Eocene, pp. 245–268.

Tectonophysics (1990). The eastern margin of Eurasia, vol. 183.

California

Aubouin, J. et al. 1986. Traits généraux des cordillères de l'ouest des USA. *BSGF*, p. 741.

Ernst, W. 1984. Californian blue schists, subduction and the significance of tectono-stratigraphic terranes. *Geology* 12: 436–440.

Lagabrielle, Y. et al. 1986. The Coast Range ophiolites (N. California): possible arc and back-arc basin remnants and their relations with the Nevadan orogeny. *BSGF*, pp. 981–999.

Roure, F. and M. Sosson. 1986. Late Jurassic collision between a composite exotic block and the North American continent: a model for the Cordillera building. *BSGF*, pp. 945–959.

Roure, F. et al. 1986. Evolution comparée des bassins franciscains de Californie, de l'Orégon et de Basse Californie (USA et Mexique). *BSGF*, pp. 931–944.

Intercontinental Collision Chains

General

Tectonophysics (1988), 146 (1–4): 1–231.

Wezel, F.C. (ed.). 1986. The origine of arcs. *Developments in Geotectonics.* Elsevier, 567 pp.

Himalaya-Tibet

Armijo, R. and P. Tapponnier. 1989 Late Cenozoic right-lateral strike-slip faulting in southern Tibet. *J. Geophys. Res.* 94 (83): 2787–2838.

Armijo, R. et al. 1936. Quaternary extension in southern Tibet. *J. Geophys. Res.* 91: 803–872.

Bassoullet, J.-P. et al. 1984. L'orogène himalayen au Crétacé supérieur. *Mém. Soc. Géol. Fr.*, 147: 9–20.

Chalaron, E. 1994. Modélisation numérique et signature géologique des interactions entre tectonique, érosion et sédimentation dans l'avant-pays himalayen. Mém. Géol. Alpine, Grenoble 23, 132 pp.

Geodinamica Acta, Paris (1987), vol. 1, fasc. 4–5.

Geological Evolution of Tibet. Phil. Trans. Roy. Soc. London, (1988) ser. A, 327 (1594): 413 pp.

Guillot et al. 1995. Un nouveau témoin du métamorphisme de HP dans la chaîne himalayenne: les éclogites rétromorphosées du dôme de Tso Morari. *C. R. Acad. Sci. Paris* 320: 931–936.

Makovsky, Y. et al. 1996. Mid-crustal reflector truncating the India-Asia suture and magma beneath the Tibetan rift-system from wide-angle seismic data of project INDEPTH. Abstr. 10th Himalaya Workshop, Flagstaff, pp. 91–92.

Mascle, G. et al. 1990. La formation de l'Himalaya. *La Recherche* 21: 31–39.

Mattauer, M. 1983. Subduction de lithosphère continentale, décollement croûte-manteau et chevauchements d'échelle crustale dans la chaîne de collision himalayenne. *C. R. Acad. Sci. Paris* 296: 481–486.

Molnar, P. and P. Tapponnier. 1975. Cenozoic tectonics of Asia: effects of a continental collision. *Science* 189: 419–426.

Nelson, K.D. et al. 1996. Initial results of INDEPTH II (1994) deep crustal profiling in S. Tibet. Abstr. 10th Himalaya Workshop, Flagstaff, p. 110.

Tapponnier, P. et al. 1986. On the mechanics of the collision between India and Asia. In: *Collision Tectonics* (Coward and Ries, eds.). Geol. Soc. Spec. Publ., no. 19, pp. 115–157.

Tectonic Evolution of the Himalayas and Tibet. Phil. Trans. Roy. Soc. London (1988), ser. A, 326 (1589): 325 pp.

Tectonics of the Western Himalayas. Geol. Soc. Amer. (1989), Prof. Paper no. 232, 320 pp.

Tectonophysics (1987). Deep-seated processes in collision zones, no. 134.

Zhao, W., K.D. Nelson and INDEPTH Project team. 1993. Deep seismic reflection evidence for continental underthrusting beneath S. Tibet. *Nature* 366: 557–559.

Variscan Chain, Europe

General articles

Bard, J.-P. et al. 1980. La chaîne hercynienne d'Europe occidentale en terme de tectonique de plaques. CR 26th CGI, Paris (C6). *BRGM, pp.* 233–246.

Matte, Ph. 1983. Two geotraverses across the Ibero-Armorican Variscan arc of Western Europe. In: *Profiles of Orogenic Belts.* Amer. Geophys. Union, Geodynamic series 10: 53–81.

Matte, Ph. 1986. La chaîne varisque parmi les chaînes paléozoiques peri-atlantiques: modèle d'évolution et position des grands blocs continentaux au Permo-Carbonifère. *BSGF*, pp. 9–24.

Matte, Ph. 1986. Tectonics and plate tectonics model for the Variscan belt of Europe. *Tectonophysics* 126: 329–374.

Thematic aspects: ophiolites and high-pressure metamorphism

Bouchardon, J.L. et al. 1989. Ecologites in the French Paleozoic orogen: geodynamic significance. *Tectonophysics* 169: 317–332.

Cabanis, Br. and G. Godard. 1987. Les éclogites du pays de Léon (NW du massif armoricain). *BSGF*, pp. 1133–1142.

Girardeau, J. et al. 1986. Cinématique de mise en place des ophiolites et nappes cristallophylliennes du Limousin. *BSGF*, pp. 849–860.

Ménot, R.P. et al. 1988. Les associations magmatiques leptyno-amphiboliques du massif de Belledonne. *BSGF*, pp. 917–926.

Quinquis, H. and P. Choukroune. 1981. Les schistes bleus de l'île de Groix dans la chaîne hercynienne: implications cinématiques. *BSGF*, pp. 409–418.

Also *see:*

Earth Plan. Sci. Let. (1988) 88: 82–92.

Schweiz. Mineral. Petrogr. Mitt. (1986) 66: 229–258.

Schweiz, Mineral. Petrogr. Mitt. (1987) 67: 273–284.

Thematic aspects: Variscan flyschs and Variscan front

Franke, W. and W. Engel. 1986. Sédimentation synorogénique dans la chaîne varisque d'-Europe. *BSGF*, pp. 25–34.

Raoult, J.F. 1986. Le front varisque du nord de la France. *Rev. Géogr. Phys. Géol. Dyn.* 27 (3–4): 247–268.

Raoult, J.F. and F. Meillez. 1986. Commentaires sur une coupe structurale de l'Ardenne selon le méridien de Dinant. *Ann. Soc. Géol. Nord* CV: 97–109.

Late Variscan phases

Arthaud, Fr. and Ph. Matte. 1975. Les décrochements tardi-hercyniens du SW de l'Europe. Géometrie et essai de reconstitution des conditions de la déformation. *Tectonophysics* 25: 139–171.

Burg, J.-P et al. 1994. Syn. to post-thickening extension in the Variscan belt of W. Europe. *Géol. de la France.* BRGM 3: 33–52.

Prost, A. and J.-F. Becq-Giraudon. 1989. Evidence for mid-Permian compressive tectonics in Western Europe. *Tectonophysics* 169: 333–340.

Alps

General and western Alps

CR 26th CGI Paris, 1980:

An Outline of the Geology of Switzerland. Spec. Mém. BRGM and Comm. Suisse de Geol., 334 pp.

Introduction à la géologie du sud-est de la France. Spec. Mém. BRGM, 249 pp.

Coward, M.P. et al. (eds.). 1989. *Alpine Tectonics.* Geol. Soc., London, spec. publ., no. 45, 450 pp.

Debelmas, J. 1983. La formation des Alpes. *La Recherche* 150: 1542–1552.

Lemoine, M. 1984. La marge occidentale de la Téthys ligure (les Alpes). In: *Les Marges Continentales Actuelles et Fossiles autour de la France* (G. Boillot et al., eds.). Masson, Paris, pp. 157–248.

Tricart, P. 1984. From passive margin to continental collision: a tectonic scenario for the Western Alps. *Amer. J. Sci.* 284: 97–120.

Deformation, metamorphism

Escher, A. et al. 1988. *Coupes Géologiques dans les Alpes Occidentales Suisses.* Mém. Géol., Lausanne, no. 2.

Frey, M. 1978. Progressive low-grade metamorphism in Central Swiss Alps. *J. Petrol* 19: 95–135.

Hunziker, J. et al. 1989. Alpine thermal evolution in the Central and Western Alps. In: *Alpine Tectonics* (P. Coward, ed.). Geol. Soc. Amer. Spec. Publ. no. 45, pp. 353–367.

Merle, O. and B. Guillier. 1989. The building of the Central Swiss Alps: an experimental approach. *Tectonophysics* 165: 41–56.

Platt, J.-P. 1986. Dynamics of orogenic wedges and the uplift of high-pressure metamorphic rocks. *Geol. Soc. Amer. Bull.* 97: 1036–1053.

Steck, A. 1994. Structures de déformations tertiaires dans les Alpes Centrales. *Ecl. Geol. Helv.* 77: 55–100. Also *see: Ecl. Geol. Helv.* (1990) 83: 603–627.

Deep structures

Ménard, G. 1988. Structure et cinématique d'une chaîne de collision: les Alpes occidentales et centrales. Thèse, Grenoble, 268 pp.

Ménard, G. and F. Thouvenot. 1984. Ecaillage de la lithosphère européenne sous les Alpes occidentales. *BSGF*, pp. 875–884.

Nicolas et al. 1996. General conclusions of the ECORS-CROP Alpine survey. Mém. no. 170, pp. 99–104.

Soc. Geol. France (1990). *Deep Structure of the Alps* (F. Roure et al., eds.). Mém. no. 156, 350 pp.

Soc. Geol. France (1996). *The ECORS-CROP Defined Seismic Traverse.* Mém. no. 170, 113 pp.

Spakman, W. 1990. Tomographic images of the upper mantle below Central Europe and the Mediterranean. *Terra Nova* 2: 542–553.

Eastern Alps

Clar, E. 1973. Review of the structure of the Eastern Alps. In: *Gravity and Tectonics* (De Jong and Scholten, eds.). J. Wiley & Sons, New York, pp. 253–270.

Dietrich, V. 1983. Evolution of the Eastern Alps: the traditional concept and modern plate tectonics. *Ann. Geophysicae* 2 (2): 155–160.

Flügel, H. and P. Faupl (eds.). 1987. *Geodynamics of the Eastern Alps.* Deuticke, Wien.

Soc. Geol. France (1966). CR réun. extraord. dans les Alpes autrichiennes. *CRSGF* 11: 413–472.

Tollmann, A. 1989. Eastern Alpine sector, northern margin of Tethys. In: *Evolution of the Northern Margin of the Tethys.* Mém. SGF, n.s., II (154): 23–94.

Southern Alps

Aubouin, J. 1963. Essai sur la paléogéographie post-triasique et l'évolution secondaire et tertiaire du versant sud des Alpes orientales. *BSGF*, pp. 730–766.

Castellarin, A. and G.-B. Vai. 1982. Introduzione alla geologia strutturale del Sudalpino. In: *Guida alla Geologia del Sudalpino Centro-orientale* (A. Castellarin and G.-B. Vai, eds.). SG It, Bologna.

Doglioni, C. 1987. Tectonics of the dolomites (S. Alpen, N. Italy). *J. Str. Geol.* 9 (2): 181–193.

Doglioni, C. and A. Bosellini. 1987. Eoalpine and mesoalpine tectonics in the Southern Alps. *Geol. Rdsch.* 73 (3): 735–754.

Syntectonic and post-tectonic extensions of the Alps

Dewey, J.-F. 1988. Extensional collapse of orogens. *Tectonics* 7: 1123–1139.

Merle, O. 1994. Synconvergence exhumation of the Central Alps. *Geodinamica Acta, Paris* 7 (3): 129–138.

Philippot, P. 1990. Opposite vergence of nappes and crustal extension in the French-Italian occidental Alps. *Tectonics* 9: 1143–1164.

Ratsbacher, L. et al. 1991. Extrusion in the Eastern Alps. *Tectonics* 10: 257–271.

Ring, U. and O. Merle. 1992. Forethrusting, backfolding and lateral gravitational escape in the northern part of the W. Alps (M. Rosa). *Geol. Soc. Amer. Bull.* 104: 901–914.

COLLAGES AND TERRANES

General

Howell, D. 1986. Des terrains déplacés: les terranes. *Pour la Science*, pp. 18–29.

Howell, D. 1989. *Tectonics of Suspect Terranes.* Chapman & Hall, London, 232 pp.

North America

Bruns, T. 1983. Model for the origin of the Yakutat block, an accreting terrane in the NW Gulf of Alaska. *Geology* 11: 718–721.

Bruns, T. et al. 1984. The eastern Gulf of Alaska transcurrent plate boundary and the migration of terranes. *Ann. Soc. Geol. Nord,* CIII: 325–331.

Hall, Cl. 1983. Lanière d'accolement (= accreted terrane) le long de la faille de San Andreas (Californie). *Rev. Géogr. Phys. Géol. Dyn.* 24 (4): 343–350.

Jones, D. et al. 1977. Wrangellia, a displaced terrane in NW North America. *Can. J. Earth Sci.* 14: 2565–2577.

Jones, D. et al. 1982. The growth of Western N. America. *Sci. Amer.* 247 (5): 50–64.

Monger, J. 1984. Cordilleran tectonics: a Canadian perspective. *BSGF,* pp. 255–278.

Nur, A. 1984. Oceanic plateaus and accreted terranes: the view from Western N. America. *Ann. Soc. Geol. Nord,* CIII: 377–386.

Panuska, B. 1985. Paleomagnetic evidence for a post-Cretaceous accretion of Wrangellia. *Geology* 13: 880–883.

Other Regions

Charvet, J. and J. Fabbri. 1987. Vue générale sur l'orogenèse Shimanto et l'évolution tertiaire du Japon S.-W. *BSGF,* pp. 1171–1188.

Frizon de Lamotte, D. 1987. Un exemple de collage synmétamorphe: la déformation miocène des Temsamane (Rif. externe, Maroc). *BSGF,* pp. 337–344.

Mizutani, S. 1985. Tectonostratigraphic terranes in the Japanese islands and their environs in E. Asia. 3rd Circ. Pacif. Terrane conf. Geol. Soc. Austr., Sydney. Abstracts 14: 156–158.

INDEX

Printed in India